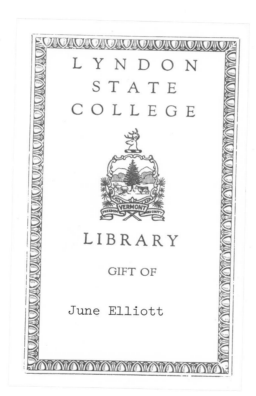

REGIONAL LANDSCAPES OF THE UNITED STATES AND CANADA

REGIONAL LANDSCAPES OF THE UNITED STATES AND CANADA

THIRD EDITION

Stephen S. Birdsall
John W. Florin

University of North Carolina at Chapel Hill

John Wiley & Sons

New York   Chichester   Brisbane   Toronto   Singapore

Copyright © 1978, 1981, 1985, by John Wiley & Sons, Inc.

All rights reserved. Published simultaneously in Canada.

Reproduction or translation of any part of
this work beyond that permitted by Sections
107 and 108 of the 1976 United States Copyright
Act without the permission of the copyright
owner is unlawful. Requests for permission
or further information should be addressed to
the Permissions Department, John Wiley & Sons.

Library of Congress Cataloging in Publication Data:

Birdsall, Stephen S.
    Regional landscapes of the United States and Canada.

    Includes bibliographical references and indexes.
    1. United States—Description and travel—1981–
2. Canada—Description and travel—1981–
3. United States—Economic conditions—1981–
4. Canada—Economic conditions—1945–   I. Florin,
John William. II. Title.
E169.04.B57 1985   917.3   84-22086
ISBN 0-471-88490-1

Printed in the United States of America

10 9 8 7 6 5

PREFACE TO THIRD EDITION

Authors face the prospect of writing a new edition with mixed feelings. On the one hand, the effort provides an opportunity to include new information and ideas and to respond to the (in the authors' opinion) reasonable criticism of the earlier edition. On the other hand, the task involves considerable effort and time. This third edition results from a number of additions, subtractions, and modifications we wished to make to our second edition.

Data from the 1980 U.S. and 1981 Canadian censuses have been incorporated throughout the book. Trends and patterns indicated by these data led to numerous substantive changes in text material. Additionally, many of the maps and tables have been updated. Tables of metropolitan area population from the most recent censuses have been added to each of the regional chapters.

We have substantially expanded our systematic treatment of both human and physical patterns without surrendering what we consider the important focus on regions and regional character. In this third edition, we now cover systematic topics in two chapters, rather than one.

"Additional Reading" lists have been added at the end of each chapter. These short lists should allow readers to pursue material covered in the chapters beyond that referred to in footnote citations. The focus in the reading lists is on relatively recent books, although some articles are included to introduce students new to geography to alternative sources of geographic literature. We have also tried to include on each list at least one item that is a more popular, or at least nonacademic, presentation of regional character.

Several other general changes have also been made. The glossary has been substantially expanded. A chapter on the South Atlantic lowlands, part of the earlier editions of the text, has been eliminated; many readers did not perceive that area as a separate region. Additional Canadian content material has been added to several of the regional chapters. And a number of the occasional end-of-chapter case studies have been changed or new ones added.

Most of what we originally wanted to do remains. That which was well received by those who wrote us from classes across both countries has been retained. No text can be all things for all users, whether instructors or students, and we make no such pretense in this one. We do hope, however, that those who read this book find as much pleasure and utility in it as we found in writing it.

Stephen S. Birdsall
John W. Florin

PREFACE TO SECOND EDITION

It is in the nature of all instructors to want to modify available texts; most instructors have their own teaching orientations and research interests. We believe that student interest and understanding is increased by interweaving factual information with the distinctive geographic approach. This is hardly a new notion to professional geographers. It seems to us, however, that many regional texts emphasize facts without offering students a sufficiently explicit geographic perspective to provide meaning to those facts. We agree that factual information is important; it has been included in even greater abundance in this edition than in the first. Our primary emphasis, however, remains on the character of a region: its "flavor" as defined on the basis of regional themes and on the specifically geographic contributions that have helped form the region's distinctiveness.

One of the ways that geographic perspective can be shown in a regional text is through the application of organizational ideas that are basic to the discipline. Several dozen fundamental geographic concepts have been fused into the regional discussions of this book. Each is presented where it will enhance students' understanding of both the region and the concept. Many of the concepts appear in more than one chapter, and most can be applied to other sections of the book as well. We encourage instructors and students to demonstrate to themselves (and to each other) the vitality of regional geography by actively "testing" the usefulness of these concepts in succeeding regional chapters.

This book can be used effectively by students with little or no previous experience with college-level geography. Most of the concepts explained in the text are defined again in the glossary at the end of the book. We hope that, by noting the ways in which these concepts are used to clarify the complexities of a region's character, students will understand the kinds of questions that interest geographers and will be inspired to ask similar questions themselves.

We believe that a regional text should be organized within a regional framework. We have chosen to define the major regions of the United States and Canada thematically as well as locationally. In doing so, we are simply making explicit the difference between a region and a territory. A territory is a portion of the earth's surface, but a region is a portion of the earth's surface that is different from other areas because of the features found there. Those who look in each chapter for a rigid topical sequence applied to a territory—for example, the physical environment, the agriculture, economy, population, urbanization, and so forth—will be disappointed. There are general similarities in the subjects treated in many regions, of course, but this does not depend on an arbitrary structuring of topics. Much of the material in each regional chapter is built around the themes that we have selected as most important in defining the region's character. The identifying themes in most chapters stem from the cultural rather than the physical environment, although the interaction of the two landscapes is usually strong. The major focus, then, is on human patterns and the interaction between these patterns and the physical environment.

Although the kind of information presented in most chapters is much the same from region to region, the thematic emphasis affects this somewhat. Some of our regions, for example, are contained within political boundaries; most are not. A few follow the general pattern of physiographic or climatic variations; others are more fully defined by the sprawl and interaction of their human geography. Some portions of the United States and Canada have been identifiable as distinct regions since the earliest stages of European development; others are only recently formed or are still in the process of development and, therefore, remain vague in outline and functional character. None of the regions that we define should be viewed as permanent. A complex pattern such as that in any large region is always in flux with the region's character and boundaries changing. It is possible that several of the regions described in this book may disappear through a merging process, whereas others may be created as subregional characteristics become stronger. Throughout the discussion we point out these possible shifts in regional alignment.

The thematic regions that we use are not spatially exclusive—their territories overlap. The continental manufacturing core, for example, overlaps spatially with the continent's agricultural core region, the Canadian national core, and the continent's major urbanized region, called Megalopolis. Northern Indiana is located in both the manufacturing and agricultural core regions and is discussed in both regional chapters, although it is the urban-manufacturing aspects that are discussed in Chapter Five and the rural small-town features that are described in Chapter Eleven.

This territorial overlap of regions takes two forms. For instance, for northern Indiana, the important characteristics of two different types of regions must be considered. Most of the familiar political unit, the state, is located well within both regions. In other instances, a territory may lie in parts of two regions primarily because the boundary between them is indistinct. In the latter example, the thematic elements of both regions blend together; they are not at all separated by a sharp line. The western margin of our primary agricultural core region, for example, falls well to the west of the eastern boundary of the adjacent Great Plains region. The area between the boundaries is peripheral to both regions but contains elements of each. This complexity is reflected in Figure 1-5, our basic regional map. Some students may find this map difficult to read because of the illustration of regional overlaps. We have tried to ease this problem by eliminating the peripheral, or "transitional," zones around each region. Students should use the regional maps at the beginning of each chapter for more specific locations. Throughout the book we make every effort to make our discussions of overlapping thematic regions especially clear as well as to emphasize the advantages of this approach.

This is a thematic regional geography of North

America. On the basis of political and cultural patterns, it treats only the United States and Canada; Mexico and the Caribbean countries are not included. The term "Anglo-America" has also been discarded because of the strong non-Anglo cultural influences in many sections of both countries. French culture in Canada and the strong Spanish influence in the American Southwest are only two examples of this cultural diversity. The British heritage is dominant in both countries, but non-British cultures have modified and enriched this heritage sufficiently to render the term "Anglo-America" unnecessarily misleading to the unwary.

Stephen S. Birdsall
John W. Florin

ACKNOWLEDGMENTS

A project of this scope cannot be the work of its authors alone. To the many friends, colleagues, and reluctant participants who encouraged and in some way assisted us with this book, we express our deep appreciation. To those among this group who find their suggestions unused or their criticism unheeded, we ask their forbearance. Authors, like parents, can be incredibly stubborn about their creations at times. J. Lewis Robinson made many suggestions for improving the Canadian content of this edition. James Anderson, Clyde Browning, Charles Kovacik, Peter Muller, Richard Pillsbury, Merle Prunty, and Jean Wheeler all reviewed chapters or larger portions of earlier editions of the text. We thank each of them. Although we can extend credit to others for many of the book's strengths, we must bear full responsibility for its weaknesses.

The staff of our publisher, John Wiley & Sons, has helped us in many ways. We are especially indebted to our editor, Katie Vignery. We also thank Lisa Heard, Jan M. Lavin, Elyse Reider, and Susan Winick, and all of the others at Wiley who performed their jobs with professional competence.

S.S.B.
J.W.F.

CONTENTS

CHAPTER ONE

THEMES AND REGIONS

This is a book about two of the largest and wealthiest countries in the world. One, the United States, is almost surely the world's most powerful nation, both economically and militarily. It has had a dominant influence on the world for decades. That position of strength has diminished somewhat in recent years, but not enough to rob the country of international preeminence. The other, Canada, even larger than the United States in land area, has only a tenth as many people. The sparseness of its population coupled with a large potential resource availability has led many to anticipate that Canada's already significant world economic influence will grow in the future.

Yet this is not a book about international power, nor is it really about the wealth of nations. Instead, our goal is to discuss and, to some degree, evaluate the geography of these two countries. Although we will look at their physical geography, our central interest is not landforms, climate, soils, or vegetation but rather the human imprint on the landscape. The people in these technologically advanced countries are the prime creators of their own environment, and it is the cultural environment that interests us most. The specific elements of this environment are diverse, and their number is almost endless, from the pattern of land use in a city, to the distribution of racial or ethnic groups, to the style of fences that farmers use to enclose their land.

This does not mean that the physical environment will, or can, be ignored. In fact, in some instances, it will hold a central role in the text. There are several reasons for this. One is that the physical environment often has a significant influence on people's activities in the landscape. One factor in the importance of New York City is certainly its location on one of the world's finest natural harbors. Southern Florida's long growing season and mild winters enable it to be a leader in the United States in the production of such crops as oranges, lemons, and sugarcane. You can surely add many other examples.

Still, Florida's mild climate does not automatically mean that it will be a supplier of oranges

1

and New York City's harbor is, taken by itself, only one of many important reasons for the city's growth. The physical environment is one of the elements that influences human activities, but it does not in itself determine those activities. In general, the more advanced the level of technology, the greater the latitude that a population has in dealing with the land. For Canada and the United States, this means that the possible variations in activity for any particular environment are great. If one simply looked at its climate, the valley around Great Salt Lake in central Utah should be a barren wasteland. The valley is, however, a productive agricultural region and the home of several sizable urban areas, largely because the people of Utah wanted this to be and had the technology to make the necessary changes.

It is obviously impossible to cover all the material that might fit into a geography of the two countries. To do so would require many volumes. We have chosen to limit the unmanageable complexity by subdividing the United States and Canada into a number of areas of distinctive character, or regions. We believe that each of these regions has a special identity that has developed out of a small number of interacting elements. We have used these elements to form the themes around which each of the regional chapters is organized.

A FEW BASIC THEMES

There are a few general cultural patterns in the two countries that are important because they serve to give the countries an identity distinct from most of the rest of the world. These identifying themes are often so common or universal within the United States and Canada that most of their residents view them as simply a part of the normal nature of things. Their influence cuts across regional and political boundaries and, in many cases, ignores major differences in the physical environment. These general themes indicate what appear to be elements that are especially characteristic of the ways that Cana-

dians and Americans have organized their national territories. In this manner, they define the character of the entire continent.

Urbanization

Citizens of the United States and Canada hold dearly to their rural roots, and there continues to be a widespread attitude that life in a rural setting is somehow good, whereas life in an urban environment is less good, if not downright bad. Millions of Americans, most of them urbanites, prefer to consider their country as a basically rural place. They seem to believe that this rurality sets them apart from Europeans and provides our two North American countries with a basic national vigor that has been lost in Europe. Few politicians seeking elective office in either country would have their chances damaged by a personal claim to a rural background.

There is no longer much justification for this view of rural dominance in either country. Although most of the countryside is rural by definition—a matter that may distort the residential image of the two countries to the casual visitor—most of the people simply do not live in most of the country. By their own definitions, about 70 percent of the residents of Canada and the United States live in urban areas. It has been predicted that by the year 2000 over 75 percent of the U.S. population will live in 30 urban regions, each numbering over 1 million people (Figure 1-1). The U.S. farm population was only 2.8 percent of the total population in 1982 and that percentage has declined steadily since the first national census in 1790, when over 90 percent of all Americans were farmers. Over 40 percent of the population of the United States lives in Standard Metropolitan Statistical Areas (SMSAs) of 1 million or more people. Nearly one-third of all Canadians live in Toronto, Montreal, and Vancouver, which are the country's three largest metropolitan areas. Nearly half of all Canadians live in one of their country's eight largest urban areas. Urban percentages in both countries have grown steadily, and no decline is foreseen.

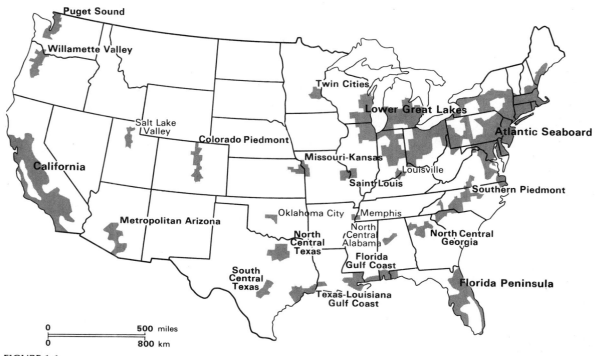

FIGURE 1-1

PREDICTED URBAN AREA EXPANSION TO THE YEAR 2000. By the year
end of this century, the major urban regions in the United States are expected to
be much more extensive than at present. The largest may have merged across vast
areas as shown.

In short, these are urban countries. Most people were born in urban areas; few of those presently in urban areas are likely to move to farm
locations. The rural-farm component of the
population, already small, continues to decline.

Several elements of urbanization will be emphasized in our discussion. Cities have a particular *form*, a particular layout. Most Americn cities have a rectangular-grid pattern, partly a result
of cultural attitudes, partly a result of a desire
for efficient transport before the automobile, and,
to a degree, simply because that pattern is an
easy way to survey the land. Modern shopping
centers are usually in the suburbs near major
highways. Within the city there is a collection of

industrial and commercial centers, residential
areas, warehouses, and so on.

Cities exist for many different reasons. They
may serve an important transportation function, such as New Orleans' role as an ocean port
or Omaha's as a railroad center. They may provide an important administrative function, as does
the Canadian federal capital at Ottawa or as does
New York City as the administrative center for
many corporations. Perhaps your city is a center
of recreation, like Las Vegas or Atlantic City.
Others, such as Hamilton, Ontario, or Detroit,
are usually thought of as manufacturing centers. Thousands of smaller towns and cities in
both countries exist primarily to provide goods

3

and services to residents of the town and a surrounding trade territory. Most cities, certainly all large ones, contain many different urban *functions*. Nevertheless, many urban places are characterized by certain dominant functions that were the reason for their initial development or for much of their early growth and that today continue to give them their special character.

The pattern of continuing and often rapid urban growth in the two countries during the last 100 years is another key element in their geography. Coupled with the increasing mobility of the urban population, this growth has stimulated a great sprawling pattern of urbanization. The total urbanized area of Canada and the United States increases at a rate of about 800,000 hectares (2 million acres) each year. This loss of rural land has generated a growing concern for the future availability of adequate amounts of good agricultural land. In some areas, the result of urban spread is urban coalescence, with the edges of different urban areas meeting and blending. As independent cities grow together, the role and meaning of urban political organization changes, and attempts may be made to create new admininstrative organizations to cut across city boundaries (see the discussion on political complexity).

Industrialization

The United States is the world's largest industrial power. No other country in the world can come close to matching the volume of output of American manufacturing, nor can any country

Urban housing built before the advent of the family automobile often consisted of closely-packed houses and walk-up apartments like these units in Boston. Afterward, housing spread out and cities exploded in area as people commuted great distances by car. (Donald Dietz/Stock Boston.)

match the variety of U.S. products. Canada, being much smaller than the United States in population, does not reach the magnitude of America's industrial output. Furthermore, with its much smaller population and less available financial capital, Canadian manufacturing lacks the diversity found in U.S. manufacturing; *economies of scale* similar to those of its southern neighbor are less possible with Canada's smaller market. Thus, Canada depends more on imported manufactured goods than does the United States, and many of the items it does manufacture are somewhat more expensive than their counterparts in the United States.

Still, both are highly industrialized countries. Each manufacturing job generates two or three other jobs. This *multiplier* effect emphasizes the importance of manufacturing as *basic* employment. Such employment is that which generates wealth in an area by serving the needs of people who live beyond the margins of the area. *Nonbasic* employment, on the other hand, is that which does not bring in additional wealth from outside because it is concerned primarily with serving the population of the area itself. Thus, it is an area's basic employment that is most important to long-term growth; only by bringing in wealth from outside can more activities, basic and nonbasic, be supported. Although manufacturing may be the most important of the basic industries in many cities, other activities can also be basic. For example, if a university attracts most of its student body from places beyond the bounds of the community in which it is located, the faculty have basic occupations. Police, retail sales personnel, public school teachers, and local bankers are among the many in the community who have nonbasic occupations.

A substantial part of the total employment in both countries, thus, is related to manufacturing either directly or indirectly. Furthermore, most cities were founded and experienced their major periods of growth when manufacturing was easily the primary factor in urban growth. The fundamental urban pattern, the pattern that re-flects where most people in the two countries currently live, can be viewed as a direct result of industrialization.

Substantial regional specialization in manufacturing exists in both countries. This is partly the result of variations in the availability of industrial raw materials. No one should be surprised that North Carolina is a leader in cigarette manufacturing, for the state is a major producer of tobacco, or that Quebec, with vast forests in its interior, leads both countries in pulp and paper production. The industrial pattern is also the result of industrial linkages; manufacturing concerns that produce component parts of some final product are located near each other as well as near the final assembly site to minimize total movement costs. The many manufacturers of automobile parts in northern Ohio and southern Michigan are a classic example.

Other important sources of variation may be differences in labor availability or labor skills, the quality of transportation facilities, and local political attitudes. Regions tend to specialize in the production of whatever it is that they can best produce. And with this regional specialization has come regional interdependence; few sections of either country are truly self-sufficient in manufacturing, in spite of what local pride might lead us to believe.

High Mobility

America's extensive transportation network is an important element in its high level of economic interaction. Goods and people move freely within and between most sections of both countries. Regional interdependence is great; it is made possible by these interregional flows. Relative isolation is uncommon, but it does exist; in a few cases, it may be important in a region's economy (e.g., see Chapter 7).

The populations of both Canada and the United States have high residential mobility. Nearly all Americans move more than once during their lives. On the average, nearly 20 percent change their residence in any one year. Although much of this residential migration is local

Sun City West, located near Phoenix, Arizona, has a current population of 11,000 and is projected to reach 50,000 by the end of the century. It is basically a retirement community. Recent urban expansion in both countries often ignores the grid street system that dominated American cities from the early nineteenth to mid-twentieth centuries. (Courtesy Del E. Webb Development Company.)

in nature, it does result in substantial interregional population movement. A decision to migrate is a function of the balance between negative characteistics at the origin (push factors) and positive characteristics at the destination (pull factors). The easier it is to move, the more likely this push-pull balance will be the primary influence on the migration decision.

Migration has traditionally been toward areas perceived to contain greater economic opportunity. Until the last decade of the nineteenth century in the United States and the 1910s in Canada, there was a strong westward population shift toward frontier agricultural lands. The focus of opportunity then changed and migra-

tion became oriented overwhelmingly toward urban areas. More recently, particularly in the United States, the national economy has entered what some call a *postindustrial* phase; employment growth is primarily in professions and services rather than *primary* (extractive) or *secondary* (manufacturing) sectors. Such employment is much more flexible in its location, and there has been a more rapid growth in such employment in areas that appear to contain greater amenities. This has led to rapid growth of population in the so-called Sunbelt states of the South and West and slow population growth or population decline in the Snowbelt states of the upper Midwest and Northeast (Figure 1-2).

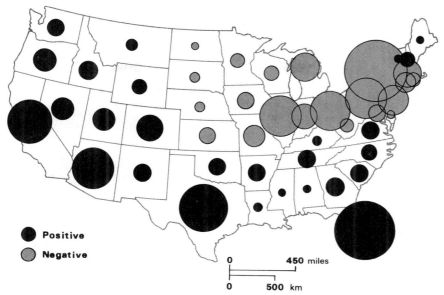

FIGURE 1-2

ESTIMATED RELATIVE POPULATION CHANGE, 1979–1989 (from R.J. Kopec). The total areas of the positive and negative circles are equal. The size of each circle is proportional to the degee of relative change for that state. Population in the South and West is expected to grow for the rest of this century, whereas that of the Northeast and Middle West will decline.

Resources: Abundance and Dependence

The United States and Canada are among the world's most important suppliers of raw materials of many types. Foodstuffs are the key raw material export of both countries. Over half of all world food exports of the 1981–82 crop year were from the United States, including 52 million tons of corn, 48 million tons of wheat, and 25 million tons of soybeans. Total U.S. grain and soybean exports rose from 56 million tons in 1970–71 to an estimated 131 million tons in 1981–82. The United States exports a majority of its wheat crop, 40 percent of its soybeans, and over one-quarter of its corn. The United States is the leading international exporter of most of the world's major grains, including wheat, corn, and rice. It also leads in the export of such products as soybeans, tobacco, and cotton. Agricultural sales accounted for almost one-fifth of all U.S. exports in 1981. Canada ranks a distant

second to the United States in the tonnage of food exports, most of it wheat.

In 1981 the United States had a total trade deficit of $28 billion. This resulted from a net deficit of $50 billion from the sale and purchase of manufactured goods and minerals and a surplus of $22 billion from the sale and purchase of agricultural products. The United States during that period purchased a total of $19 billion in agricultural products on the international market and sold some $41 billion in return. Both Canada and the United States grow enough food in a normal year, measured in terms of calories, to feed more than twice their combined populations. About 25 percent of the total land in row crops in the United States produces export crops; in Canada the percentage is even higher. The surplus moves onto the international market.

Canada, in particular, is also an important exporter of a variety of other raw or semimanufac-

The massive Ford Motor Company assembly plant at River Rouge in Detroit employs many thousands of workers. The complexity and intensity of manufacturing is a basic element in the economic character of both countries. (Ford Motor Company.)

tured materials, such as petroleum, pulp, lumber, iron ore, natural gas, copper, and nickel. The United States is easily the most important buyer for most of these raw materials, although other countries also purchase large amounts. For example, Japan buys much of the forest products exported from the west coast of Canada. The United States is able to satisfy much (but not all) of its gigantic demand for industrial raw materials with domestic sources, an indication of its large production of raw materials. The United States does have the potential to be a major supplier for a few nonagricultural raw materials in-

ternationally and is the world's leading exporter of coal.

Although the populations of both countries are predominantly urban, the extraction of natural resources from this abundant base does require a large nonurban labor force. Furthermore, particularly for agriculture, the development of these resources often involves a substantial land area. As a result, the relationship between the physical environment and human adaptations to that environment are clearly visible in many of the landscapes of both countries. Governments play an important role in this relationship by establishing controls on land use and agricultural production and by regulating the development of many resources.

High Income and High Consumption

Canada and the United States are rich countries. Their populations earn a great deal of money, and they spend it. The demands of the populations and of the economies themselves generate massive consumption.

There are substantial regional variations in the income within both countries (a topic that will be treated in several chapters), but as national units, both countries have per capita incomes that rank among the highest in the world. Income levels in the United States are somewhat higher than those in Canada; indeed, the average U.S. citizen does have a somewhat higher standard of living than the typical Canadian. Still, the difference is often between two or three cars or one or two televisions.

The economies of both countries have grown

increasingly energy dependent. Their high national incomes are achieved through high levels of worker productivity. High productivity levels require a greater use of machines. And modern machines are fueled by inanimate energy sources. High mobility also implies a high use of energy resources. High income spread somewhat evenly among a large share of the population will generate high product demand. All this increases energy consumption. North Americans presently consume nearly a third of the world's total energy production, and the absolute amount used continues to increase.

As mentioned, the United States also has grown increasingly dependent on imported resources, with Canada a principal supplier. The United States now imports more than one-third the petroleum it consumes, an increasing share of its iron ore and natural gas needs, nearly all of its tin and aluminum, and large quantities of many other mineral ores. There are few industrial raw materials in which the country can still claim self-

The Chesapeake and Ohio railyards and associated coal docks at Newport News, Virginia. Coal is brought here by rail from the coalfields of the central Appalachians and then sent by ship to points along the east coast of the United States and throughout the world. (Daily Press-Times Herald Photo/Newport News, Virginia.)

sufficiency. Coal is the most notable exception. Although Canada is less dependent on foreign sources, it too purchases from other countries materials not available domestically plus a substantial minority of its needed petroleum resources. Canada will become self-sufficient in petroleum when its domestic resources are better developed and when it completes a pipeline to move surplus western supplies to the eastern part of the country.

High income also affects the diets of Canadians and Americans. They eat far more meat products and have a substantially more varied diet than most of the world's population. Beef and dairy production are, therefore, especially important in the agricultural economy. Al-

though both countries are major food exporters, both must turn to foreign suppliers to obtain products not grown at home. This is especially true of Canada, whose climate does not allow production of the variety of foodstuffs that can be grown in the United States.

The high consumption that goes along with high incomes has a number of important manifestations. Travelers are bombarded with a constant series of ads, signs, carefully designed buildings, and the like, all of which urge them to part with at least a small portion of their money. The resources of the two countries are consumed at a rapid rate in the attempt to satisfy their populations' appetites for manufactured goods and specialized foods. There re-

FIGURE 1-3

MEDIAN HOUSEHOLD INCOME, 1979. Higher incomes in the Northeast and much of the far West (especially in suburban counties) are counterbalanced by lower incomes across much of the South and portions of the country's midsection.

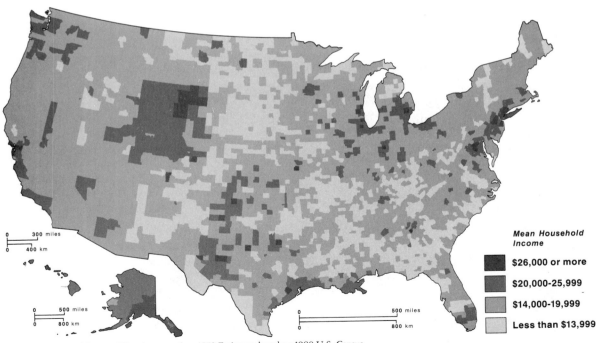

Mean Household Income

- $26,000 or more
- $20,000-25,999
- $14,000-19,999
- Less than $13,999

Source: International Data and Development, Inc. 1979 Estimates based on 1980 U.S. Census.

cently has been an increased recognition and concern that continental reserves of many materials are definitely limited. The implications of high consumption in a finite environment and of high consumption's natural partnership with high income will be among the most important, and most difficult, issues facing the United States and Canada during the next few decades.

In spite of this high overall income, problems of poverty remain important to both countries (Figure 1-3). Many rural areas of French Canada and parts of the rural Atlantic Provinces have area income averages that are little more than half that found in Toronto or Vancouver. Much of the rural U.S. South, Mexican-American and American Indian areas of the Southwest, sections of Appalachia, and nonwhite ghettos in most American cities have populations with average incomes far below the national average. The persistence of these poverty areas is a major problem for the two countries because the poverty represents both a real economic loss to the national economies and a large number of people who are not sharing in the abundance of their countries.

Environmental Impact

One consequence of the combination of high consumption with resource abundance and dependence has been strong disruption of the physical environment. Resources seldom can be removed from the natural landscape without some visible (or invisible but still important) impact. The increased severity of such environmental impacts has enlivened the age-old argument between development and conservation. The argument recently has stimulated greater governmental intervention in both processes as federal and state/provincial governments attempt to establish a middle ground between these two divergent viewpoints.

The governmental problem frequently is described as one of establishing regulations that provide some appropriate and meaningful level of environmental protection without raising extraction costs so much that the development of

a necessary resource is impossible. As domestic resources become increasingly scarce and their costs of extraction and production increase, the importance of this conflict also will grow. The entire field of environmental impact promises to be one of the major regulatory battlegrounds during the next several decades.

Political Complexity

Both the United States and Canada have complex political structures, with *jurisdiction* over an activity or territory divided among many different decision-making bodies, some elected and some appointed. At the national level, Canada has a parliamentary government after the British model, whereas the United States separates its executive and legislative branches. The Canadian federal system provides for substantially more provincial independence than the U.S. Constitution allows individual states. The current demand by Quebec for greater autonomy within Canada would be impossible to accept in the United States, but not impossible in Canada.

Below the state and provincial level, the two countries have rather similar political structures, and their complexity presents a major problem in the effective and efficient distribution of governmental services. Counties, townships, cities, and towns are all governed by their own elected officials. Many special administrative units, some appointed and some elected, oversee the provision of specific services, such as education, public transportation, and water supply. The resulting administrative pattern is often nearly impossible to comprehend because there may be many overlapping jurisdictions that provide one service or another in a given area.

One problem is that a jurisdictional unit, once created, seldom disappears and seldom is willing to surrender any of its powers. It is rare for political units to join their efforts and form administrative organizations under a single new structure as populations grow and merge spatially. Thus, certain service inefficiencies increase over time, but the spatial pattern of administration remains the same (see Figure 4-9).

Americans value what they perceive as their rural or small town background. The essence of that perception is captured by small New England villages such as East Cornith, Vermont. (Vermont Travel Division.)

Variety of Cultural Origins

Finally, both Canada and the United States have grown from diverse cultural backgrounds. Much of this diversity remains evident within the two countries. The most obvious example of the differences that can continue as a result of varied backgrounds is the distinction between British and French cultural areas in Canada. Canada is officially bilingual, and the government of the country follows a program that stresses this bicultural development.

There are a number of obvious cultural differences within the United States. One example is the important contribution of blacks to the national culture and particularly to that of the South. Another example is the distinctive culture region that has developed in the Southwest with an admixture of Hispanic Americans, American Indians, and Anglos. Another might be the contribution of Chinese in such cities as San Francisco and New York. Several other examples will be treated at some length in the text, for this cultural diversity is an important element in the distinctive character of many places in both countries.

American landscapes are often a creation of the culture of the people who live there rather than the physical environment. The houses strung out along the main road, the large size of the church, and the long, narrow farm lots all help identify this as a village in Quebec. (George Hunter/Miller Services.)

REGIONS

One notable geography text, written decades ago, began with the observation that, "Hell is hot." This was not a veiled threat to those students who might treat the material in the book too lightly. Nor was it the author's suggestion that the book would have a strong religious tone. Rather, he was making a significant, yet obvious, comment about one of his culture's more widely accepted regions. Although substantial differences of opinion existed concerning its exact location, and many argued that it did not exist at all, most could agree on its general position. There were many different evaluations of the nature of its environment, although nearly everyone accepted the idea that intense heat was a necessary element. In this case, all the available information about the region came not from individual experience but from written materials, the words of others, folktales, and beliefs. Even though they personally had not experienced the place, most people were willing to state firmly their evaluation of the general nature of

13

that region and of the quality of the experience to be had there. In short, it was a region in much the same way that the sections of the United States and Canada that we are discussing are regions.

A *region,* as we use the term, is very much a mental construct. One person may choose to divide an area into one set of regions; another may regionalize the same area in a different way. In part, this may be due to different goals or different information. A district sales manager who wishes to determine the precise borders of the sales regions of the people working for him or her would probably create a pattern quite unlike that of a botanist interested in regional variations in tree types. Similarly, a regionalization based on agricultural patterns might bear little resemblance to one based on manufacturing activities. One set of geographers, when subdividing the United States and Canada into regions, may well create a map noticeably different from that formed by a second group of geographers simply because each group selected different elements as most influential to their patterns.

These various reasons for regionalizing and the alternative attitudes expressed by the regionalizers obviously can lead to the creation of many different maps of any given territory. No one regional scheme is necessarily "better" than another. If it meets the requirements of its creators, if those requirements were thought out reasonably, and if the map is executed correctly and factually, then the regionalization must be perfectly acceptable. Geographers use regions as a neat system of categorization, a way of organizing a large or complex set of facts about places into a more compact meaningful set of areas. As with any categorization, the regions are satisfactory if they identify understandable patterns in the facts and if they help clarify the complex patterns.

To geographers, a region can be either *nodal* or *uniform, single feature* or *multifeature.* A nodal region is characterized by a set of places connected to another place—a focus (or node)—by lines of communication or movement. The places that make up this region, therefore, are associated with each other because they share a common focus. The circulation area of a newspaper is an example of such a region, as is a "milkshed"—the set of places surrounding a city that send their raw milk production to the markets of that city. Each of these separate places in the region may be quite diverse in appearance, but because they are tied together, they constitute a single nodal region.

In comparison, a uniform region is a territory with one or more features present throughout the area and absent or unimportant (or much diminished) beyond the territory. The entire area where the cottonwood constitutes a significant part of the total tree cover may be such a single-featured, uniform region. Or a uniform region may represent some characterization of the total environment of a territory, including both its physical and cultural features. These quite general and frequently subjective regions are meant to reflect distinctive mixtures of all elements of the landscape. It is this last type of region that we use for the general structure of this book.

As we said earlier, there is an almost endless variety of elements that could be considered in the creation of these regions. In fact, however, our perception of the nature of a region, of the things that together shape its personality, is based on a relatively small group of criteria. In each major section of the United States and Canada, we have tried to identify the one or two underlying themes that reflect ways in which the human population has interacted (with itself or with the physical environment) to create distinctive regions. The most important identifying themes for a region may vary greatly from one region to another. It is impossible to speak of the Southwest of the United States without a focus on aridity and water erosion, of the North without its cold winters, or of the Northeast without cities and manufacturing. The key element that establishes a total uniform region, then, is not how that section compares with others on a predetermined set of variables, but how a certain set of conditions blend there. It is not possible to create such regions by simply amassing a large

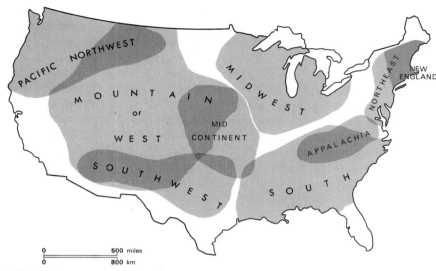

FIGURE 1-4

PERCEIVED REGIONS OF THE UNITED STATES. Is this set of perceptions
from students in Ohio likely to match patterns arising from groups in other parts
of the country? How might they differ?

set of statistics and then grouping areas and places
on the basis of relatively similar numbers, proportions, or averages. To follow such an approach ignores the personality of a place. It is
that personality that gives meaning to the regions we use in the text.

Students in a geography class at Ohio State
University were asked to identify those parts of
the United States that had meaning to them as
regions and to locate these regions on a map
(Figure 1-4). What emerged most clearly was that
certain sections of the country were well defined
in the minds of most of the students, with the
result that these sections were named and located consistently by the students. Such sections
as the Southwest, New England, and the South
were almost always indicated, and their borders,
broadly stated, were similar. Other parts of the
country, such as the Middle Atlantic states, were
often excluded from the regionalization entirely.

With minor variations, this pattern of identification would be repeated by individuals from

most areas of the country. Such places as New
England and the South are regions with firm
personalities, ones that most Americans recognize and can at least loosely define. As with Hell,
the South exists with remarkably consistent
boundaries for almost all Americans, even though
opinions about its character may vary widely.
Similarly, to most Canadians, British Columbia
and the Maritimes conjure up definite impressions, although residents of Nova Scotia and
Ontario may differ outrageously in the specifics
of these impressions.

An elaboration of the idea of a uniform region suggests that every such region is composed of several distinct zones. Somewhere within
the region there is a *core area* in which the major
identifying features are most obvious. This is
usually a well-accepted area that is clearly a part
of the region, even by the most stringent application of the selection criteria. Somewhere toward
the margins of the region, these identifying features grow fuzzy or less distinct; hence, we grow
less sure of the appropriateness of, or have

15

greater disagreement over, the specifics of our regional assignment. For example, although nearly everyone would accept central Alabama and Mississippi as a part of the South, opinion varies as to what to do with southern Illinois or the Louisville area or much of northern Virginia. Boundaries between multifeatured regions are seldom sharp or well defined on the landscape. Regions blend into each other, and travelers are often well into one region before realizing that they have left another region. The boundary between such regions is usually a transition zone, not a sharp line.

It is, therefore, often difficult to map the boundaries of a multifeature region. If we were to view regions simply as a form of categorization, we might convince ourselves that we must create regions that occupy mutually exclusive portions of the earth's surface and are represented equivalently by all of the data for that portion—that is, every place would be a part of one and only one region and would be characterized in the same way as all others. We feel that the arbitrary nature of such an approach is inappropriate and sometimes misleading.

Within this book, regions have been presented largely as though they are distinct territorially, even though they are not. The "feeling" of a region we wish to present is a function of place, but it is also (and perhaps more important) a function of the subject theme chosen. Therefore, for example, the intense urban character of Megalopolis is discussed in Chapter 4, but the aspects of manufacturing that affect New York, Philadelphia, and Boston and other manufacturing core cities are presented in Chapter 5. There are two important aspects of regional "feeling" in the region usually called the Midwest—the urban-industrial and the rural-agricultural. Both are important enough for us to treat each separately in some detail. Therefore, students who wish to have a complete picture of North American geography should be aware of the two major themes of the Midwest, the two aspects of Megalopolis, the three or four pertinent phases of the West, and so on.

Rigid regional boundaries do not fit the landscape of the United States and Canada. A given portion of the continent's territory may be occupied by parts of two or more regions, but the boundaries of many regions may also be fairly broad transitional zones. These zones are areas that contain many of a region's characteristics, even though their diminished strength indicates the area is beyond the edge of the region itself. At times, these zones mark an area where the mix of characteristics is so subtle or complex that it is difficult to assign the area to any one region. Parts of the margin between the Deep North and the Great Plains are examples of this, as are sections of the transition between the Deep North and the Deep South. Therefore, students should be forewarned (1) that the regions we discuss in this text are complex, (2) that because of this complexity, the regions are not exclusive territories—they may overlap other regions, and (3) that the outer edges of the regions blend gradually into the neighboring areas and should not be viewed as hard-and-fast lines.

Regional boundaries and regions themselves are not static. Settlement patterns shift, society develops signifcant new technological abilities, and political patterns are altered; regions reflecting these patterns may expand, contract, appear, or disappear. A regionalization of North America for 1492 would have been quite different from one for 1776, 1865, or 1985. There is no reason to believe that the pattern for 2085 will be at all similar to that for 1985.

An examination of the regions that we have created for this text indicates a subdivision that should be generally recognizable (Figure 1-5, see pages 18 and 19). Again, this is because many of the regions of the two countries are widely accepted. Other regions represent combinations that are normally not expected. For example, consider the Bypassed East, a combination of the Atlantic Provinces of Canada and northern New England. Most casual observers firmly lump all of New England into one region. This is largely a historical carry over, reflecting the long-term identification of the states of New England as a separate region with strong cultural cohesion. But there have been great changes in southern New England in recent decades because of heavy immigration and urbanization. Still, the popular image of the New England Yankee and small towns with village greens has not changed.

Several of the regions closely follow political boundaries. The reason for this in Hawaii is obvious. California is separated from most of its adjacent landscape because of its leadership role in changing the culture of America and its state-wide political "solutions" to its local resource problems. Megalopolis has been defined traditionally along county lines, and we choose to follow that approach. It does serve to make the collection of statistics about these regions much easier, for nearly all information bases follow standard political boundaries.

Finally, it is significant that only one of the regional boundaries follows the international border between the two countries. Few international borders are as loose as this. Communication and transportation across the border is frequent, and economic and settlement changes influence both sides of the boundary. It should not be supposed, however, that the Canadian-American border has no meaning. Although the Great Plains region extends easily into both countries, differences in early settlement, governmental land policies, and general economic conditions result in noticeable differences in the plains landscape in the two countries. Canada and the United States are remarkably cooperative, and their settlement histories are similar, but there are many differences between them.

As we mentioned, each of the regional chapters is developed around one or a few basic themes (Table 1-1, see page 20). Most of these regional themes are drawn at least indirectly from the basic themes of the entire book. In certain regions, the expression of some themes is stronger or clearer than others. The themes are intended to provide an explicit basis for treatment of information about the region, although, in many chapters, it will not be difficult to identify elements of national or continental geography.

Additional Readings

Callwood, June. *Portrait of Canada*. Garden City, N.Y.: Doubleday, 1981.

Canada. Surveys and Mapping Branch, Geography Division. *National Atlas of Canada*, 4th ed. Toronto: Macmillan, 1974.

Doster, Alexis, et al. (eds.). *The American Land*. Washington, D.C.: Smithsonian Institution, 1979.

Hart, John Fraser, (ed.). *Regions of the United States*. New York: Harper & Row, 1972.

Jordan, Terry J. "Perceptual Regions of Texas." *Geographical Review*, 68 (1978), p. 293.

McCann, L. D., (ed.). *Heartland and Hinterland: A Geography of Canada*. Scarborough, Ont.: Prentice-Hall Canada, 1982.

Peirce, Neal R., and Jerry Hagstrom. *The Book of America: Inside 50 States Today*. New York: W. W. Norton, 1983.

Reed, John S. "The Heart of Dixie: An Essay in Folk Geography." *Social Forces*, 54 (1978), pp. 925–939.

Robinson, J. Lewis. *Concepts and Themes in the Regional Geography of Canada*. Vancouver: Talon Books, 1983.

U.S. Geological Survey. *National Atlas of the United States of America*. Washington, D.C.: U.S. Geological Survey, 1970.

Watson, J. Wreford. *The United States: Habitation of Hope*. New York: Longman, 1982.

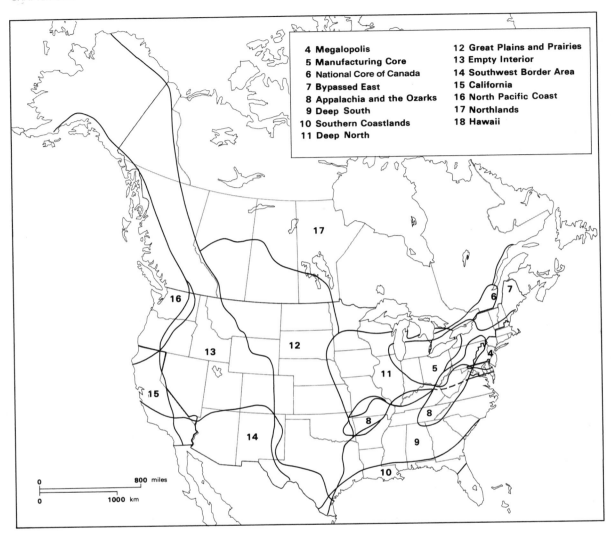

FIGURE 1-5
REGIONS OF THE UNITED STATES AND CANADA. Separated by theme but
overlapping territorially, the regions used in this book reflect the complex variations that characterize the geography of these two countries.

TABLE 1-1
Regional Themes

Chapter	Chapter Title	Themes
4	Megalopolis	Urban growth and sprawl, urban problems
5	North American Manufacturing Core	Industrialization, how regions change
6	The National Core Region of Canada	Federalism, cultural pluralism
7	The Bypassed East	Relative isolation, regional economic problems
8	Appalachia and the Ozarks	Resource dependence, governmental intervention
9	The Deep South	Culture
10	The Southern Coastlands	Amenities and economic resources, development conflict
11	The Deep North	Agriculture's cultural and economic foundations
12	The Great Plains and Prairies	Environmental perception, human/land interaction
13	The Empty Interior	Settlement in a difficult environment (environmental impact)
14	The Southwest Border Area	Ethnic diversity
15	California	Perception of amenities, population redistribution, environmental dependence
16	North Pacific Coast	Regional independence
17	Northlands	Settlement in a difficult environment, resource extraction, governmental intervention
18	Hawaii	Ethnic diversity, isolation, environmental impact

CHAPTER TWO

GEOGRAPHIC PATTERNS OF THE PHYSICAL ENVIRONMENT

Some landscape characteristics are most easily discussed when they are used to highlight particular regional themes. Such characteristics are often reflected in the regional thematic organization of this text. For example, the basic pattern of manufacturing in the United States and Canada is discussed in Chapter 5 and that of religion is identified in Chapter 3. Some subjects are developed in several contexts. Ethnic diversity, for example, is a theme in Chapters 5, 6, 10, 14, and 18; agriculture is discussed in almost all of the chapters.

Nevertheless, several important geographic patterns are not a focus of any single region. They are so broad that they must be dealt with independently of the regional structure. Most elements of the physical environment—topography, climate, and vegetation—are important almost everywhere, although there are regional variations even in this. Because our regions are expressions of human landscape patterns, we consider the physical environment only one of several factors that may help explain the char-

acter of human geographic distributions in the United States and Canada. Even though the physical environment may be the most important element in some regions, we have chosen to discuss its broad patterns in this chapter.

THE PHYSICAL ENVIRONMENT

Topography

Compared to a continent such as Europe, where a traveler can encounter very different landforms within a small area, the major topographic regions of Canada and the United States tend to be large (Figure 2-1). If you select your route correctly, it is possible to drive for several days through areas with little variation in landforms.

The major topographic features of the continent tend to extend north-south across the two countries. The interior of the continent is a vast, sprawling lowland that stretches from the Gulf of Mexico to the Arctic. At times close to 5000 kilometers (3000 miles) wide, this lowland en-

FIGURE 2-1
PHYSIOGRAPHIC REGIONS (adapted from Fenneman).

compasses well over half the entire land area of the continent. Geomorphologists (scientists with an interest in landform development) place this expanse of flat land and gently rolling hills in three different physiographic regions—the Atlantic and Gulf coastal plains, the interior lowland, and the Canadian (or Laurentian) Shield. Some split the interior lowland into two parts, the Great Plains and the interior plains. These regions may be best differentiated by the processes that led to their appearance and by their subsurface (bedrock) materials. However, there are only moderate differences in the visual appearance of the landforms in each of these three major divisions. More significant in differentiating the regions, at least to the untrained eye, are variations in climate and in the cultural landscape.

The Atlantic and Gulf coastal plains reach north along the east coast of the United States as far as the southern margins of New England. Underlying this area are beds of young, soft, easily eroded rock deposited in recent geologic time as shallow seas lapped back and forth across the land. These low plains actually extend well out under the current ocean surface to form a Continental Shelf, which in places extends as much as 400 kilometers (250 miles) beyond the shore.

The coastal plains are generally flat and often swampy. For example, a geomorphology field trip, conducted annually near Baton Rouge, Louisiana, visits a hillside where the students can observe the boundary between one rock bed and the one above, a difference in elevation of little more than 1 meter (3 feet). In many locations across the coastal plains, farm fields are bounded by drainage ditches 5 or more feet deep to handle runoff from heavy rains and to lower the water table. These are examples of a very flat environment.

Northward, the interior of the continent is occupied by the central, or interior, lowland. Although noticeably hillier than the coastal plains, there is almost no rough terrain in the lowland. This region is like a saucer, turned up at the edges and covered with a deep series of sedimentary rocks. The rocks are older and more durable, can withstand erosion more effectively, and hence contribute to a more varied topography than those of the coastal plains. These sedimentary beds are generally quite flat and most topographic variation in the interior lowland is the result of local erosion or, in the north, of glacial debris deposited during the Ice Age. Still, to observers accustomed to a mountainous landscape, the overwhelming topographic impression of the lowlands is a bland, slightly modified flatness.

The geologic structure of the Great Plains differs little from that of the interior plains. The sedimentary beds dominate and are nearly horizontal, although in the north, they are broken by some eroded domes, most notably the Black Hills of western South Dakota. The boundary between the Great Plains and the interior plains is marked by a series of low escarpments that indicate the eastern edge of the sediments of the Plains. The sediment cover creates a gentle eastward slope from an elevation of 1750 meters (5500 feet) at the foot of the Rockies to 650 meters (2000 feet) at the section's eastern edge. One result of the regularity of this gradient in the U.S. Great Plains is a parallel west-to-east stream drainage pattern.

The character of this massive lowland area has had a number of important influences on the economic and settlement history of the United States. In addition to the vast agricultural potential it provides, fully half of the country can be crossed without encountering any significant topographic barriers to movement. This has facilitated the integration of both this region and the distant West into the economic fabric of the country. Nearly all of the interior lowland is drained by the Mississippi River or its tributaries. This drainage pattern also assisted regional integration by providing a transport and economic focus for the land west of the Appalachians. To many of its residents, this region is the heart of America—vast, economically powerful, productive, separated from the outside influences felt to be so common along the coast. These residents may overstate their case, but it is hard to deny that this is the region that most clearly represents many of the popular images of America.

North and northeast of the central lowland lies the Canadian Shield. At over 2.8 million square kilometers (1 million square miles), it is the two countries' second largest physiographic region. Here, old, hard crystalline rocks lie at the surface. Father south in the lowlands, similar rocks are covered by the sedimentary beds deposited under the sea that once filled the midsection of the continent. A long history of erosion has worn down the surface of the Shield into a lowland of small local relief.

The Shield, more than any other North American physiographic region, has had its landforms remolded and shaped by massive continental glaciers during the last 1 million years. Vast sheets of ice, probably originating on either side of Hudson Bay, have repeatedly crept outward before melting back. At their greatest extent, the massive sheets of ice covered nearly 13 million square kilometers (5 million square miles), two-thirds of the total land area of the continent. These continental glaciers covered most of Canada and Alaska east of the Rocky Mountains and the Coast Ranges, and they reached southward to approximately the present valleys of the Mississippi and Ohio rivers.

Major landscape changes were caused by this moving ice cover. The ice could pluck rocks weighing many tons off the surface and carry them great distances. Preexisting drainage patterns meant little to the glaciers, and old patterns were disrupted greatly by the passage of the ice. The massive weight of the ice, up to 1.6 kilometers (1 mile) thick in places, gradually depressed the land on which it rested. Much of the Shield is still "rebounding," slowly rising in response to the removal of the ice. Ice melt along the peripheries of the glaciers created major rivers and cut broad new pathways to the sea.

Glaciation scoured much of the Shield's surface. Today the soil cover of the region remains thin or nonexistent. The heavily disrupted drainage pattern dammed many streams with debris and led others into the area's labyrinth of lakes and swamps rather than to the sea. Central and northern Minnesota, for example, called the "Land of 10,000 Lakes," is part of the southern lobe of the glaciated Shield that extends into Minnesota, Michigan, and Wisconsin. Massive boulders called *erratics*, often many feet in diameter, are strewn across the landscape, resting where they were dropped by the glaciers.

The landscape impact of continental glaciation is also found far south of the Shield. Southward, where the ice was not as thick and its force correspondingly less, the glaciers were diverted or channeled by higher elevations. For example, the ice was blocked in central New York by the highlands south of the Mohawk River. Narrow probes, however, did push up the valleys of streams tributary to the Mohawk, gradually broadening and deepening them. As the ice began to melt around the glaciers' peripheries faster than it was replaced by new ice, the glaciers "retreated" northward. They rested for a long period where they had piled up against the highlands before sufficient melt occurred to pull the ice away from the margins of the hills. During this period, as the ice melted, the debris that the glaciers had collected was deposited along their margins. These deposits created natural dams across the tributary streams. Today the deep, narrow Finger Lakes of New York State fill these glacially enlarged valleys and form one of the country's truly beautiful landscapes.

All along and beyond the southern edges of the glaciers, deposition replaced erosion as the prime result of glaciation. Large areas of the interior lowland are covered by a mantle of *glacial till* (rocks and soil dropped by the glaciers). The till covers the land to depths varying from a meter or less to more than 100 meters (330 feet). Where the glaciers remained stationary for long periods of time, higher hills, called *moraines*, were created. In the East, Staten Island, Long Island, Martha's Vineyard, Nantucket, and southern Cape Cod are end moraines that mark the farthest major extension of glaciers toward the southeast. The landscape south of the Great Lakes is laced with long, low, semicircular moraine ridges and other glacial deposits.

As the ice retreated, massive lakes were created along the glacial margins as the ice blocked normal drainage patterns. On the northern Great Plains two huge lakes, Agassiz and Regina, together covered an area larger than today's Great Lakes. With continued glacial retreat, these lakes mostly disappeared. Their existence is now marked by the former lake bed, a flat area covering parts of Manitoba, Saskatchewan, and North Dakota. The St. Lawrence River, the outlet for the Great Lakes, was dammed for a time by the ice; the flow from the lakes was diverted southward into the Mississippi and Susquehanna rivers.

This is a composite satellite photograph of the U.S. coast from New York City to Norfolk, Virginia. The excellent harbor facilities, seen here in Chesapeake Bay, Delaware Bay, and New York harbor, have played an important role in the economic growth and urbanization of this region. (NASA.)

Sea level was significantly lower during periods of widespread glaciation. This lowered the *base level* of many rivers and, thus, fostered increased erosion by those streams. Furthermore, many of these stream valleys extended well into what is today the ocean because of the lower sea level. Along with many others, the Susquehanna and Hudson rivers cut much deeper valleys during this period. As the ice retreated and sea level rose, the ocean filled these deepened valleys. Two of the world's finest harbor areas were formed in this way: New York Bay with the deep Hudson River and the protective barriers formed by Staten Island and Long Island; and Chesapeake Bay, the drowned valley of the Susquehanna River and some of its major former tributaries, such as the Potomac and the James rivers.

In the East, the coastal plains are gradually squeezed against the coast northward along the ocean by the Appalachian Highlands until the lowland disappears entirely at Cape Cod. From there northeastward to Newfoundland, the coastal landscape is a part of the northern extension of the Appalachian system. The Appalachians separate the seaboard from the interior lowlands along much of the eastern United States. These are old, eroded remnants of what were once much higher mountain ranges. Today they are less than half the height of their counterparts in the interior west. Mt. Mitchell, in North Carolina, the tallest peak in the east, is 2036 meters (6684 feet) high.

These mountains are important, despite their relatively low elevations. The two regions dominated by these peaks and ridges, the Bypassed East and Appalachia, have had a settlement and economic history profoundly affected by the mountains. Soils in most parts of the Appalachians and the Bypassed East are shallow, and as Robert Frost once noted, rocks are often a farmer's most dependable crop. The steep slopes, difficult to farm under any circumstances, are totally unsuited to modern agricultural practices that emphasize mechanization. Large-scale urban or industrial growth is cramped by the small local lowlands. Early settlers found the Appalachians from the Mohawk River in New York southward to northern Alabama to be a surprisingly effective barrier to western movement; there are few breaks in the mountains' continuity. Inadequate transportation remains a major regional problem.

The Appalachians, although generally rugged throughout, do contain several noticeably different kinds of landforms. South of New England, the eastern Appalachians are an intricate pattern of folded linear ridges and valleys. In the west, the mountains offer a jumbled landscape that was created as many streams cut into the region's level-bedded sedimentary rocks. In Newfoundland and the Atlantic Provinces, massive bubbles of *igneous* rock created huge *domes* that have been heavily eroded by water and ice. These domes are most obvious in the highlands of central New Brunswick and on Cape Breton Island in Nova Scotia. New England's landscape appears to represent a little of both patterns, with perhaps a dash more of the combination from the southern Appalachians. These variations will be discussed more thoroughly in the appropriate regional chapters.

If the physical landscape of North America from the Great Plains eastward can be characterized as a gently rolling terrain punctuated by mountains in a few places, the West is a land of mountains and of sudden, great local changes in elevation. Few parts of the West are beyond sight of mountains. Instead of punctuating flatness, the mountains form and mold the landscape, dominating all of its other elements. The physiography again is arranged in a series of three large north-south trending bands, with the Rocky Mountains on the east separated from the mountains and valleys of the Pacific coastlands by a series of high, heavily dissected plateaus. On the map, these landscape regions create the impression of great homogeneity. In fact, there is great visual diversity in this section of the continent.

Starting in the east, the Rocky Mountains generally present a massive face to the Great Plains, with peaks occasionally rising a mile or more above the plains. The Front Range of central Colorado is a fine example of such a sudden change in elevation. Elsewhere, as in south-central Wyoming, the Rockies almost seem not to exist at all. Travelers passing through Wyoming on Interstate 80 must search to find the moun-

The West in both countries is a land of dramatic landscapes. Few places from the Rocky Mountains westward are beyond the sight of mountains. This view is in the northern Cascades of Washington. (National Park Service.)

tains through which they are passing. Really a western peninsula of the Great Plains, the Wyoming Basin has allowed millions of travelers to circumvent the more rugged sections of the Rockies. In the northern Rockies in Idaho, the north-south linearity of most of the region's mountains is replaced by massive igneous domes irregularly eroded into a rugged, extensive series of mountain ranges that contain the largest remaining area of wilderness in the United States outside Alaska. Farther north still, parts of the Canadian Rockies, carved by alpine glaciation during the Pleistocene, contain some of the most beautiful mountain scenery on the continent.

The high plateaus of the interior West are also varied in their origin and appearance. The southernmost subsection, the Colorado Plateau, is a series of thick beds of sedimentary rocks rising more than 1000 meters (3300 feet) above the lowlands' elevation. The basin and range subsection is characterized by interior drainage, with many of the area's usually small or intermittent streams ending their courses in the area's lakes and dry sinks. Much of the region has been covered with a thick mantle of alluvium washed down from the mountains. To the east-west traveler, the area appears to be occupied by a series of narrow mountains and barren lowlands in boring repetition.

The Columbia Basin has been filled by repeated lava flows to a depth of more than 1000 meters (3300 feet). Rivers, both past and present, have eroded into the rock. The resultant landscape is similar to that of the Colorado Plateau, although the stepped appearance of the eroded sedimentary rocks of the Colorado Plateau is missing.

After a brief disappearance along the international border, the plateaus reappear farther north. They gradually widen northward, encompassing the valley of the Yukon River in Alaska. There is a complex of lava plateaus, low mountains, and river plains in British Columbia and the Yukon Territory. In comparison, much of central Alaska is a broad flat, lowland that is poorly drained.

Along the Pacific Coast, the pattern of complexity within apparent simplicity is repeated. In the conterminous United States (i.e., excluding Alaska and Hawaii), the region seems to consist of two mountain chains separated by a discontinuous lowland. In southern California, the Coast Range is fairly massive, with peaks reaching 3000 meters (10,000 feet). From there to the Oregon border, the mountains are low and linear, seldom rising above 1000 meters (3300 feet) in elevation. This also is the major *fault zone* of the state and a region of frequent earthquake activity. Along the California-Oregon border, the Klamath Mountains are higher, more extensive, and much more rugged—much like the Idaho Rockies. Except for the Olympic Mountains in Washington State, the Coast Ranges in the rest

of Oregon and Washington are low and hilly rather than mountainous. In Canada, the Coast Ranges are called the Insular Mountains as they extend along the coast of British Columbia.

The interior lowlands along the coast—the Central Valley of California, the Willamette Valley in Oregon, and the Puget Sound lowland in Washington and British Columbia—are the only extensive lowlands near the coast. Filled with relatively good soils, these lowlands have supported much of the Pacific Coast's agriculture. The Central Valley, especially, has some of the flattest land to be found anywhere in the United States.

East of the lowlands are the Sierra Nevada and the Cascade ranges. The Sierra Nevada appears as though a massive section of earth was tilted upward in what is called a *fault block,* with the sharpest exposed face toward the east. Although the western approaches into the Sierra Nevada are reasonably gentle, on the eastern side, the mountains rise in some places more than 3000 meters (10,000 feet) in just a few linear miles. Volcanic activity was important in the formation of the Cascades. Some of the continent's best known volcanic peaks, such as Mt. Ranier and Mt. St. Helens in Washington, are found there. As the Cascades continue into Canada, they are renamed the Coast Mountains.

Climate

Climatic patterns are a result of the interaction of three geographic controls. The first is *latitude*. The earth is tilted on its axis with reference to the plane of its orbit around the sun. As it makes its annual revolution around the sun, first the Northern Hemisphere, then the Southern Hemisphere, is exposed to the more direct rays of the sun. During the Northern Hemisphere's summer, higher latitude locations have longer days, with far northern points experiencing a period of continuous daylight. Conversely, daylight periods during the winter months are shorter at higher latitudes. Places north of the Arctic Circle receive no daylight at all at midwinter. Thus, latitudinal variation in temperature is minimized during the summer when the heating of the sun's more direct rays at lower latitudes in the United States is partially balanced by the longer period of daylight in northern Canada or Alaska. During winter, more southerly locations have both longer days and exposure to more direct rays than northern places. Canadians and residents of the U.S. Northeast can journey to Florida in January and find temperatures averaging 15° C or 20° C (roughly 30° F or 40° F) warmer than those they left behind. Floridians making the reverse trip in summer would find temperatures only slightly (perhaps 6° C [10° F]) lower than those in Florida. Thus, seasonality is great in the midlatitudes and the high latitudes, and warm to hot summers are widespread.

The second control is based on the relationship between land and water. Land tends to heat and cool more rapidly than water. In a tendency called *continentality*, nonequatorial places far from large bodies of water experience greater seasonal extremes of temperature than do coastal communities. Parts of the northern Great Plains and the Prairie Provinces average annual temperature ranges close to 65° C (150° F); annual differences of as much as 100° C (180° F)—from 50° C (120° F) to −50° C (−60° F)—have been recorded in some locations. The larger the landmass, the greater is this annual range in temperature. North America is a very large landmass.

The converse effect occurs at *maritime* locations, especially on the west coasts of continents in the midlatitudes. These locations have much smaller temperature ranges as a result of what is called a "maritime influence." Vancouver, British Columbia, for example, has an average annual temperature range of only 15° C (27° F). Summer and winter extremes are moderated by the movement onshore of prevailing westerly wind systems from the ocean. Horizontal and vertical ocean currents minimize seasonal variations in the surface temperature of the water. The moderated water temperature then serves to curb temperature extremes in the air mass above the surface.

Proximity to large water bodies also tends to have a positive influence on precipitation levels, with coastal locations receiving generally higher amounts. The reason for this should be obvious; large water bodies provide greater levels of evaporation and, thus, increase the amount of moisture in the atmosphere. That, in turn, increases the possibility of precipitation. There are, however, notable exceptions to this general rule. An example is the dry coast of southern California. Moisture-producing weather systems that develop over the north Pacific and then move southward are blocked during much of the year by a large stable *air mass* that maintains the hot, dry weather of that coastline. The Arctic coastline of Canada and Alaska is even drier. Here the long winters minimize atmospheric humidity. Very cold, stable air dominates the Arctic for long periods, reducing still further the likelihood of snowfall.

The third prime geographic influence on climate is topography. Most obvious is the relationship between elevation and temperature, with higher elevations cooler than lower elevations. The influence of topography can be broader, however, because of its effect on wind flow. If a major mountain chain lies astride a normal wind direction, the mountains force the air to rise and cool. Precipitation falls on the windward side, and the lee is dry. The wettest area in North America is along the Pacific coast from Oregon to southern Alaska, where moisture-laden winds strike mountains along the shore. Average annual precipitation is more than 200 centimeters (80 inches) throughout the area, and in some places exceeds 300 centimeters (120 inches) (Figure 2-2). Mountains also can reduce the moderating effects of maritime conditions on temperature, as happens in the interior of the Pacific Northwest in the United States. Within the United States and Canada, the Western Cordillera (mountain mass) confines West Coast maritime climatic conditions to that coast. Some of the greatest variations in both precipitation and temperature to be found across a small distance anywhere on the continent exists between

The close relationship between climate and vegetation is well illustrated by the altitudinal tree line found especially on the higher mountains of the West. Above the often sharply identified line, climatic conditions are too cold for tree growth. (John Florin.)

the west and east sides of parts of the Coast Ranges. The aridity of the central and northern interior West is due in large part to the barrier effect of the north–south trending mountain ranges of the West.

East of the Rockies, the topographic effect on precipitation eventually disappears, partly because the eastern mountains are much lower and partly because much of the weather of the interior is a result of conflict between two huge air masses that are unimpeded, one flowing northward from the Gulf of Mexico and the other flowing southward out of Canada. The contact of these two different air masses creates what are often violent displays of weather in the region.

This illustrates a fourth major and complex influence on climate, the impact of air mass characteristics and wind systems. The weather of most of the two countries is affected markedly by the confrontation between polar continental air masses (usually cold, dry, stable) and tropical maritime air masses (warm, moist, un-

stable). The former push farthest south in winter, whereas the latter extend farthest north in summer. Most parts of North America are subject to a generally westerly wind flow that tends to move most weather systems eastward. The continental climate of the interior is, thus, pushed onto the east coasts of the two countries. Coastal New England and the Atlantic Provinces, for example, have a climate with much greater average seasonal variations in temperature than do similar latitudinal locations along the west coast of Canada and the United States.

The interaction of all these climatic controls creates a pattern of climatic regionalization that is decidedly different between east and west (Figure 2-3). In the east, the principal element in climatic variation is temperature; in the west, it is precipitation. In the east, the divisions between the climate regions are based largely on the length of the growing season—the period from the average date of the last frost in spring to the first frost in fall—and on the average summer maximum temperature or winter min-

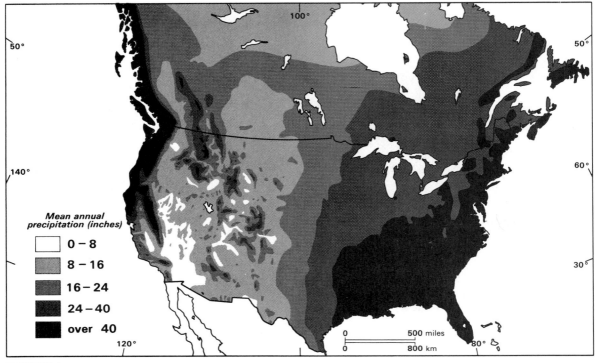

FIGURE 2-2

PRECIPITATION. From the Rocky Mountains westward precipitation is strongly influenced by topography, whereas to the east most precipitation is generated by storm systems. One result is a pattern of dramatic local variation in the West compared with broad areas of similar moisture conditions in the East.

imum temperature. In the west, average annual precipitation is the key, although moderated temperatures are, as mentioned, an important aspect of the marine West Coast climate. To be sure, in the east the more northerly areas are generally drier; in the west they are colder. Still, in the east the key is temperature and in the west it is precipitation. In the east, the major influence on climatic variation is latitude; in the west it is topography. Many climatic margins are indistinct. It is often difficult to tell exactly where one influence is less important and another becomes more important. In general, the western influences reach beyond the Rockies onto the higher, drier western portions of the interior grasslands.

Vegetation

Botanists speak of something called *climax vegetation*. It is defined as the assemblage that would grow and reproduce indefinitely at a place given a stable climate and average conditions of soil and drainage. For most of the inhabited portions of North America today, that concept has little meaning. The "natural" vegetation, if it ever existed, has been so totally removed, rearranged, and replaced that it seldom is found now. In the Southeast, for example, much of the land is presently covered by pine trees. The original mixed broadleaf and needleleaf forests were cut and replaced by the economically more important needleleaf forests. The grasses of the plains and prairies of the interior are now mostly Eu-

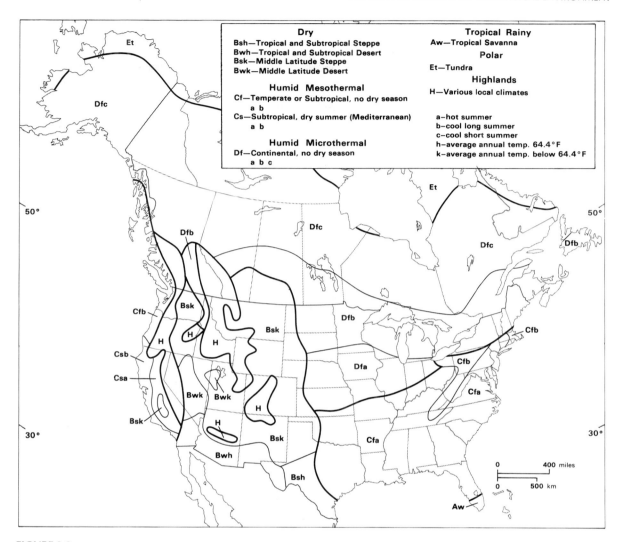

Dry
Bsh—Tropical and Subtropical Steppe
Bwh—Tropical and Subtropical Desert
Bsk—Middle Latitude Steppe
Bwk—Middle Latitude Desert

Humid Mesothermal
Cf—Temperate or Subtropical, no dry season
  a b
Cs—Subtropical, dry summer (Mediterranean)
  a b

Humid Microthermal
Df—Continental, no dry season
  a b c

Tropical Rainy
Aw—Tropical Savanna
Polar
Et—Tundra
Highlands
H—Various local climates

a—hot summer
b—cool long summer
c—cool short summer
h—average annual temp. 64.4°F
k—average annual temp. below 64.4°F

FIGURE 2-3

CLIMATE REGIONS (after Koppen). Climate results from the combination of the long-term pattern of moisture and temperature conditions.

ropean imports. Their native American predecessors are gone either because they offered an inferior browse for farm animals or simply because they could not withstand the onslaught of modern humanity and its imported weeds. Most of what climax vegetation remains in the two countries is in the mountain west and the north. Logging and settlement expansion continue to reduce even that small remnant.

Still, vegetation regions are worth mentioning for at least two reasons. First, the area association between climatic and vegetative regions clearly indicates the central importance of climate on vegetation. Second, even with all the human changes, areas that once were forested continue to have substantial tree cover today, and areas that were grasslands still have few trees.

There are several ways of creating vegetation

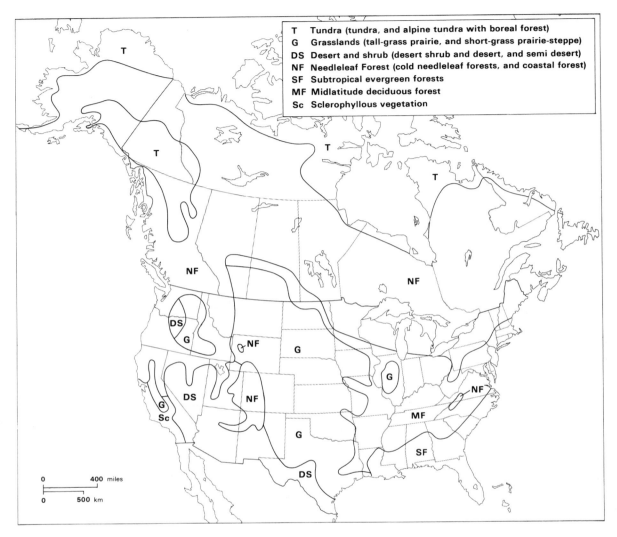

FIGURE 2-4
VEGETATION REGIONS. Notice the similarity between vegetation patterns and
the map of climate regions.

regions. Perhaps the simplest is to divide an area
into three broad categories—forest, grasslands,
and scrublands (Figure 2-4). Forests once covered most of the East, the central and northern
Pacific Coast, the higher elevations of the West,
and a broad band across the interior North. Forests of the Pacific Coast, the interior West, the
North, and those of a narrow belt in the Deep

South were all needleleaf and composed of many
different trees. Much of the Ohio and lower
Mississippi river valleys and the middle Great
Lakes region were covered by a deciduous
broadleaf forest. Between the broadleaf and
needleleaf forests were belts of mixed needleleaf and broadleaf forests.

Grasslands cover much of the interior lowlands, including nearly all of the Great Plains from Texas and New Mexico to central Alberta and Saskatchewan. This is an area of generally subhumid climate where precipitation amounts are not adequate to support tree growth. An eastward extension of the grasslands, the Prairie Wedge, reached across Illinois to the western edge of Indiana. Precipitation in this area is clearly adequate to support tree growth. Although there is disagreement about why the Prairie Wedge existed, fires set by the American

Indians to improve buffalo browse probably had a major impact.

Scrublands usually develop under dry conditions. They are concentrated in the lowlands of the interior West of the United States, with some minor extensions into Canada. Actual vegetation varies from the cacti of the Southwest to the dense, brushy chaparral of southern California and the mesquite of Texas. The scrubland has expanded into some parts of the drier southwestern sections of the grasslands as overgraz-

A desert landscape rich in a variety of cacti, such as the majestic saguaro intricate cholla, and rope-like ocotillo in the scene from southern Arizona—so much a part of the image of the interior West after countless western movies and television shows—is, in fact, uncommon in the United States. (Alan Pitcairn/Grant Heilman.)

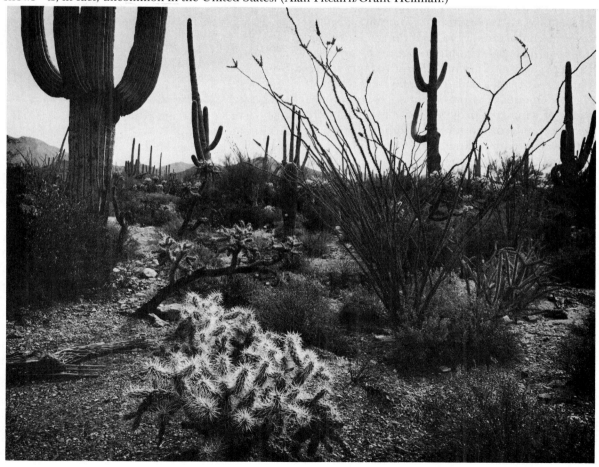

ing destroyed the fragile grass cover (see Chapter 13).

The tundra of the Far North is the result of a climate that is too cold and too dry for the growth of other vegetation. The tundra vegetation does not fit well into our three-part classification. Grasses, lichens, and mosses dominate the region's vegetation. Tundra exists in small areas far southward into the United States, where climatic conditions at high elevations are inhospitable to tree growth. Northward, the *tree line* is found at lower altitudes until, eventually, the latitudinal tree line is reached.

Soils

Soil is a mixture of weathered rock material and organic matter that has been shaped and altered by the physical environment. The soil of a place owes its characteristics to such things as the parent rock material, climate, topography, and decaying plants and animals. Hundreds of different types of soil result from the interaction of these elements. Any particular soil is unique because of its mix of properties (such as color and texture) and composition (including such conditions as organic content and the action of soil colloids). Even so, general soil regions can be specified.

Colloids are small soil particles. Their properties and influences on soil are complex and important. Soil acidity (or alkalinity), for example, is a result of the alteration and integration of soil colloids. Acid soils are characteristic of cold, moist climates; alkaline soils typically are found in dry areas. Most soils of the major agricultural zones of the eastern United States and Canada are moderately to strongly acidic. Lime must be added to neutralize that acidity before these soils can be used to produce most row crops.

Color is perhaps the most obvious soil property. The red clays of Georgia, the black soils of the Red River Valley in Manitoba—these colors may be the first characteristic to catch an observer's attention. A dark color usually indicates an abundance of organic materials; red, the presence of iron compounds. Generally, however,

The soil of a place is not an unlimited resource. The quality of these different soil horizons differs substantially, with the greatest amount of nutrients commonly concentrated in the upper horizons. (USDA—Soil Conservation Service.)

color is a result of the soil-forming processes. For example, the pale-gray soil of the northern needleleaf forest results from the *leaching* of organic matter and minerals from the soil's surface layer.

Soil texture refers to the proportion of particles of different size in the soil. Sand is the coarsest measure of soil texture, silt is intermediate, and clay is the finest. Soil texture depends largely on the composition of the parent material. Texture is important because it determines a soil's ability to retain and transmit water. Soils called "loams" contain substantial proportions of each of the three particle grades and are considered best. They are fine enough to hold moisture yet are not so fine that they cannot easily take up water.

Where soils have developed in place for a sufficiently long period, they take on a characteristic layered appearance. These layers are called soil *horizons*. The uppermost, the A-horizon, is the zone from which certain colloids, chemical compounds, and other matter has been removed; below that is the B-horizon, the zone of accumulation of the matter removed from the A-horizon. The red color of the clay of the Georgia Piedmont, for example, appears because coloring materials other than iron have been removed from the A-horizon, a result of the substantial precipitation of the area. The lowest soil zone, the C-horizon, is the parent material. A thin O-horizon at the surface, the zone of active addition of organic matter, can also be identified in many soils.

There have been many attempts to classify soils. Soil scientists usually prefer a genetic classification, one that focuses on the nature of soil creation. At present the most widely accepted in the United States is the Department of Agriculture's soil classification. It is often called the Seventh Approximation because it is the department's seventh classification in a gradually evolving process. It is based on the identification of precise soil horizons that develop under particular environmental conditions. The map of soil regions based on the Seventh Approximation (Figure 2-5)—indeed, any regionalizaton of soils—must be viewed with some question, for it hides the tremendous complexity that can exist in small areas. The regions simply indicate the most important soil category for the area.

Aridisols gain their name from *arid*. These soils of dry climates are low in organic content and have little agricultural value. *Spodosols* develop in cool, moist climates. In Canada and the United States, spodosols are largely the soils of the northern needleleaf forests. They are quite acidic and low in nutrients; they are of agricultural value only for acid-loving crops, such as potatoes or blueberries. *Tundra soils* are associated with a cold, moist climate. The soil is shallow, frequently water saturated and with a subsurface of perennially frozen ground (see Chapter 17). Tundra soils also have little agricultural value. *Highland soils* are little developed and agriculturally worthless.

Mollisols are grassland soils of the semiarid and subhumid climates of the midlatitudes. They are characterized by a thick dark brown to black A- and B-horizon, loose texture, and high-nutrient content. They are among the most naturally fertile soils in the world and produce most of North America's cereals.

Alfisols are second only to mollisols in agricultural value. They are soils of the midlatitude forest and the forest-grassland boundaries. They are very much "middle" soils in a climatic sense. They are located in areas moist enough to allow for the accumulation of clay particles in the B-horizon (their characteristic identifying feature), but not in areas so moist as to create a heavily leached or weathered soil.

Alfisols are divided into four subcategories, each with its own characteristic climatic association. *Boralfs* are soils of the boreal needleleaf forests of central Canada. They are usually thin, acidic, and have little agricultural potential. *Udalfs* are soils of the deciduous forests of the American Middle West and southern Ontario. Somewhat acidic, they are, nevertheless, highly productive when lime is used to reduce the acidity. *Ustalfs* are found in warmer areas with a strong seasonal variation in precipitation. In the United States, they are most common in Texas and Oklahoma. They are highly productive if irrigated. *Xeralfs* are soils of a Mediterranean climate of cool, moist winters and hot, dry summers. Found in central and southern California, they too are highly productive.

Ultisols represent the ultimate stage of weathering and soil formation in the United States. They develop in areas with abundant precipitation and a long frost-free period. Particle size is small, much of the soluble material and clay has been carried downward. These soils can be productive but high acidity, leaching, and erosion are often problems.

Another soil class, *entisols*, is nowhere sufficiently predominant to be found on the soils map,

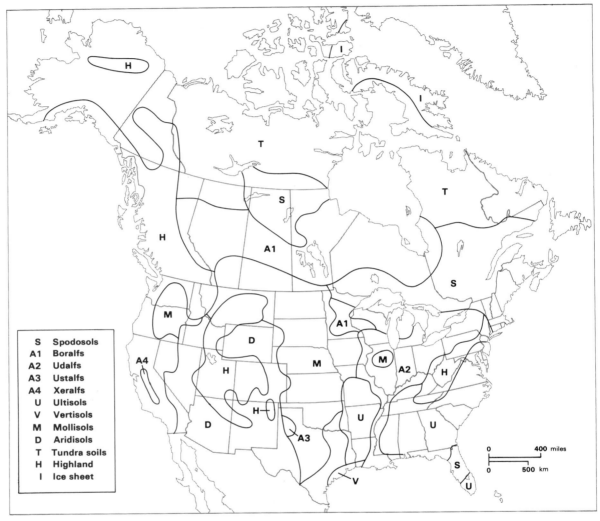

FIGURE 2-5
SOIL TYPES (based on the Seventh Approximation from the U.S. Department of
Agriculture). This map is very generalized. The elements that together influence
soil type are so complex and can be so localized that a single small field or city
block often contains several different soils.

but it is important locally. These are recent soils,
too young to show the modifying effects of their
surroundings. They are widely scattered and of
many types, from the Sand Hills of Nebraska to
the alluvial floodplains of the St. Lawrence and
Mississippi river valleys. The agricultural potential of entisols varies, but the alluvial floodplain
soils, drawn from the rich upper layers of upstream soils, are among the continent's most
productive.

MINERAL RESOURCES OF NORTH AMERICA

There is a distinct association between the location of minerals that meet the needs of heavy manufacturing and the land's subsurface rock structure. Each of the three major forms of rock is capable of containing a type of mineral economically useful to humans, although this potential is not always realized. Sedimentary and metamorphic rocks are the most prevalent rock substructure and are more likely to contain minerals of broad utility than are igneous rocks, the third category.

Sedimentary rock is the result of the gradual settling of small solid particles in stationary water. For example, if a shallow sea was located adjacent to an arid landscape subject to occasional rainstorms, sand particles would be washed into the sea and spread across its bottom by water currents and the force of gravity. As this process continued—for a length of time perhaps five hundred times that since Imperial Rome's legions marched or a length five thousand times that since the United States of America was conceived—each layer of sand would press down on the layers beneath it, squeezing and solidifying the sandy mass that had been deposited only a few thousand years before. When this seabed was raised and folded into mountains by shifts in the earth's crust, the method by which at least some of the rocks were formed would be betrayed by the presence of layers of sandstone. A wide variety of other sedimentary rocks were also formed in this general fashion from various kinds of parent material and dependent on local conditions.

About 300 million years ago during what earth historians call the Carboniferous period of the Paleozoic era, conditions present in most existing land areas were such that unusual sedimentary sequences were created. Heavily vegetated and thick, swampy regions were drowned and covered with another layer of sediment. In some cases, the organic matter came to be represented in liquid form, trapped between folds of impermeable rock and eventually drawn off as petroleum. Most of these petroleum deposits are found in conjunction with another by-product of the period—natural gas. In other cases, the organic matter became solid layers of coal that were sometimes only inches thick but occasionally found dozens of feet thick.

In North America, vast regions are underlaid by sediments formed during the Carboniferous period. These areas where coal, oil, or natural gas might be found are located in the interior and Great Plains physiographic provinces (Figure 2-1), sections of the Gulf coastal plain, the Arctic rimland, and in folded and broken form along the western margins of the Appalachian Highlands and into the eastern Rockies. Together these areas of mineral fuel potential comprise almost all of the United States between the Appalachians and the western mountain ranges and, in Canada, much of the Prairie Provinces and on northward to the Arctic.

Large deposits of mineral fuels have been identified across extensive portions of these sedimentary lowlands (Figure 2-6). The most important coal deposits on the continent have, in fact, been mined in the more rugged Appalachian field, not in the lowland interior. In spite of irregular surface topography, the coal seams are generally easy to work. Most seams are between 1 and 9 feet thick, and in many areas are close to the surface. Both factors facilitate the use of machinery for extraction. Mines throughout this nearly continuous field in eastern Kentucky, West Virginia, and western Pennsylvania were the earliest to be brought into production, and they continue to supply almost 70 percent of the nation's coal needs. And the reserves that remain in this field still comprise about 20 percent of the total reserves in the United States (Figure 2-7). An additional reason that this field has been developed so heavily is the high quality of bituminous coal found there. The coal is low in impurities and can be turned into coke, the form appropriate for use in iron and steel manufacture.

Much of the remaining coal mined in the United States has been obtained from the East-

FIGURE 2-6
MINERAL FUELS EXCEPT COAL. North America is favored with abundant
mineral fuels resources. The two countries also have a large demand for fuels,
which generates a constant search for new sources. Major new petroleum and natural gas fields are being sought off the eastern seashore, in northern Canada and
the Arctic Ocean, and in the interior West.

ern Interior Field. This large coalfield underlies
most of Illinois and extends into western Indiana and western Kentucky. Although some of
the Eastern Interior Field's coal has been used
in iron and steel production, its higher sulfur
content has restricted most use to heating and
electric-power generation. Recently, this sulfur
content drew the attention and ire of those concerned with air pollution. A by-product of burning this coal is the noxious gas, sulfur dioxide.
The consequent acid rain threatens to diminish
the quality of downwind lakes in southern Can-

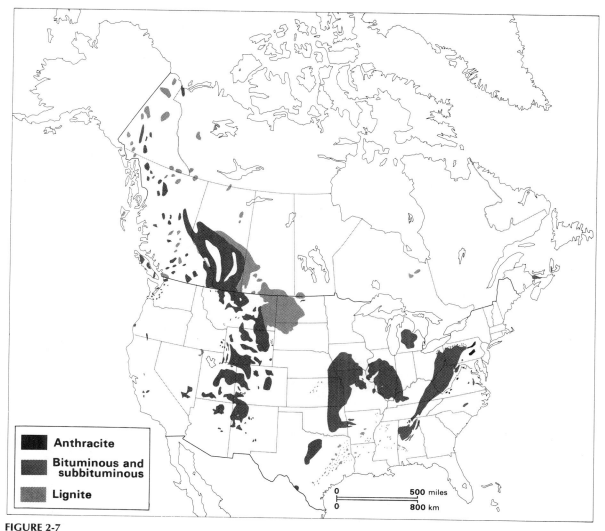

Anthracite

Bituminous and subbituminous

Lignite

0   500 miles
0   800 km

FIGURE 2-7

MAJOR COAL FIELDS. Canada and the United States have a coal reserve that would last hundreds of years at present use levels. However, the increased cost of alternative fuels should result in a large increase in coal production over the next several decades.

ada and the northeastern United States and has been cited as the cause of serious vegetation damage throughout the higher elevations in the Appalachians. The Western Interior Field is also large, located under Iowa and Missouri with a narrowing extension southward into eastern Oklahoma. The coal found in this field is of slightly poorer quality than that found in the eastern fields and has only recently begun to be mined. There are also many small and a few large bituminous deposits scattered through and along the eastern margins of the Rocky Mountains.

Extensive deposits in Wyoming and Montana have come into production in the last decade. There are also several extensive fields of *lignite* ("brown coal") in the northern Great Plains and the Canadian western prairies.

Although petroleum and natural gas are not formed in quite the same manner as coal, their genesis is similar. Scattered deposits of petroleum and natural gas are found through the Appalachian coalfield, including the first commercial well in the United States at Titusville, Pennsylvania. Southern Illinois and south-central Michigan produce some petroleum, as do scattered sites across the northern Great Plains and the northern Rockies. Easily the most important petroleum fields, however, have been those in the southern plains, along the Gulf coast, and in southern California. One great arc of producing wells is located along the full length of the Texas and Louisiana coasts. Another slightly broken arc extends from central Kansas south through Oklahoma and westward across central Texas to New Mexico. Between and beyond these two large areas lie two more fields of great importance, the East Texas Field and the Panhandle Field in northwest Texas. Separate from these fields but also of major importance are those located in southern California. In the mid-1960s, exploitation of deposits of petroleum and natural gas was begun along the north Alaska slope and in the Arctic fringes of Canada; the development of significant deposits in Alberta has recently increased the economic influence of that province within Canada.

The broad pattern of mineral fuel deposits in North America, then, is one of extensive deposits of coal south of the Great Lakes and west of the middle Appalachians and of large petroleum and natural gas fields southwest of the major coal reserves, stretching from the middle Great Plains south to the Gulf coast. Scattered, and occasionally significant, deposits are located west and northwest of the major reserves.

Metamorphic rock is formed in a quite different manner than sedimentary rock. Under the tremendous pressure exerted through the grad-ual deformation of the earth's crust, the internal structure of previously formed rocks can be metamorphosed (changed). So great is the pressure exerted over thousands of years and so great is the heat generated that the very molecular structure of the rocks is altered. This transformation indicates why metallic minerals in economically extractable quantities are located most often in areas of metamorphic rock. In North America, therefore, the primary deposits of metallic ores are located in the three major zones of metamorphic rock: the Canadian Shield, portions of the Appalachian Highlands and the eastern Piedmont, and in sections of the western mountain ranges (Figure 2-8).

Many of the mining sites for early exploitation of the metallic minerals on the Canadian Shield were located near the margins of the Shield. The pattern of mineral production follows a long arc extending from the North Atlantic and the St. Lawrence River *estuary* across the Great Lakes and northward to the Arctic Ocean. This arc defines the edge of the Shield itself in the broad horseshoe shape it exhibits around Hudson Bay. As the interior of the Shield began to be developed in recent decades, the arc has blurred. It now includes the cluster of iron deposits on the Quebec-Labrador border at Burnt Creek and Carol Lake; a tremendous concentration of minerals on the Quebec-Ontario border, including the nickel-rich region at Sudbury; and such self-explanatory places as Cobalt, Ontario, and Val D'Or (Valley of Gold), Quebec. The arc continues on both sides of Lake Superior: in northern Michigan, Wisconsin, and Minnesota with copper and iron; in Ontario, with iron at many sites. Then, it swings north through Flin Flon (copper, lead, zinc) to such likely places as Uranium City, Yellowknife (gold), Port Radium, and Coppermine.

A second zone of metamorphic rock is located along the eastern Appalachian Mountains. Iron mines at Wabana, Newfoundland, supply the ore for a small steelworks. Copper and iron were important minerals found locally by New England colonists. The most important centers in this

FIGURE 2-8

METALLIC MINERALS. The metamorphic rocks containing metallic minerals in
the two countries arc around the interior lowland and its fossil fuels.

zone still producing are Ducktown, Tennessee (copper), and Birmingham, Alabama, where iron ore can be mined at the southern end of the Appalachian chain. This narrow, linear band of metamorphic rock remains only moderately important as a source of metallic minerals. Most of the small deposits were depleted as demand continued to grow and transportation eased shipment from more distant locations with more easily worked deposits. But these eastern minerals were significant during the early years of industrial growth in the United States.

A third and extensive region of metallic minerals is formed by the western mountains. Scattered deposits of gold and silver, a few of them rich, drew prospectors and mining companies to isolated locations scattered from south of the Mexican border to central Alaska—and provided Hollywood writers with a setting for countless (even if mythical) thrillers. Of great industrial importance are the large deposits of copper, zinc, lead, molybdenum, and recently, uranium found in this western region. When this list of major deposits is swelled by adding the minerals found in smaller deposits, such as tungsten, chromite, maganese, and many others, we can appreciate the great flexibility this diversity gave to the growing industrial economies in the United States and Canada.

It should not be assumed that the two countries' industrial requirements were met fully by the tremendous variety of minerals found in these three zones of metamorphic rock. A few minerals needed by modern industry have not been located in North America in sufficient quantities to satisfy domestic needs (e.g., tin, manganese, and high-grade bauxite for aluminum). In addition, the growth of industrial capacity has been matched by a growth in demand for many minerals. Demand now often exceeds the diminishing domestic supply of these nonrenewable resources. Few other countries, however, ever equaled, or even approached, the original quantity and diversity of metallic minerals and mineral fuels located in the United States, especially when complemented by the vast reserves in

Canada. This abundance of minerals has been critical in assisting the development of the immense North American manufacturing-industrial complex (Chapter 5).

Additional Readings

Atwood, Wallace W. *The Physiographic Provinces of North America*. Boston: Ginn, 1940.

Bennett, Charles F. *Conservation and Management of Natural Resources in the United States*. New York: John Wiley & Sons, 1983.

Bryson, Reid, and F. Kenneth Hare. *World Survey of Climatology, vol. 2, Climates of North America*. New York: Elsevier, 1974.

Dixon, Colin J. *Atlas of Economic Mineral Deposits*. Ithaca, N.Y.: Cornell University Press, 1979.

Falconer, Allan, et al. *Physical Geography: The Canadian Context*. Toronto: McGraw-Hill Ryerson, 1974.

Farb, Peter. *Face of North America: The Natural History of a Continent*. New York: Harper & Row, 1963.

Gersmehl, Philip J. "Soil Taxonomy and Mapping." *Annals of the Association of American Geographers*, 67 (1977), pp. 419–428.

Hare, F. Kenneth, and Morley R. Thomas. *Climate Canada*. Toronto: John Wiley & Sons Canada, 1974.

Hunt, Charles B. *Natural Regions of the United States and Canada*. San Francisco: W. H. Freeman, 1973.

Kuchler, A. W. "Potential Natural Vegetation of the Conterminous United States." (Special Publication no. 36.) New York: American Geographical Society, 1964.

Mitchell, Bruce, and W. R. Sewell (eds.). *Canadian Resource Policies: Problems and Prospects*. Toronto: Methuen, 1981.

Steila, David. *The Geography of Soils*. Englewood Cliffs, N.J.: Prentice-Hall, 1976.

Thornbury, William D. *Regional Geomorphology of the United States*. New York: John Wiley & Sons, 1965.

Visher, Stephen S. *Climatic Atlas of the United States*. Cambridge: Harvard University Press, 1954.

CHAPTER THREE

FOUNDATIONS OF HUMAN ACTIVITY

C anada and the United States had a small, dispersed native population at the time of initial European discovery. The American Indian and Inuit (or Eskimo) populations totaled perhaps 1 million people, four-fifths of them in what came to be the United States. These peoples usually were organized in small tribal units. By the time of European arrival in North America, there was great diversity among American Indian cultures. Several hundred dialects were spoken along the coast of California alone. And the diversity was more than linguistic. The Pueblo, who lived in what is now New Mexico and who were probably influenced culturally by the Aztecs to the south, resided in permanent towns and constructed irrigation ditches. At the same time, the Piutes of the Great Basin lived in temporary thatch dwellings and pursued a semi-nomadic existence based on available wild edible vegetation and small game. The Eskimos, who were the most recent of the pre-European arrivals, shared close cultural ties with Eskimos in Greenland and Siberia.

Although American Indians did represent a real barrier to the expansion of European settlement at times, for the most part their impact was minimal. Many died of imported European infectious diseases, such as smallpox or measles, before they experienced direct contact with the Europeans themselves. Occasionally, the Indians made important contributions to the arriving Europeans, especially during the first decades of settlement. Still, the cruel fact is that, in general, they simply were ignored as much as possible. Most often, as time went on, they were killed or shunted off to reservations somewhere to the west, where it was hoped they would stay out of the way. As the settlement frontier moved westward, so did the Indians and their reservations. There was almost no integration of the Indians into the European settlement pattern, something that was common in portions of Latin America. The Canadian record on relations with the Indians was—and is—a little better than that of the United States, but neither country deserves any humanitarian awards for their treat-

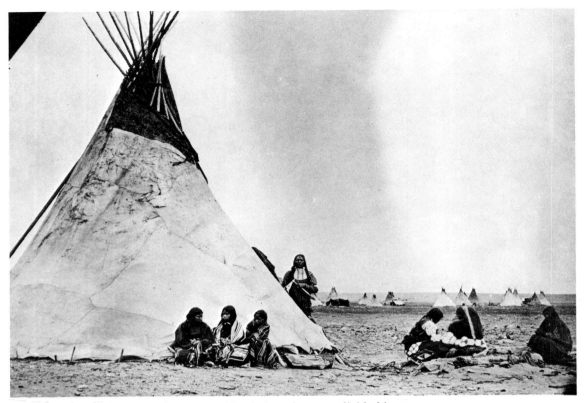

The American Indians of pre-European North America were divided into many, many different tribal and cultural groups. Here, for example, are scenes from (above) the northern Great Plains (of a tribal group actually heavily influenced already by the introduction of the horse) and (right) the coast of British Columbia. (Smithsonian Institution National Anthropological Archives. Photo by William S. Soule.)

ment of the previous inhabitants of the two countries.

European Settlement

What was easily the greatest long-distance migration in human history followed in the wake of the retreating American Indians. Although it is impossible to state precisely how many people entered what is now the United States and Canada from Europe and, to a much lesser extent, from Africa, a reasonable estimate would place the figure at somewhere close to 60 million.

The overall pattern of source areas of Euro-

pean immigrants to North America is complex, but a few generalizations can be made. Most early immigrants came from northwestern Europe, primarily the British Isles (Table 3-1). At the time of the first national census of the United States in 1790, more than two-thirds of the white population of the country had an ethnic origin on the British Isles, with Germans and Dutch next in importance. Nearly all of the French migration to Canada came early in the European settlement history, mostly in the seventeenth century. The total number of French who made the trip was small, less than 15,000. They have in-

(Courtesy of the American Museum of Natural History. Photo by Edward Curtis.)

creased hundreds of times over in the ensuing 300 years, a striking example of human capacity to increase in numbers under reasonably good conditions.

Immigration into North America slowed during the period between 1760 and 1815. This was a time of intermittent warfare in Europe and North America as well as on the Atlantic Ocean. Between about 1815 and the start of World War I in 1914, the volume of immigration tended to increase with each passing decade. Periods of war or widespread economic depression were exceptions, marking times of reduced movement. At the beginning of this period, which lasted a century, annual migration was perhaps 10,000. In 1913, nearly 1 million immigrants entered the United States and another 400,000 streamed into Canada.

The United States passed its first major legislation to restrict the volume of immigration in the 1920s. Canada followed suit a decade later. These enforced limitations coupled with the depression of the 1930s and a war in the 1940s cut immigration to a fraction of its annual high in 1913. Since 1945, the number of arrivals has increased somewhat. Today the United States typically receives roughly 450,000 legal immigrants annually; Canada perhaps 125,000. An even larger number of illegal aliens probably enter the United States each year (see Chapter 14).

There has been a great change in the major sources of immigrants to the United States with the passage of time. During the period from 1815 to 1913, the home-country locations of migrants from Europe gradually shifted southward and

45

TABLE 3-1
Immigration into the United States and Canada

Decade	United States Number in Thousands	Principal Sources	Canada Number in Thousands	Principal Sources
1820s	129	Ire., Br.		
1830s	538	Ire., Ger.		
1840s	1427	Ire., Ger.		
1850s	2815	Ire., Ger.	253	
1860s	2081	Ger., Br.	156	
1870s	2742	Ger., Br.	329	
1880s	5249	Ger., Br., Scand.	850	
1890s	3694	E. Eur., It., Ger.	373	
1900s	8602	E. Eur., It., Russ.	1401	Br., U.S.
1910s	6347	E. Eur., It., Russ.	1859	Br., U.S., Russ.
1920s	4296	Can., L. Am., It.	1273	Br., U.S., Ire.
1930s	699	Can., Ger., It.	251	U.S., Br.
1940s	857	L. Am., Can., Br.	429	Br., E. Eur.
1950s	2300	Ger., Can., L. Am.	1540	Br., It., U.S.
1960s	3212	L. Am., Can., Asia	1375	Br., It., U.S.
1970–1980	4371	L. Am., Asia	1589	Br., U.S., Hong Kong

SOURCE: M. C. Urquart, ed., *Historical Statistics of Canada* (Toronto: Macmillan Co. of Canada, 1965); Statistics Canada, *Canada Year Book*, 1965, 1981; W. Zelinsky, *Cultural Geography of the United States* (Englewood Cliffs, N.J.: Prentice-Hall, 1976); *Statistical Abstract of the United States* Washington, D.C.: U.S. Government Printing Office, 1970–1981.

eastward. For the first half of the period, most migrants continued to come from northwestern Europe. There was a decline in the percentage from Britain, but an increased number arrived from Ireland and Germany. By late in the nineteenth century, these migration origins became relatively less important. In the 1880s, there was a major outflow from Scandinavia, followed in subsequent decades by streams of people from southern and eastern Europe. In 1913, well over four-fifths of all immigrants were from these areas of Europe, especially Italy, Austria-Hungary, and Russia.

The reasons for this southeastward shift are based on the impact within Europe of the diffusion of the Industrial Revolution. Beginning in the eighteenth century in the British Isles and the Low Countries across the English Channel, it spread southeastward across Europe during the following 150 years or so. With industriali-

zation came a rapid rise in population size as mortality declined. The economy shifted to manufacturing, urbanization increased, and there was a proportional decline in the agricultural population. The growth in the demand for urban labor did not match the increase in the potential labor force; thus, there were many willing emigrants.

The destination of these immigrants within North America was largely the result of the distribution of existing economic opportunity. It has been suggested repeatedly that migrants chose areas that were environmentally similar to their European homes. The substantial Scandinavian settlement in Minnesota and the Dakotas is indicated as a case in point. There may be some small truth in this, but it was more important that those states happened to represent the principal settlement frontier at the time of major Scandinavian immigration. Did Germans

The immigrants who have come to North America over the past 150 years have added to the diversity of the culture of the two countries while at the same time integrating into that which was already there. This family is arriving in New York City around 1910. (Lewis W. Hine photo from the Library of Congress.)

the use of blacks as slave labor are the most obvious examples (see Chapters 9 and 14). Much of the early heavy Irish immigration in the late 1840s and 1850s was a panic move by a destitute population held in virtual serfdom by British landowners. The migrants were fleeing starvation that resulted from massive failures of the country's principal food crop, the potato. These Irish had no money to move beyond the cities of the East Coast and became the first major poor immigrant ethnic population in many of America's rapidly growing cities. French Canadians moving into the northeastern United States or Mexicans into the Southwest or, more recently, Cubans into the Miami area have all chosen locations close to their origins, although not necessarily because they lacked money to move elsewhere.

Non-European Settlement

Until recently, most immigrants into Canada and the United States were from Europe. The major exception was black settlement in the American South. Imported as slave labor for the region's plantations, this was a small part of the large movement of Africans to the Caribbean Basin, the northeast coast of South America, and the American Southeast. Next to the European exodus, this was probably the second largest long-distance movement in human history. Perhaps 20 million left Africa. It is believed that fewer than 500,000 blacks came into the United States, although accurate statistics are not available. Most who came probably arrived from the islands of the Caribbean rather than coming directly from Africa. At first, slaveholding was legal throughout the United States; in practice, however, the great majority went to the plantation South. The 1790 census indicated that fully 20 percent of the American population was of African origin. There was little African immigration after that date, and the percentage of the population that was black declined slowly but steadily until well into the twentieth century.

move to Texas because it reminded them of home? Unlikely. Did southern Italian immigrants find Boston or Newark much like the farms and villages of their Mediterranean origin? Certainly not. For the most part, the mosaic of ethnic patterns on the continent is the result of a movement toward opportunity—opportunity first found most often on the agricultural settlement frontier and then in the cities.

There are several notable exceptions to this pattern of essentially free choice of location. Restrictions of American Indians to reservations and

Immigration laws severely limited the possible source areas of immigrants after the 1920s. Es-

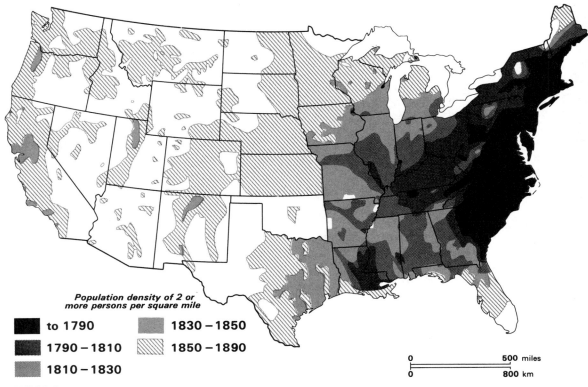

Population density of 2 or more persons per square mile

■ to 1790   ▨ 1830 – 1850

▦ 1790 – 1810   ▧ 1850 – 1890

▨ 1810 – 1830

0 ————— 500 miles
0 ————— 800 km

FIGURE 3-1

EXPANSION OF THE SETTLEMENT FRONTIER IN THE UNITED
STATES. For three-quarters of the 375 years since its initial settlement, the
United States was in a period of active frontier expansion. Many have suggested
that this long history was critically important in shaping American culture and
ideals.

sentially, except for a large quota for other
Western Hemisphere countries, most migrants
had to come from Europe. After 1881, Chinese
immigration was excluded entirely from the
United States, and the Chinese also were banned
by Canada in 1923. Far more liberal immigration laws were passed by both countries in the
1960s. Coupled with a decline in interest on the
part of most Europeans to move to North
America, this has meant that most immigrants
to the United States no longer come from Europe. Typically, only about one-quarter of the
immigrants are from Europe, somewhat more
from Asia, and the rest from the Americas. Most

Canadian immigrants are still from Europe, although the non-European component is growing.

Settlement Expansion

The European settlement of North America
started in the east and moved westward (Figures
3-1, 3-2). That is a simple, obvious statement.
Yet it is the single most important element in
understanding the European occupation of the
two countries. Settlement did develop first in the
east, and it did gradually extend westward. Although there are certainly some interesting var-

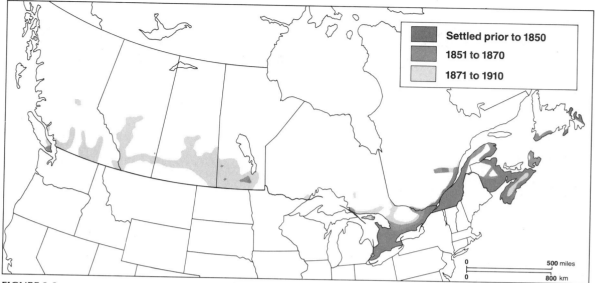

FIGURE 3-2

EXPANSION OF THE SETTLEMENT FRONTIER IN CANADA. The Shield was a major barrier to the north and to westward expansion in Canada. The St. Lawrence lowland and southern Ontario thus developed as the dominant Canadian settlement focus.

iations in the trend—places that were settled earlier or later than might have been expected—in general, you can predict the date of earliest settlement at any given place merely by knowing how far it is from the East Coast.

The first settlements were small, clinging to the ocean and looking more toward Europe than toward the land that crowded in about them. When settlement pushed tentatively away from the ocean, it still followed the waterways, for they offered trade pathways to the coast and an important link to Europe as well. Thus, the British settled the indented coastline of the Chesapeake Bay and its tributaries or they spread a thin band of settlements along the rugged coastline of New England; the Dutch moved up the Hudson River from New Amsterdam (New York); and the French gradually settled the banks of the upper St. Lawrence. To break away from the water was in many ways a break from Europe. That step was taken slowly and with trepidation.

Settlement expansion in both countries began slowly, then gradually picked up speed as time passed. For example, in the United States during the first 150 years after the beginnings of permanent European settlement—until about 1765—Europeans had moved westward only as far as the eastern flanks of the Appalachians. Within a century after that, the frontier had reached the Pacific Ocean, and by 1890 the U.S. Bureau of the Census was able to announce that the American settlement frontier was gone entirely. Perhaps this increasing speed was a reflection of the law of physics that states that an object once in motion tends to stay in motion. More likely, it resulted from a reorientation in attitude away from Europe toward North America. By the early nineteenth century, an increasing number of Americans viewed the occupation of the continent as their manifest destiny. The land laws of the country, culminating in the Homestead Act of 1862, became increas-

Both countries adopted grid systems of land subdivision to facilitate the survey and settlement of their large interior lowlands. The result is the regular, rectangular pattern of fields and roads so familiar to residents of the territory. (George Hunter/National Film Board-Phototèque.)

ingly expansionist. This made it much easier to obtain land along the settlement frontier. Also, as the country expanded in size and as the annual population increase grew—the result of both natural increase and immigration—there were more people who hoped to improve their lot by moving westward. The expansionist policies of both Canada and the United States led to an almost constant state of underpopulation and surplus production. More people were needed for the opportunity that was available, but both countries already were producing more raw materials than they needed. To Europe, both countries were economic colonies, providing raw materials and buying finished products.

Settlement moved westward onto the Great Plains more rapidly in the United States than in Canada. Whereas the Mississippi River and its many tributaries offered easy routes in the interior of the United States, movement west beyond the Great Lakes in Canada was primarily overland. Furthermore, and more important, American settlers found a vast expanse of excellent agricultural land with a generally good climate for crop production that stretched from the western margins of the Appalachians well into the Great Plains. Canadian settlers, on the other hand, faced some 2500 kilometers (1500 miles) of land nearly worthless for agriculture that extended from about 150 kilometers (95 miles) north of Lake Ontario to just east of the site of present-day Winnipeg, Manitoba. This is a portion of the glacially scoured Canadian Shield with its thin, rocky soils. Although most of the Great Plains of the United States were settled between 1860 and 1890, the prairies of Canada were not opened to settlement until near the end of the nineteenth century. Even then, many of the early settlers were Americans who had moved northward across the border.

In the eastern half of the United States, about as far west as Kansas and Nebraska, settlement expanded westward in a generally orderly fashion. To be sure, advances were more rapid along certain transportation routeways, such as the Ohio River, and slower in other places. Some areas were bypassed by the early settlers in favor of more promising lands farther west. At times, this created a hollow frontier. For example, parts of the Adirondack Mountains of New York remained unoccupied while the principal line of settlement advance pushed across the Mississippi River. A somewhat similar pattern of even expansion developed in southern Canada along the St. Lawrence River and the lower Great Lakes.

Elsewhere, in northern Canada, in the United States and Canada from the Rocky Mountains westward, and in Alaska, this even pattern of settlement expansion did not occur. These lands were characterized by generally poor agricultural potential. Much of this broad area was either too dry, too hot, or too cold for farming. Rugged topography hampered transportation and further limited agricultural development. Settlement congregated in areas that offered an identifiable economic potential. For example, agriculture developed in small areas of good soil and adequate precipitation, such as in the Willamette Valley of Oregon. Mineral resources offered another settlement impetus, as in the case of the mining centers of north-central Ontario. Other sites developed to service the transportation lines that extended across these empty areas. Whatever the specific reason for growth, the result was a pattern of point settlement scattered across an otherwise nearly unpopulated landscape.

Population Distribution Today

The 1981 census of population in Canada revealed a population of about 24.3 million. In 1970, the United States had a population of 204 million, and by 1980, its population exceeded 226 million. This 10:1 ratio has been maintained for about a century. Canada had a 1981 population density of about 2.4 people per square kilometer (6.3 per square mile), whereas the conterminous United States (i.e., excluding Alaska and Hawaii) had a 1980 density of roughly 210 people per square kilometer (77 people per square mile).

Neither country has anything approaching an even population distribution (Figure 3-3). Three principal zones of population can be identified. First, a primary zone fills a quadrant defined approximately by the cities of Montreal, Chicago, St. Louis, and Washington, D.C. This is the traditional population core of the United States and Canada. Most of the larger cities in the two countries are found in this zone: 60 percent of the Canadian population lives in southern Quebec and Ontario; 8 of the 12 most populous states in the United States are here. It is the area of earliest major population growth in both countries and long their most advanced sections economically. Fine natural routeways and many excellent harbors along the Atlantic shore have been augmented by one of the densest transportation nets in the world (Figure 3-4). Some of the continent's best agricultural lands plus rich mineral resources are either within the region or around its periphery. Now in decline relative to some other parts of the continent, this region is still North America's vital core.

Wrapping around the southern and western margins of the core and extending westward to the eastern sections of the Great Plains in the United States, there is a secondary zone of population. Much of the best agricultural land of the country is in this zone. With relatively minor exceptions, this area is amenable to agricultural development, and the greatest part of its potential agricultural lands are farmed. Most of the area is populated although densities are gener-

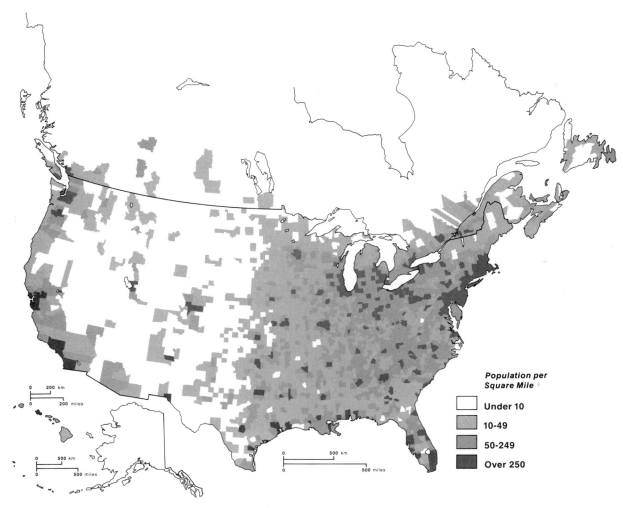

FIGURE 3-3
UNITED STATES AND CANADA POPULARION DENSITY, 1980. Most geographers would argue that we have the technological ability to work with any physical environment, and humans can live anywhere on the earth's surface. Nevertheless, notice how areas of generally higher population density are similar to those of greater precipitation on Figure 2-2.

ally much lower than those found in the core. Cities are spaced more widely and more evenly in this zone than in the core, and they are primarily service and manufacturing centers for the regional economy. As a group, these cities have less significance nationally than the cities of the core.

Finally, a peripheral population zone fills the land from the Great Plains westward. The pattern of population and economic growth at lo-

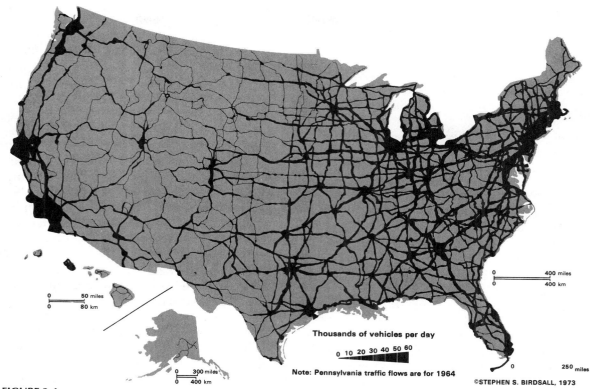

Thousands of vehicles per day

0 10 20 30 40 50 60

Note: Pennsylvania traffic flows are for 1964

©STEPHEN S. BIRDSALL, 1973

FIGURE 3-4

HIGHWAY TRAFFIC ON PRIMARY ROADS, 1969–1970. Air, rail, and highway movement all focus on the urban centers of the country.

cations of special potential in an otherwise limited region continues to dominate. This is a pattern that developed early in the region's settlement history. Although some areas are now densely populated—notably California's San Francisco Bay area and Los Angeles Basin and the Fraser Delta-Puget Sound Lowland in British Columbia and Washington State—most of the land remains sparsely populated.

Canada, beyond the southern Ontario and Quebec core, does not fit easily into this scheme. The Canadian population is split into about four major populated areas, each separated from the others by extensive empty areas. If we view the Quebec-Ontario core as a single unit, it is separated from the settlement of the Prairie Provinces by the Canadian Shield in western Ontario

and eastern Manitoba. Otherwise, the farmlands of the Prairie Provinces might be viewed as a part of the secondary zone of population that we saw in the United States. The Prairie Provinces are themselves separated from the coastal population of British Columbia by the Coast Ranges and the Rocky Mountains. This western region is similar to the peripheral zone in the United States. To the east, the population of the Atlantic Provinces is scattered widely along an indented coastline and in a few interior lowlands. It is connected to the rest of the country by a thin, tenuous line of settlement along the southern shore of the St. Lawrence River and overland into New Brunswick. Taken together, these zones include well over 95 percent of the Canadian population.

Each of these four population clusters is pressed against the southern margins of the country. In fact, most Canadians live within 150 kilometers (90 miles) of the country's southern border. They have often had closer economic and cultural ties with the areas south of the border in adjacent parts of the United States than with other parts of Canada. This has been particularly true in the past. Rising Canadian nationalism plus an effort by the Canadian government to improve interregional communications has lessened this exterior association somewhat.

Population Redistribution-
Mobility Patterns

Population redistribution is a pervasive aspect of the geography of the two countries (almost one American in five changes his or her residence in an average year). Substantial internal redistribution plus continued immigration has maintained a high potential for rapid regional change. The pace of population increase is used frequently as a measure of local or regional well-being. If a regional population is growing rapidly, the region's economy and stature are viewed as sound; if the population is declining, grave concern usually is expressed about the future.

"Go West, young man," that famous admonition attributed to Horace Greeley, has long characterized one element in the internal migration patterns of these two mobile populations. In fact, the center of population in the United States (defined as the intersection of two lines that equally divide the country's population north-south and east-west), located southeast of Baltimore at the time of the first national census in 1790, has moved westward during each ensuing decade. By 1980, it was found southwest of St. Louis, west of the Mississippi River for the first time. There is solid evidence to suggest that this westward movement has slowed, although it is by no means over.

It is possible to divide the mobility history of both countries into three periods. First came the period of east-to-west movement, then one from rural to urban areas, and, finally, the present

period, when most long-distance movement is between metropolitan areas. If the U.S. population has moved westward with every decade, it has urbanized in an equally unvarying fashion. Whereas only about 5 percent of the population could even loosely be defined as urban in 1790, about three-quarters of the country's population was urbanized by 1980. The 1981 Canadian census classified about 76 percent of all Canadians as urban.

These statistics reflect not only a relative decline in rural population, but also an absolute decline in farm population. For example, between 1960 and 1982, the farm population of the United States fell by over one-half; from more than 15 million to an estimate of just over 6 million. Although not as dramatic, Canada's farm population has also decreased substantially.

The movements from east to west and from rural to urban America were both clearly in response to the perception of economic opportunity. First, more and more farmlands became available in the two countries as the settlement frontier pushed westward. Then, there was a tremendous surge in urban employment generated by the technological advances of the Industrial Revolution. To emphasize a point, both of these trends developed together throughout the nineteenth century and continued into the twentieth century in Canada. The westward movement was relatively more important at first, whereas the urbanward movement had gained ascendency by the end of the period. And once North Americans were predominantly urbanites and economic opportunities were also urban based, variations in these opportunities ensured that most subsequent population migration would occur between metropolitan areas.

U.S. population statistics for the 1970s suggest that a fourth major mobility period is at hand. Areas that had long experienced no change or even declining population size suddenly are growing. The South is a prime example. For more than a century, it was a region of outmigration as southerners were drawn to the opportunities located in the cities of the North and the West. Recently, however, it has become a growth region; far more people are entering the region than are leaving it. Between 1970 and 1980 every state of the South (and every state from the Rockies westward) had a population growth rate above the national average. To the north and east of these states, only Vermont, New Hampshire, and Maine did not have population-change rates below the national average. The Bureau of the Census predicts that this pattern of population change will continue through the remainder of the twentieth century.

Perhaps this shift should not come as a complete surprise. Many observers of the American economy have suggested that the United States is becoming, or has become, the world's first post-industrial country. That is, the major growth areas in the national economy are not in manufacturing. Instead, they are in what are called *tertiary* and *quaternary occupations*. These are, respectively, occupations that provide services and those that manipulate and create information. The number of Americans employed in manufacturing has increased only slightly during the past two decades, whereas tertiary and quaternary employment has boomed. Much of what increase there has been in manufacturing employment has been in the production of high-value, lightweight products, such as electronic components.

It is suggested that this employment shift has freed many employers and their workers from traditional locational ties. In the past, the economic advantages offered by locations within the continent's core region were the most important considerations in attracting population. High-value light industry can presumably locate almost anywhere. High-speed, inexpensive communications and transportation allow many tertiary and quaternary employers to locate in any one of many possible locations. People, or at least more and more people, may live where they want, not where a specific economic function dictates they live. The Sunbelt states of the South and West seem to provide the kind of residential environment that Americans want. Outdoor recreation and mild winters—these evidently represent the goals of many Americans today. Now that more and more can live where they want and find employment as well, they are moving to the South and the West.

The socioeconomic nature of the migrants presently leaving and entering the South seems to support these comments. For both blacks and whites, the average immigrant into the South is better educated and better paid than the average outmigrant. This suggests that the new arrivals are there in response to a growth in the number of high-quality jobs available to them and that fewer are moving in the traditional pattern by seeking opportunity in the industrial centers of the North.

Recent data suggest the possibility of what some have called a rural renaissance in both countries. In Canada, between 1971 and 1981, the rural population grew at a rate that was much more rapid than that for the urban population after having remained essentially stable for two decades. Similarly, areas distant from major urban centers in the United States have experienced a surge of growth in the 1970s. This new migration does not represent some type of return to the land movement. Although many thousands have left the cities over the last decade to create for themselves a personally more satisfying rural lifestyle, they represent a small percentage of all movers. Most of those moving to rural areas have nonrural occupations. In census terminology, they are rural nonfarm residents. They have moved to smaller communities in response to new job opportunities or they live in rural areas and commute to urban employment.

URBANIZATION

The United States and Canada are urban countries. This has been a fact of national life since World War I. The two countries' urbanizing trend is significant because it involves a spatial rearrangement of population, but it also represents a change in the type of space occupied.

By residing in urban locations, North Americans live in an environment quite different in character from that affecting most people several generations ago. The differences are more than technological. Rural and urban environments are different. The former is characterized by individual isolation created by distance; personal isolation in the latter is more likely a result of self-imposed or socially prescribed psychological distance. Economic activities in rural areas tend to be extractive and individual, whereas in cities, most economic activities are joint efforts at manipulating materials, service opportunities, or ideas. And the rural environment offers a lifestyle that retains at least a degree of self-sufficiency, whereas the urban environment contains people entirely dependent on each other for virtually every aspect of their lives. The highly interdependent urban environment clearly has a spatial organization distinct from that typical of rural areas.

Spatial organization is the way a set of activities, each of which takes place at a separate location, is linked together by the interactions between them. Urbanized areas possess a more intense and complex spatial organization than rural areas. In large cities, the variety of activities is greater than in small towns, and the intensity of spatial organization varies accordingly.

North America possesses many urban areas that have grown in population and extent. In a few instances, the spatial growth was so great and the size of the core cities so large that major urban areas merged and formed clusters of cities. Geographers and others began to treat these clusters as something quite different in the history of urbanization. The group of large cities extending from Boston, Massachusetts, to Washington, D.C., along the northeastern U.S. coast is the clearest example (see Figure 1-1). Another group of urban areas—more widely dispersed and containing central cities that are not as large as those in the Boston-Washington, D.C., complex—is found along the southern margin of the Great Lakes. Milwaukee and Chicago anchor this region in the west and Buffalo and Pittsburgh do the same in the east. South-

ern California, from San Diego to San Francisco, is offered by some observers as yet another set of urban areas that will be merged by the end of this century. Others do not expect much spread in southern California beyond the current San Diego-Santa Barbra cluster considering the urban resource limitations of California (see Chapter 15). Much of east coastal and central Florida is expected to be urbanized by the year 2000, and numerous small clusters of cities repeat the pattern in other sections of the country. The one cluster of coalescing urban areas that overshadows all of the others, however, is the urbanized northeastern seaboard of the United States (Chapter 4).

Each major cluster of urban places and the isolated towns and cities, too, exist where they do for identifiable reasons. We might ask: Why do cities come to be located where they do? In a simplified and general way, we can say that most of the largest urban places have developed where transportation routes connect with each other. Quite often it is the land-water connection that is important. Other factors matter as well: hinterland quality, proximity to alternative transportation, security, and even healthfulness of the local environment may be important. However, where goods and people must transfer from one form of transportation to another, there are a great many opportunities to process, exchange, manufacture, repackage, sell, and buy goods. Cities, in other words, are usually located where they are because of certain economic advantages associated with movement to and from their location.

Think for a moment about the major urban centers of the United States and Canada (Figure 3-5). Many of the largest are located adjacent to navigable water. Some are on a seacoast or large estuary: Boston, New York City, Baltimore, Quebec, Miami, Tampa-St. Petersburg, Mobile, San Diego, Los Angeles, San Francisco, Seattle, and Vancouver, British Columbia, are examples. Others are on naturally navigable waterways, such as Chicago, Milwaukee, Minneapolis-St. Paul, Detroit, Cleveland, Buffalo, Hamilton,

FIGURE 3-5
THE LARGEST URBAN AREAS IN CANADA AND THE UNITED STATES.
Most of these large urban areas in the two countries are at highly accessible ocean
or inland waterway locations.

Toronto, Montreal, Philadelphia, New Orleans, Memphis, St. Louis, Kansas City, Cincinnati, Pittsburgh, Louisville, and Portland, Oregon. Still others are on rivers or channels that have been modified extensively just to give the cities water access: Houston and Knoxville are examples. There are exceptions to this water orientation, of course, such as Atlanta, Calgary, Denver, and Dallas-Ft. Worth, but these, too, were on early transport routes of some kind.

The specific reasons for any particular city's growth and why it grew where it did are often divided into aspects of urban *site* and urban *situation*. The site of a city is its immediate physical environment—that is, the characteristics of the landscape in which the city is located. Pertinent information includes the topography across which the city spreads—Is it flat or rolling? Is it on a plateau or in a valley?; its soil characteristics— Are large areas of poorly compacted river soils present? Are there good water-bearing sands?; drainage—Are there swampy areas? Does it have a river for water supply or water transportation?; the depth of the bedrock; and so on. Many

57

times such site characteristics help account for the difference between rates of urban growth for competing cities.

Urban situation is the set of factors associated with the position of the city relative to other places. Situation is sometimes referred to as *relative location*. Several aspects of situation are proximity to other centers—short distances might make trade easier; position between two areas that produce different products—this may provide the opportunity to serve the middleman function in exchange; and the productivity of land using the city as a regional focus. In urban geography, situation is used to describe the locational features of a city that link it to other places and form the relations the city has with other cities, other regions, or locations of potential and actual activity. In other words, a city's situation reflects the support it receives through the way it is associated with other places. In some ways, situation is defined by a city's *accessibility* to regions that contribute to the city's growth. Thus, situational factors are related much more exclusively to the process of urban growth, whereas site factors, as we have just seen, are related to both the initial appearance of settlement and its growth.

PATTERNS OF REGIONAL CULTURE

The territorial space of Canada and the United States has been neatly subdivided into political units—2 countries, 50 states, 10 provinces and 2 territories, over 3000 counties in the United States and hundreds in Canada, and many thousands of smaller units. These units mark and bound our political activity. They are far less reliable, however, as indicators of the spatial expression of culture. Although some of our political subdivisions do have substantial meaning as culture regions (Quebec is perhaps the best example), most do not.

The term "culture," and the concept behind it, have long been debated by anthropologists, geographers, and others. No universally acceptable definition is possible. Basically, culture is our assemblage of learned behavior. Its complexity and durability sets humans apart from other animals. A common culture often represents the strong bond that ties together the people of a country, an ethnic or social group, or a region. In turn, it is the group's heritage—its shared set of experiences—that is of greatest importance in the creation of that cultural togetherness. Culture is transmitted through symbolic means, such as language, not through biological means. Thus, a common language is an important element in the maintenance of a culture or culture region. Some have argued that one of the greatest strength's of the United States is that it is the world's largest and most populous country joined geographically and socially by a common language.

The Nine Nations of North America

Geographers and others have published literally hundreds of books and articles attempting to identify and delimit North American culture regions. Most of these studies have focused on a small territory or region (e.g., see Meinig's "Mormon Culture Region" in Chapter 13) or on a single or small set of cultural elements, such as house types; word usage—Is it pop, soda, or coke? Is it a sofa, couch, or davenport?; or food habits—a good North Carolinan can subdivide the state by regional styles of barbecue. An interesting and readable exception is Joel Garreau's *The Nine Nations of North America* [1] (Figure 3-6).

Garreau argues, as we have, that North American culture regions are very different from North American political regions.

> Forget the maze of state and provincial boundaries, those historical accidents and surveyors' mistakes. . . .
>
> Consider, instead, the way North America really works. It is Nine Nations. Each has its capital and its distinctive web of power and influence. . . .
> These nations look different, feel different, and sound different from each other, and few of their

---

[1] Joel Garreau, *The Nine Nations of North America* (Boston: Houghton Mifflin, 1981).

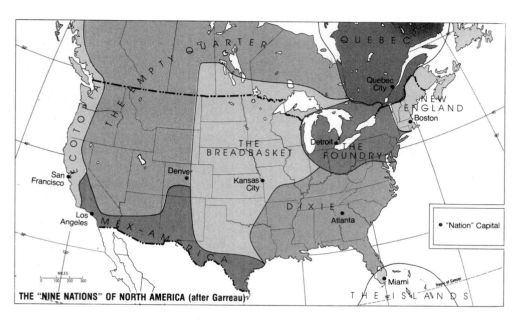

FIGURE 3-6

THE NINE NATIONS OF NORTH AMERICA. This regionalization suggests
the interrelated influences of culture, economy, and the physical environment.
(From Joel Garreau, *The Nine Nations of North America* (Boston: Houghton Mifflin,
1981.))

boundaries match the political lines drawn on current maps. Some are clearly divided topographically by mountains, deserts, and rivers. Others are
separated by architecture, music, language, and
ways of making a living.[2]

Some might question whether these are culture regions. Mountains, deserts, water—what do
these have to do with culture? In addition, is the
connection between culture and economy really
that close? Still, it can be argued that in any culture the tie to the land—the interaction between
the human and physical environments—is important. Similarly, the economy of a place often

provides a powerful cultural influence. Must we
not associate the urban economic structures of
Detroit and Pittsburgh or the wheat fields of
Kansas with their regions' cultures, just as surely
as we associate plantation agriculture with the
antebellum South?

Garreau's New England, which is quite close
in territorial extent to our Bypassed East, is (he
suggests), "the poorest of nations," yet it prides
itself on being "the only really civilized place in
North America" (p. 16). The differences between Canada's Atlantic Provinces and New
England in the United States are minimal: "the
differences between the two regions are quantitative, not qualitative. The Maritimes are more
cold, more poor, and, remarkably, even more

---

[2] Ibid, pp. 1–2.

59

beautiful [than New England]" (p. 17). A difficult but beautiful environment, an old settlement history by American standards, a shared history of British settlement, a perhaps uncommonly strong regional identity and regional pride, and poverty bred at least partly by isolation—these are this region's special ingredients.

Quebec, as mentioned earlier, is perhaps the most distinctive of North America's culture regions. The French in the province—guaranteed a separate cultural identity by the Canadian confederation and increasingly intent over the last decade on maintaining that independence—have voiced their cultural independence by electing a provincial government with a primary platform position emphasizing provincial rights and support of the French cultural tradition in Quebec (see Chapter 6).

The Foundry, as described by Garreau, is the industrial heartland of the two countries. Large, gritty manufacturing cities here face an uncertain future as heavy industry declines in importance. A resultant special concern is for the future of the American economy, a concern that springs from a fear that the Foundry may not share in the benefits of that future. A European ethnic hodgepodge reflects the diverse migration into the region from 1850 to 1915. Large central-city black populations throughout the Foundry are a result of the twentieth century migration northward by southerners in search of opportunity. Traditionally strong support for trade unionism and the concept of the working man in heavy industry as an ideal are also considered regional characteristics.

Garreau's Dixie is perhaps the most universally recognized of his nations. The impact of the plantation economy is still obvious. Its population is largely of either African or northwest European (especially British) origin. Antagonisms between the groups remain a problem, although changing attitudes and federal legislation have done much to reduce racial segregation. Poverty remains a regional concern. Yet Garreau calls this an emerging nation characterized by dramatic change. The focus of this change is

manufacturing. The result is a declining economic disparity between the South and the rest of the country and (perhaps) a blurring of the region's distinctiveness.

The Islands is perhaps the most original of Garreau's nations. He argues that Caribbean and Latin American migration to south Florida, most of it dating only from the late 1950s, has turned Miami into the capital city of the Caribbean Basin (see Chapter 10). It provides a commercial opening for the United States into that basin. Much trade with the Caribbean passes through the port of Miami. Not all of this trade is legal. It has been suggested that the international drug trade has introduced billions of dollars into the Florida economy and that this money is important in helping finance the state's building boom. The *Miami Herald* carries more news about the Caribbean Basin and Latin America than any other U.S. newspaper. South Florida, because of its southward connections, has become a center of creativity, change, and opportunity.

The Breadbasket is the conservative agricultural core of the continent. Interestingly, Garreau also sees it as "that North American nation most at peace with itself. It is the nation that works best. Based on the most prodigiously successful agriculture the world has ever known, the Breadbasket has built an enviable, prosperous, renewable economy" (p. 331). He also points out that "Stability, here, is a virtue . . . the Breadbasket has become the ratifier of what constitutes a truly mainstream continental idea" (p. 336). Most of its people are not farmers, but agriculture is a cornerstone of the economy. The virtues of farming and rural life are held in generally high esteem, even by city folk. It is also the most homogeneously Anglo of any of America's regions.

The Empty Quarter is less obviously a culture region than any of the other eight. Its obvious character is emptiness—a lack of population. The visitor to the region is more often impressed by that lack of obvious habitation and the accompanying, often dramatic, appearance of the physical environment than by the human im-

There has been a significant Cuban presence in southern Florida for a long time, but the influx of immigrants from that island country since the late 1950s has transformed large sections of the major urban areas in the region. This street corner in Miami provides vivid proof of the Latinization of the metropolitan area.

pact on the land. Indeed, the central political issues in the region center around the physical environment—water and how to use it, the development of mineral and metal resources, and public versus private ownership of land. Except for the American Indian population, this is the territory of America's last frontier. Much of the Empty Quarter's population is recently arrived; 1980 data show that 30 percent of Colorado's population did not live in the state 5 years earlier—for New York City, a similar statistic was less than 9 percent. Growth, change, and a focus on the physical environment perhaps best characterize the Empty Quarter's culture.

A Hispanic community of substantial size and territorial extent occupied portions of Mexam-

erica well before the region was acquired by the United States in the 1840s (see Chapter 14). Although Anglos are the dominant population in most of the region today, the Hispanic influence remains strong; substantial legal and illegal migration, relatively high Hispanic fertility rates, and growing interaction across the international border insures that the Hispanic influence will remain so. This is in many ways a plural society, with Spanish-language newspapers, radio stations, churches, stores, and billboards serving the needs of that considerable portion of the Hispanic population that lives largely within their own cultural community. Still, cultural interactions, however casual, are substantial and increasing. The Anglo visitor is struck by the widespread use of Spanish, whereas the resident Anglo often simply accepts it as part of the way things are.

Finally, Ecotopia is that narrow, cool and moist band of land along the Pacific coast from near San Francisco to southern Alaska. It looks to industries of the future (or, at least, the recent past), like computer chips, and finds the industry of the Foundry irrelevant, clinging to a firm belief that it has found the best of all worlds. It has a tradition of independent thinking and a willingness to consider new ideas (or, at least, many of its people so believe). Oregon, for example, has a recent history of sending men to the U.S. Senate who act as constant thorns in the side of their own party. Quality of life, as something that is somehow separate and better than just income, is considered important. Recreation is often individual and out of doors.

Garreau's Nine Nations represent just one approach to how the culture of North America can be regionalized. As noted in Chapter 1, there are many different regionalizations that can be created from one set of data. Garreau's analysis is certainly influenced by elements that cannot be considered as entirely cultural. His patterning is worth considering because of his emphasis on the complexity of regional patterns, on the notion that the elements critical to the existence of one culture region may be unimportant to another culture region, and on the dynamic nature of culture and culture regions.

Also (perhaps obviously) not all portions of a given culture region necessarily fit the defining characteristics of that region. Are the wealthy suburbs of eastern Long Island really a part of the Foundry? Is industrial Oakland, California, part of Ecotopia? Is Hartford, Connecticut, affected by the difficult climate, isolation, and poverty of New England? The answer to all of these questions is probably no even though each place is located within the larger culture region. A lesson in this is that it is important to retain the scale at which a region is described when considering the accuracy of the description.

American Religion

Of the individual elements of American culture, one of the most interesting and telling is religion (Figure 3-7). This is largely because it is an essentially conservative culture trait. If we were to ask most Americans why they are members of a particular denomination, the answer would either be because their parents were members or that they had always been members of that group.

A number of the larger Christian churches were brought to North America by European migrants. The distribution of these denominations closely matches the areas where those various migrant groups and their descendants form a large part of the population. For example, German and especially Scandinavian settlers carried their Lutheran church to the northern Great Plains and northwest portion of the Deep North. Hispanics in the Southwest; southern and eastern Europeans in the Northeast, Middle West, and in most large cities everywhere outside the South; and French Acadians who were forced to leave eastern Canada—they migrated in substantial numbers to southern Louisiana—all help explain the widespread distribution of Roman Catholicism in much of the country outside the South and Utah.

The United States has also been a place of active denominational creation. This is the result of a combination of a number of elements. One is the lack of an association between church and state and the relative wealth of American society (operating an independent denomination can be expensive) is another. A third element is the desire for national or regional identity. Several denominations, such as the Episcopalians (formerly part of the English Anglican church), were created at the end of the American Revolution in reaction against any continued dominance by Britain. The Civil War led to a similar north-south split in the country. Presbyterianism in the United States is divided into several denominations as a result of a post-Civil War split in which each subdenomination felt that the others did not adequately reflect its particular interests.

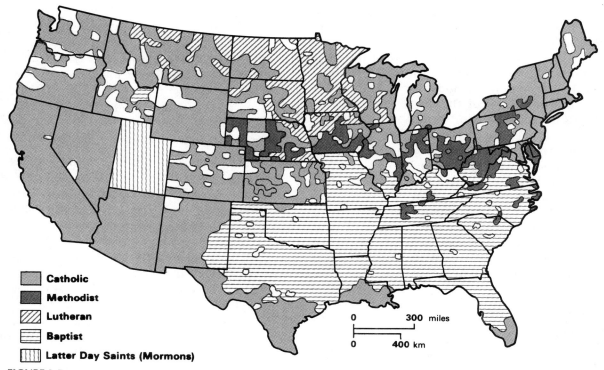

Catholic

Methodist

Lutheran

Baptist

Latter Day Saints (Mormons)

0   300 miles

0   400 km

FIGURE 3-7
PREDOMINANT RELIGIOUS AFFILIATIONS IN THE UNITED STATES
(adapted from *Churches and Church Membership in the United States, 1971* by Douglas
W. Johnson, Paul R. Picard, and Bernard Quinn. Copyright National Council of
Churches, 1974; published by Glenmary Research Center, Washington, D.C.) International migration into the United States (Catholics and Lutherans), internal
migration (Mormons), and the development of regional identity (Baptists) are
among the influences on the religious geography of the country.

Another explanation has been the creativity
of American religion. Individuals would establish their own churches. Or congregations or
groups of congregations would leave a denomination to form a new one because of disagreements over such questions as biblical interpretation or church administration. Many of these
native American churches follow a conservative
theology. Many were founded in the South. Although often today large in size, they are nowhere numerically dominant and, hence, are not
on the map.

One native American church that is on the map
is the Church of Jesus Christ of Latter-Day Saints,
commonly called Mormons. Founded in upstate
New York in the mid-nineteenth century, it was
gradually carried westward by its followers in
search of an isolated place to settle and follow
their beliefs (see Chapter 13). They eventually
chose Utah. Today most residents of Utah are
Mormons. Mormons are procreative—they believe in large families. One result is that average
completed family size (the total number of children ever born to an average family) in Utah is

the highest in the United States, standing at about 3.2 children. The average for the country is near 2.0 children.

Southern Baptists (the territory on the map dominated by Baptists is almost entirely Southern Baptist in denomination) are an interesting mixture of several of the above explanations. Baptism was brought to North America by early European migrants as a nonestablished church seeking freedom of worship. Although founded in the South, it did not become the dominant regional church until after the Civil War. During the last third of the nineteenth century, it became almost the religious expression of Southern culture, and it was easily the dominant regional church. One measure of whether a community is culturally part of the South surely must be the existence in it of at least one Southern Baptist church.

Additional Readings

Abler, Ronald, et al. *A Comparative Atlas of America's Great Cities: Twenty Metropolitan Regions.* Minneapolis: University of Minnesota Press, 1976.

Bryant, Roderick, and Kevin McQuillan. *Growth and Dualism: The Demographic Development of Canadian Society.* Toronto: Gage Publishing, 1982.

Canada Yearbook. Ottawa: Statistics Canada. (Annual)

Gastil, Raymond. *Cultural Regions of the United States.* Seattle: University of Washington Press, 1979.

Gertler, Len, and Ron Crowley. *Changing Canadian Cities: The Next 25 Years.* Toronto: McClelland & Stewart, 1977.

Greenlie, Barrie B. *Spaces: Dimensions of the Human Landscape.* New Haven, Conn.: Yale University Press, 1981.

Halvorsen, Peter L., and William M. Newman. *Atlas of Religious Change in America: 1952–1971.* Washington, D.C.: Glenmary Research Center, 1978.

Koegler, John. *Canada's Changing Landscape: Air Photos Past and Present.* Toronto: Douglas Fisher, 1977.

Population Reference Bureau. "U.S. Population: Here We Are; Where We Are Going." (Population Bulletin No. 37.) Washington, D.C.: Population Reference Bureau, 1982.

Rooney, John F., et al. (eds.). *This Remarkable Continent: An Atlas of United States and Canadian Societies and Culture.* College Station: Texas A & M University Press, 1982.

U.S. Bureau of the Census. *Statistical Abstract of the United States.* Washington, D.C.: U.S. Government Printing Office. (Annual)

Vale, Thomas R., and Geraldine Vale (eds). *U.S. 40 Today: Thirty Years of Landscape Change in America.* Madison: University of Wisconsin Press, 1983.

Zelinsky, Wilbur. "North America's Vernacular Regions," *Annals of the Association of American Geographers,* 70 (1980), pp. 1–16.

CHAPTER FOUR

MEGALOPOLIS

In 1961, a French geographer published a monumental study of the highly urbanized region located in the northeastern United States.[1] Professor Jean Gottmann's book was the outgrowth of 20 years of research on an area extending from southern New Hampshire and northern Massachusetts along the eastern seaboard of the United States to Washington, D.C., and its northern Virginia suburbs (Figure 4-1). He argued that this was a "very special region." Megalopolis, as he named it, suggested to Gottmann a type of urban region that differed from anything that had existed before. This new region was worthy of our attention because of its importance in North America, but it could also provide an indication of the future for a number of similar areas elsewhere in the world.

Megalopolis was formed by the gradual coalescence of large, independent metropolitan areas. As the population of many of the large central cities grew, the effects of this growth spilled over into the surrounding rings of smaller places. The larger suburbs made their own contributions to the total urban sprawl. The outer fringes of the resultant polynodal (many-centered) metropolitan regions (called *conurbations*) eventually began to merge with each other. Along this northeastern coast of the United States, there were a sufficient number of important centers of growth relatively close to each other to lead to the extensive urbanized region that is Megalopolis.

The dominant theme of Megalopolis is its "urbanness." Throughout the region, urban activities dominate the lives of its inhabitants, urban services provide in varying degrees for the millions who live in this region, and urban forms—dense patterns of streets and buildings, industrial centers, retail and wholesale clusters, governmental complexes—are visible almost everywhere. If one were to drive along Interstate Highway 95 between Boston and Washington,

---

[1] Jean Gottmann, *Megalopolis: The Urbanized Northeastern Seaboard of the United States* (New York: Twentieth Century Fund, 1961).

FIGURE 4-1
MEGALOPOLIS.

urban features constantly assail the senses. Office and apartment buildings, small shops and mammoth shopping centers, factories, refineries, residential areas, gas stations and hamburger stands by the thousands, warehouses to store temporarily the goods brought by ship, rail, and truck—all this and more can be viewed along the 800-kilometer (500-mile) route through Megalopolis.

But Megalopolis is more than just a collection of urban forms tied together by a web of transportation lines. It also contains many green spaces. Some are parks or other semiorganized recreation areas, but much of the green space still is devoted to agricultural uses. Within the counties defined as Megalopolis, substantial areas are farmed. Farmland in Megalopolis comprised 37.7 percent of the total land area in 1959. Although this proportion had declined to 28.1 percent by 1969 and 26.7 percent by 1978, there remained almost 3.6 million hectares (8.8 million acres) devoted to farming within a principally urban region.

Clearly, Megalopolis is a complex region. It is composed of at least five very large, but distinct, metropolitan areas and many smaller cities between and around these large urban zones. As Professor Gottmann put it:

In this area . . . we must abandon the idea of the city as a tightly settled and organized unit in which people, activities, and riches are crowded into a very small area clearly separated from its nonurban surroundings. Every city in this region spreads out far and wide around its original nucleus.[2]

In spite of the mixed character of Megalopolis, it is its massive urban presence that makes this region so important in the United States. Of the country's 38 metropolitan areas exceeding 1 million population in 1980, 7 were located in Megalopolis. Another 34 city regions within Megalopolis had at least 100,000 inhabitants each,

and the counties defined as Megalopolis contained a total of 41,114,747 people at the 1980 census. This was no less than 17.5 percent of the total U.S. population. When it is realized that Megalopolis comprises only 1.5 percent of the area of the United States, we may begin to gain a sense of the continental importance of this region. Even though there was significant outmigration from the region during the 1970s, a movement that received considerable attention, the proportion of national population that remains in Megalopolis is still extremely high.

There is much more that makes Megalopolis important than the large number of people who live there. The region also possesses an unusual economic quality. Average per capita income is quite high here. Many of the almost 17 million persons employed in Megalopolis in 1980 worked at jobs that were higher paying than that of the average U.S. worker. Transportation and communication activities are prominent in Megalopolis, partly a result of the region's coastal position. Greater than 40 percent of all commercial international air-passenger departures have Megalopolitan centers as their origin. And almost 30 percent of North American export trade passes through the six main ports of Megalopolis. The statistics could go on, but the point is clear. This has, indeed, been and continues to be a "very special region" in North America.

THE LOCATION OF MEGALOPOLIS

Megalopolis is unique in North America. But why has this particular portion of the continent developed as it has? Whenever a geographer is asked such a question—why a region developed the set of characteristics that distinguish it from other regions—the first aspect of the region considered is usually its location. In some cases, location is of secondary importance, but as we argued in Chapter 3, many times it is primary. What is it about this region's location that has helped generate the tremendous urban complex that is Megalopolis? Although Megalopolis

---

[2] Ibid., p. 5.

MAJOR METROPOLITAN AREA POPULATIONS IN MEGALOPOLIS, 1980	
New York, N.Y.-N.J.	9,120,346
Philadelphia, Pa.-N.J.	4,716,818
Washington, D.C.-Md.-Va.	3,060,922
Boston, Mass.	2,763,357
Nassau-Suffolk, N.Y.	2,605,813
Baltimore, Md.	2,174,023
Newark, N.J.	1,965,969
Providence-Warwick-Pawtucket, R.I.-Mass.	919,216
Hartford, Conn.	726,114
Allentown-Bethlehem-Easton, Pa.-N.J.	635,481
New Brunswick-Perth Amboy-Sayreville, N.J.	595,893
Jersey City, N.J.	556,972

is not a single city, it is worthwhile to consider the site and situation of this vast urban region for indications of the region's origins and growth.

Site Characteristics in Megalopolis

Many of the site characteristics of Megalopolis are visible in the region's outline. Occupying a coastal position in North America, the eastern margin of Megalopolis is deeply convoluted. Peninsulas jut into the Atlantic. Islands are scattered along the coast, some large enough to support communities. Bays and river *estuaries* penetrate the land mass throughout the region in a kind of mirror image of the land's penetration of the ocean. This interpenetrated coastline brings more land area close to the ocean and, in this way, provides easier access to cheap water transportation than would a straighter coast. This access came to be one of the major supports of Megalopolitan growth.

Proximity to water, however, does not always equal accessibility. Megalopolis also straddles some of the best natural harbors in North America. To be sure, such cities as San Francisco, Seattle, Vancouver, and Mobile are built around excellent natural harbors. But the harbors at Boston, New York, Providence, Rhode

Island, and—to only slightly less degree—at Baltimore are excellent *and* close to one another. Although Philadelphia is actually on the Delaware River, dredging maintains the port's ocean access. Numerous smaller harbors, such as those at New Haven and Groton, Connecticut, New Bedford, Massachusetts, and Portsmouth, New Hampshire, also are located along the coast of Megalopolis. In this region, proximity to water clearly is joined with many reasonably close sites that provide the means of transfer across the land-water boundary.

Why so many good harbors located close to one another? Is it coincidence or is there an explanation? Because Megalopolis is not a small region, anything that affected the whole region must have been of considerable magnitude.

As discussed in Chapter 2, the northern half of Megalopolis was covered by the most recent of the series of continental glaciers. This glaciation contributed to the excellence of Megalopolis's harbors in several ways. Sea level was much lower during the ice ages because a tremendous amount of water was held in the massive ice sheets. As the ice cover began to melt, large river spillways were formed, and the erosive power of the rivers cut them deeply into the flat coastal plain across which they rushed. When sea level rose, the lower river valleys were "drowned" to form estuaries and, thus, extended the ocean toward the land's interior. In this manner, the Chesapeake and Delaware bays are the drowned lower courses of the Susquehanna and Delaware rivers, respectively. With the navigability of the lower Hudson River enhanced as well, these waters formed some of the harbors that were later to prove useful in the development of Megalopolis.

The other major contribution is more specific to one or two locations. As mentioned before, the large amounts of soil, stone, and other debris scraped up by the expanding glaciers were deposited as moraines when the ice front retreated. One such terminal moraine is the southern arm of Cape Cod. The rise in sea level and deposition of sandy soil by ocean currents sweeping north along the east coast formed the distinctive north-reaching "forearm" of Cape Cod. An even larger separate ridge of surface material was left just south of what is now the coast of Connecticut. This latter moraine developed as an island when the seas rose, and it too was widened by deposition from the ocean. The island was not made so wide, however, that it could be called anything except Long Island.

Long Island has enhanced the harbor quality of New York in two major ways. First, from the point of view of the ocean, one end of Long Island is tucked into the harbor mouth formed by the Hudson River without actually blocking it.

The length of coastline available for port facilities, already significant along the Hudson is increased considerably. Second, from the viewpoint of the land, a large, fully developed harbor often creates an extensive urban structure and space is needed as the city grows. Urban growth was somewhat restricted to the west of the Hudson in New Jersey by tidal marshes and the erosion-resistant ridge of the Palisades. To the east of the Hudson lies only a narrow finger of land, Manhattan Island, bounded on the north by the small Harlem River and on the east by Long Island Sound and its extension, New York's East River. Beyond the East River is Long Island, a

There are many fine harbors in Megalopolis, but none can match New York's. With the Hudson River on the west and the East River on the east of Manhattan and the confluence of these rivers in the foreground also "protected" from ocean currents and storms, the city is a focus of interaction at the highest level of intensity. (Skyviews.)

flat to slightly rolling land without the barrier marshes of New Jersey. New York's boroughs of Brooklyn and Queens developed early at the western tip of Long Island, and the island offered a great deal of room for further urban expansion toward the east. In this way, both the harbor and the port city that grew around it benefited from the presence and orientation of Long Island, a landform that owes its existence to glacial deposition.

Although Megalopolis possesses many high-quality harbors, there are few other site characteristics that contributed so positively to Megalopolitan development. The climate is not exceptionally mild, although the summers are generally of sufficient length and wetness to support farming in the areas near the ports. Soils are variable, with the good soils inland from Baltimore and Philadelphia better than most found closer to New York. From New York to Boston and beyond, the thin and stony soils are generally only fair, with the exception of the notably superior soils in the Connecticut River Valley. The land resources of Megalopolis, although adequate, are hardly exceptional.

The general topographic features of Megalopolis south of New York do provide additional urban site benefits. If you were to travel inland from the Atlantic coast, a sequence of distinct physiographic regions would be crossed. A flat coastal plain is succeeded by a rolling, frequently hilly landscape called the piedmont. Further westward, you would encounter the more mountainous terrain of the Appalachians. The mountains do not parallel the coast but approach and eventually merge with the coast in New England.

Although the change from piedmont to mountains is a more obvious shift in topography, the break between the piedmont and the coastal plain may be more significant to Megalopolis. The irregularly rolling relief of the piedmont is underlain by very old, very hard rocks. This surface has resisted erosion enough to maintain the level of the piedmont above that of the coastal plain. Wherever rivers flow off the

piedmont, therefore, a series of rapids and small waterfalls are formed. Because these falls are found along a line tracing the physiographic boundary, this boundary has been named the "fall line."

Early European settlers found the fall line to be a hindrance to water navigation but an obvious source of water power. Settlements developed along the fall line, as far inland as possible while still possessing access to ocean shipping. In addition, because the fall line was often the head of navigation, goods brought inland or those to be exported had to be unloaded at these locations for transfer to another mode of transport. As we pointed out earlier, because the goods were handled in the transshipment anyway, it was clearly advantageous to break the large, bulky shipments of ocean transport into smaller loads. In many cases, manufacturing processes could be applied here as well. Thus, fall-line sites operated as *break-in-bulk* points, and the characteristics of these sites provided early support for urban growth. Within Megalopolis, fall-line cities, such as Washington, Baltimore, Wilmington, Philadelphia, and Trenton, trace the piedmont-coastal plain boundary. Only New York among Megalopolitan cities has water access beyond the fall line via the glacially eroded Hudson River Valley.

Situation Characteristics of Megalopolis

To consider aspects of Megalopolis's urban situation, it is necessary to take an even broader view than that used when looking at site characteristics. One of the most important aspects of Megalopolis's situation is the region's location relative to Europe. The coastal portions of the New England and Middle colonies were readily accessible to trading ships sailing between Europe and the New World. It was not coastal position that was critical here—that is a site characteristic. This region was also on, or very near, the most direct sea route between Europe and the productive plantations of the Caribbean colonies and southern America, at least on the homeward voyage. If we view the North Atlan-

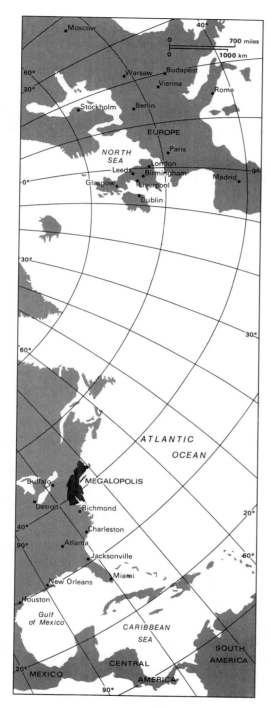

FIGURE 4-2

GREAT CIRCLE ROUTE BETWEEN THE CARIBBEAN AND EUROPE. The high quality harbors of what came to be Megalopolis were located close to the most direct sea route from the Caribbean and northwestern Europe.

tic as the focus of a map, we can see the nearly straight-line (or great-circle) route between the Caribbean Sea and the British Isles (Figure 4-2). The dominant wind patterns and ocean currents also supported this shipping route. Megalopolitan ports, therefore, were convenient stopping places and contributed actively to the transoceanic trade that grew rapidly during the eighteenth and nineteenth centuries.

Ships stopped at New York, Boston, and the other Megalopolitan ports because there was business to be transacted, there were goods to be purchased for local consumption and manufacture, and there were goods to be sold that could be found in few other places. Sugar and molasses were brought from the Caribbean to be distilled into rum. Cotton was transported from South Carolina to be woven into textiles. The obvious profitability of sea trade (or sea piracy) encouraged the growth of shipbuilding. The nearby resources, adjacent to Megalopolis and accessible to its population, were part of Megalopolis's situation and contributed to its early growth.

Critical to this growth was the location of the core cities relative to the interior of the continent. Philadelphia and Baltimore grew rapidly because each was the focus of a good-size and good-quality agricultural region. Southwestern New Jersey, southeastern Pennsylvania, northern Maryland, and much of the Delmarva Peninsula possessed good soils, among the best in Megalopolis. Access routes to the interior were constructed early and helped support the growth of the trading functions of these cities. Inland from Boston, in New England generally, the soils are too shallow and rocky and the terrain too rolling for this land to be considered good farming land. Agriculture was not as important a contributor to Boston's growth as it was to that of Philadelphia and Baltimore, although Boston did have the resources nearby to support the fishing and shipbuilding industries. The New England hills were covered with hardwood and pine forests ideal for ship construction. Also accessible were the very productive fishing banks off the New England, Nova Scotia, and Newfoundland coasts and farther south in the rich Chesapeake Bay area.

The importance of accessibility in evaluating a city's situation, however, is most apparent in the case of New York. New York's local agricultural resources are better than those of Boston but not as fine as those of Philadelphia or Baltimore. Long Island had some good land for farming but little of exceptional quality was found nearby on the mainland. New York's chief advantage lay in its position at the head of the best natural route through the Appalachian Mountains. The Hudson-Mohawk river system, later amplified by the Erie Canal and railroads, provides access to the Great Lakes. The Great Lakes, in turn, provide access to the broad interior of the continent. As settlement density and economic activity on the interior plains increased, some of the goods produced were shipped downriver to St. Louis or New Orleans, but large amounts were carried to the urban cores of Megalopolis. The city that benefited most from this growing trade was the one with the greatest natural access to the interior: New York.

Megalopolis as the Continental Hinge

It should be clear by this point that Megalopolis and the city cores that dominate it possess real advantages for growth. Although none of these cities has great quantities of what are traditionally thought of as essential resources, such as high-quality agricultural land or mineral deposits, they do share a set of resources that might be called *accessibility resources*. Such resources would be features of a place that facilitate movement into or out of that place. Good harbors, good routes to the interior, a location between regions conducting trade—all are accessibility

resources, and all are definitive features of Megalopolitan centers.

As trade grew between Europe, the Caribbean area, and the North American mainland, cities formed around the best harbors along the East Coast. Small-scale manufacturing began to appear in the larger port cities from Baltimore northward. The mutually reinforcing relationship between manufacturing and trade stimulated growth in both activities. As industry grew, the demand for labor increased, drawing immigrants from northwestern Europe or diverting large numbers of workers from farming, and the populations of these cities grew as well. Banks and other financial institutions underwrote investments in local manufacturing and shipping. Service activities, wholesale and retail business, centers of information and control, all grew and supported further urban expansion.

The greatest growth occurred within the four large port cities in Megalopolis. Each of these cities was the focus for an expanding region. The richness of each region and each city's accessibility to it were critical to the way that growth proceeded. Boston was the largest city among Megalopolitan core cities through 1750. Its advantages were those that could dominate early development—proximity to the most densely settled colonies, easy access to rich fishing banks, the shipbuilding industry, and the trade that accompanied growth in shipping. Philadelphia and, to a much lesser extent, Baltimore, enjoyed growth stimulated by early development of the rich agricultural possibilities in their general region and by the cultural and political activities focused on Philadelphia during the last half of the eighteenth century. By 1760, Philadelphia had reached a population of 18,756, exceeding that of Boston (15,631). As the land west of the Appalachians began to be developed, however, New York's preeminent site and situation began to be felt. The 13,040 people in New York in 1756 had increased to 98,373 by 1810, surpassing the population of every other city in the young nation (Figure 4-3).

The most unusual feature of this region is not

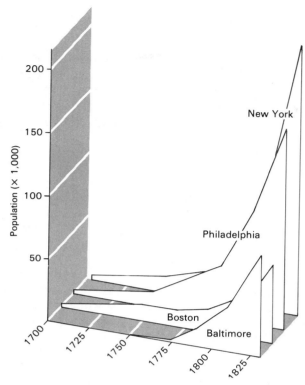

FIGURE 4-3

POPULATION GROWTH OF SELECTED U.S. CITIES, 1700–1830. The early growth of Boston and Philadelphia indicated their importance before the American Revolution, but New York surpassed both in size by the beginning of the nineteenth century.

the fact that these cities grew—the site and situation advantages were present—but that four such large cities (later to become five with the addition of Washington, D.C.) continued to grow in close proximity to one another. Washington is unique, of course, because even though it too is located on the fall line, its growth has followed directly from expansion of the national governmental structure. The other four cities, along with many smaller cities along the Megalopolitan axis, depended largely on economic stimulation. So great was national growth during the nineteenth century and so strong were the linkages

between the interior and these four ports that none of the four were able to wholly absorb the flow of goods to any of their neighbors and competitors. By the turn of this century, the combined economic resources of this region had reached the continental proportions described earlier. Remember that few of these economic resources are based any longer on local features. It is, instead, the set of accessibility resources that has come to be the most important. The flow of goods, people, and ideas enter Megalopolis from overseas and are absorbed, changed, and sent on toward the interior. Other goods, people, and ideas are transmitted through Megalopolis in the opposite direction. It is this dual set of flows that led Professor Gottmann to describe Megalopolis as the "hinge" of the continent, for this axis of urban activities functions much like a long pivot for North America's exchange with Europe, Africa, and western Asia.

THE URBAN ENVIRONMENT

The recognition that there is some unity to the many functions and forms of the component parts of the territory confirms the identification of Megalopolis as a distinct region. In a general way, the commercial "hinge" function indicates a similarity among sections of Megalopolis even if the various cities within the region do not coordinate their commercial efforts. The major parts of this region have all grown in spite of proximity, partly because the tremendous volume of goods transferred across the "hinge" has been much greater than the capacity of any single port. These port cities also have shared a pattern of growth because each has sought competitively a larger share of national export and import trade, although no one city has an absolute advantage of position over any of the others. Throughout Megalopolis, however, it is the urban form and urban functions that provide the most significant regional unity to the territory. It is the almost continuous—occasionally overwhelming—urbanness of Megalopolis that is its primary characteristic.

Any North American student will have a feeling for what is described as urban character. Images of tall buildings, busy streets, crowded housing, and industrial plants are accompanied by an impression of cultural opportunity—theaters, symphony orchestras, art museums, and large libraries—and, perhaps, by impressions of deterioration—dilapidated structures, traffic congestion, and air pollution. All of these, and more, are present in the metropolitan cores of Megalopolis. This character is also found in any major city in the United States and Canada and in most large cities around the world. What distinguishes Megalopolis, Professor Gottmann argued, is that the urban characteristics have spread out so far from their core cities in this region that these urban regions have begun to merge with one another in a process of *metropolitan coalescence*. It has proceeded in Megalopolis beyond a stage found elswhere, partly because the degree of merging is more extensive but primarily because there are so many large cities within a relatively small total region. Megalopolis is in this way a kind of gigantic laboratory in which intensively urban patterns and peculiarly urban problems can be observed developing on a large scale. It is the most urban region within an urbanized continent. Examination of its components helps set the tone for study of the other regions of Canada and the United States either through comparison or by contrast.

Population densities in Megalopolis remain high, averaging almost 295 persons per square kilometer (760 per square mile) in 1980. This average density has dipped slightly since the 300 per square kilometer (777 persons per square mile) in 1970, but it continues well above the 1960 densities of 266 per square kilometer (688 per square mile) in spite of national population redistribution trends.

Of course, the population densities in the region are not the same from place to place. Some peripheral countries have populations that average 1 person per 10 to 20 (or even more) acres. Settlement density increases as a city is approached, and high densities are reached near the city core. In New York City, for example, 1980 densities exceeded an average of 223 persons per hectare (90 persons per acre), which amounts to more than 22,300 persons per square kilometer (58,000 persons per square mile). This general pattern occurs with considerable spatial regularity. There are variations from city to city in the rate of density increase as one approaches each metropolitan center and in the height of the density peak, but the pattern repeats itself again and again. What other features of city organization are associated with this consistent density pattern? What does this association tell us about Megalopolis? How is the pattern changing, and what are the implications of these geographic changes for urban development elsewhere in North America?

Most geographers would agree that cities are fundamentally a consequence of economic organization in space. Cities do have political foundations. They do constitute a form of social clustering. Some cities have been founded or have grown because of religious stimulation. When someone decides to move or to relocate their business into a city rather than a nonurban setting, however, it traditionally has been the basic economic advantages of such a choice that have dominated the decision. So pervasive are these advantages that large numbers of people live close to one another, often closer than they would prefer, and tolerate many negative consequences of this dense clustering to participate in the benefits of city organization.

An increasing number of urbanites have sought to minimize some of the disadvantages of city living by moving their residences to suburban locations. Others have moved even farther out, well beyond the suburban fringe, into a zone referred to as exurbia. From small towns, areas of converted vacation homes, and rural estates, exurban commuters travel each day (or each week) to distant workplaces. But this spread of population has not eliminated the disadvantages of clustering, only shifted the population faced with them. It has also resulted in the spread of workplaces for those living in the region; a smaller

The very great demand for land is dramatically portrayed in the abrupt contast between New York's Central Park, preserved for recreation, and the tall, densely packed buildings crowding the park's borders. (Allen Green/Contact.)

proportion of a sprawling metropolitan population still find it necessary to enter the central city for work. The spread has also created a wide variety of new problems, many of which are geographic. Before examining these new problems and their implications, it is useful to have clearly in mind the essential features of the urban landscape of Megalopolis.

Components of the Urban Landscape

Interaction. In general, the cost of moving something is directly proportional to the distance it must be moved (i.e., the farther the item must be moved, the more it will cost to move it). Cities exist as clusters of activity because it has been advantageous to minimize movement costs.

Because *spatial interaction* is so basic to the existence of a city, the imprint of this interaction should be apparent. The visibility of interaction depends on what is interacting. If it is the human population that moves, the lines of interaction are visible in streets, subways, bridges, tunnels, sidewalks, and parking lots. In older cities, such as those in Megalopolis, the downtowns were laid out during the period of travel by foot and horse-drawn carriage; therefore, they may have only about 35 percent of their total central areas devoted to the visible features of human interaction. In newer cities, developed largely after the rise of the automobile, the proportion devoted to these visible features is much higher. In Los Angeles, for example, as much as

62 percent of the city's downtown is devoted to automobile uses. The critical importance of movement from place to place is shown by these high proportions of urban land exclusively used to facilitate interaction.

Other forms of interaction may be equally important but less visible. The easy movement of information and ideas is supported in cities by the intensively developed telephone, telegraph, and teletype systems as well as other communications systems of this nature. Only 60 years ago, telephone lines were not wound into cables and placed underground. The central business districts of all large cities were crowded with above-ground crisscrossing spiderweb networks of telephone lines testifying to the intensity of communication demands. The lines are no longer visible, and it is just as well because intraurban information flow has multiplied many times during the last half-century.

The movement of people and the movement of ideas and information are the two most obvious examples of interaction within cities, but many more things move within cities and exhibit the same general pattern of most intense movement near the city core. Also hidden, and taken for granted unless something goes wrong, are such items as electricity, a fresh water supply, and sewage and garbage disposal. Until the volume of garbage reached high levels because of urban residential densities, it could be fed to animals, buried, or discarded in nearby dumps without becoming a general health hazard. In areas of rural density, individual septic tanks are often sufficient to handle a single family's sewage safely, and each dwelling can have its own well without overdrawing the underground water supply. Whereas some of these typically "urban" services have been extended into nonurban areas, the extension is recent. Even electricity was not extended into most of rural America until well after the federal Rural Electrification Act was passed in the 1930s. In many cases, it still may not be necessary in rural areas to provide the full range of services found within the cities.

An example of the tremendous intensity of

As cities grew during the late nineteenth century, communication became intense. Telecommunication technology was still crude enough to lead to scenes like this one on Broadway in New York City in 1889. The landscape gradually changed after 1900 as wires began to be laid underground. (Culver.)

urban interaction was provided by a story in the *New York Times*. In the process of constructing a new subway line, work was slowed by the intricate networks of utility lines buried beneath the pavement of Second Avenue in New York City. In one block, the crews found

one 30-inch and two 10-inch gas mains; a four-foot water main and a 12-inch water main; a box sewer three and one-half by two and one-third feet; 16 three and one-half inch ducts for electrical wires along with 43 three-inch ducts; 64 three-inch ducts for telephones and 30 four-inch ones. In addition, they found footings for the old El;

house connections for sewers, water and gas; cables for fire alarm-system boxes, cables for police call boxes and Transit Authority cables.[3]

Remember, all this was in only *one block* and found in the process of constructing a new interaction line!

The intensity of intraurban interaction is a reflection of the diversity as well as the density of activities. Movement between spatially distinct locations does not just happen. A cost must be borne because of the movement, and, by implication, a return must be expected to make the movement worthwhile. Perhaps the prime motivation for interaction is the geographic separation of supply and demand. Economically, this means that an item or a service is needed or desired at a location that cannot meet that need; the good or service must be obtained at another location, and interaction occurs. When two or more places are deficient in items that can be obtained at the other places in the set, they are said to be complementary places or to have *spatial complementarity*. When locations are complementary, interaction will take place unless they are too far apart, there is some closer place to obtain the good or service, or there is some barrier to movement.

Functional complexity. The implications of all this for city landscapes should be clear. A city is a region of considerable functional complexity. There are a great many different kinds of activities in a relatively small area. Some functions tend to cluster together, whereas others are scattered more generally across the urban region. Because a wide variety of activities are carried on in a city, interaction between functions or within zones of the same function is stimulated. When the activities are mapped, the result is a greatly mixed pattern of land use. The generalized pattern of planned land uses in Philadelphia provides a modified example of the complexity that is presently found there as well as in other major cities (Figures 4-4).

_____
[3] *New York Times*, February 18, 1973, p. 69.

Philadelphia's land area is dominated by residential and industrial uses. The waterfront is the primary industrial region. Low-cost water transportation and Philadelphia's role as a hub of international oceanic trade combine to generate the location of industry. It is also important to note the impact that this land-use pattern has on the rest of the city. Almost the entire waterfront along the Delaware River is planned for industrial uses,

FIGURE 4-4
LAND USE PATTERN IN PHILADELPHIA. Even in this somewhat simplified map, the complex intermingling of urban functions is apparent.

Residential

Industrial

Commercial, Institutional and other

Parks and Recreation

as is a large area on the lower Schuykill River. When we add the two or three linear industrial areas extending across the city, much of Philadelphia is seen to be close to some form of industrial activity. Although industrial activity within the city has declined during the last decade as firms move elsewhere or go out of business and although it is difficult to anticipate how the land uses will change, the complex pattern remains.

Masked by the generalized nature of the categories on the map is the tremendous diversity of function within each land-use type. The densely packed and intensively interacting central business district of Philadelphia is not differentiated on the map from the retail and wholesale clusters located to the north and west of the city. Large shopping centers are not indicated in a manner that is different from that of the headquarters of a large insurance company, for instance, whereas the functions themselves are clearly complementary and produce interaction flows. Other interaction flows arise from the institutional activities such as the city government, various centers of higher education, the State Hospital and the adjacent county prison farm, and the Philadelphia Naval Shipyard. This wide diversity of functions is surrounded by, interspersed with, and often interacts with the resident population of ethnically and socioeconomically diverse communities. This pattern is repeated in other major cities of the region in spite of differences in urban form.

Public services. A third aspect of the urban landscape is implied by the dense population, the complex functional mix, and the high level of interaction stimulated by these diverse activities. The concentration of population and urban activities requires supporting functions as well. These support functions are only indirectly productive, in an economic sense, but they are necessary for commerce and industry. These public-service functions are traditionally organized and operated as various branches of city government.

As the settled urban environment becomes more concentrated, by definition the services provided by the urban administrative structure also become more intense geographically. In addition to the services mentioned earlier—water, electricity, sewage, and garbage collection—cities operate more spatially concentrated services, such as police and fire protection, construction and maintenance of public movement facilities, health care, documentation of vital population statistics, and educational facilities.

There is much variation in the quality and quantity of these services from city to city—some cities, for example, extend support to cultural activities as well. A few cities have recently chosen to turn provision of some of these services over to private firms, but it is the large number of people and functions clustered in urban areas that requires the presence of such services in some form.

As the United States and Canada have become increasingly urbanized—that is to say, as a larger proportion of the total population has come to live in urban areas—many of these services have been extended far beyond metropolitan limits. In nonurban areas, nonurban governments (county, state, and occasionally federal) have chosen to provide support for public services, although to a less intensive degree. Such services are often demanded by urbanites who have moved to nearby suburbs or semirural communities because they consider these services to be essential. Thus, these small communities may be forced to change their tax structure and institute services more typically found in larger urban places.

So pervasive have these services become in large cities that an immense governmental structure has developed to administer them. In the New York metropolitan region, to take the extreme case, more than 1550 distinct administrative agencies were in operation in 1982. More on this will be discussed later, but it should be clear at this point that an extensive network of services is an important corollary to functional diversity and the interaction flows within urban areas, such as those within Megalopolis.

Accessibility. A fourth major component of the urban landscape is the high level of accessibility found there. This too is related to the other components. Accessibility is created and maintained as a public service, but one that is demanded by the great amount of interaction and shaped by the spatial arrangement of land uses. The degree of accessibility found at a given location also may affect the decision to locate an activity at that site. And, of course, limited access between two places is a constraint on the amount of interaction possible.

Easy access between most sections of an urban area has not always been as dominant a consideration in city organization and structure as it has become in recent decades. The original street plans of most cities in Megalopolis, for example, followed the "rational" and simple grid pattern popular in the seventeenth and eighteenth centuries. Square or rectangular grids were laid down as fully as local terrain allowed in New York, Philadelphia, Baltimore, Washington, D.C. (with modifications), and most of the smaller cities and towns along the eastern seaboard. Even in Boston, where the configuration of the harbor and the winding rivers that enter the harbor make the square grid awkward, portions of the city were laid out in this fashion.

During the early years of city development, when the functional mix and interaction it stimulated were less intensive than today, the grid pattern was satisfactory. As these cities grew,

The major port cities of Megalopolis continue to be important shipping and industrial centers. This view of south Philadelphia across the Delaware River demonstrates the impact of these functions. (Courtesy Delaware River Port Authority.)

however, the inadequate access provided by this street system began to become apparent. A square grid, for example, possesses frequent right-angle interactions. Because the flow of traffic is interrupted at each intersection, an increased volume of traffic leads to longer pauses at each intersection. By 1900, Baltimore and Boston had exceeded populations of 500,000, Philadelphia had reached nearly 1.3 million, and New York was approaching a population of 3.5 million. The main impact of the automobile was then still in the future, but serious congestion was already present at these cities' centers. As long as all sections of the city were subjected to the same conditions of limited accessibility, pressures to change the form of access were not great enough to have much effect. When all sections of the population had to live within walking distance of work or within reach of the trolley tracks, none suffered significantly poorer access than any other.

Following World War II, a series of changes occurred that led to the rapid areal expansion of cities and increased use of automobiles for travel. An increasing proportion of the work force in the cities began to live at distances and in residential densities that made it uneconomical for mass transit to reach them. Economically, the speed and flexibility of truck transport accelerated the diversion of short-haul freight from rail to road carriers. These changes plus many others increased demand for access within the city center, between the center and the periphery, and eventually between sections of the periphery.

Intensity of change. All of this accentuates another component of the urban landscape. This is a landscape of change. All landscapes change, but urban patterns shift and proceed through major rearrangements that are made more vivid by the intensity and complexity of urban functions. Tens of thousands of new residents enter a large city like Philadelphia or New York each year. Even greater numbers have been leaving, some to distant cities, some only into the metropolitan fringes. Structures are destroyed, new ones built. Street patterns are altered, the pattern of functions is changed, and flows of people, goods, and ideas are shifted to fit these new patterns. The changes may be observed in any major region in North America, but the changes, in some ways, actually created Megalopolis. After 1940, this region became less and less a compact series of large cities performing somewhat similar functions for the continent. Instead, increasingly, it became a single urbanized region. The transformation of Megalopolis into a "super city" is far from complete and will probably never occur because of yet other changes, but the process is more advanced here than anywhere else on the continent. The impact of these changes and the problems they have raised may well provide insight into some of the patterns that are developing elsewhere.

CHANGING PATTERNS IN MEGALOPOLIS

Perhaps the most fundamental and far-reaching change in Megalopolis during the last 40 years has been the great areal expansion of each of the major metropolitan areas (Figure 4-5). Greater New York clearly has extended its population the farthest, but the Boston, Philadelphia, Baltimore, and Washington urban areas have also grown greatly. New York had both the largest initial population and the most intensive economic concentration, but the other three port cities possessed firm foundations for growth as well (see Chapter 5). At the same time, the federal government rapidly expanded its operations. The District of Columbia was no longer of sufficient area to absorb the burgeoning civil-service population and the growing number of people needed to feed, clothe, and otherwise serve them. Urban development spilled into Virginia and Maryland.

The spread of urban population far beyond city limits had a strong impact on rural activities in Megalopolis. As pointed out earlier, substantial areas within Megalopolis have yet to become urbanized, but the cities have had a profound influence nevertheless. There are two types of

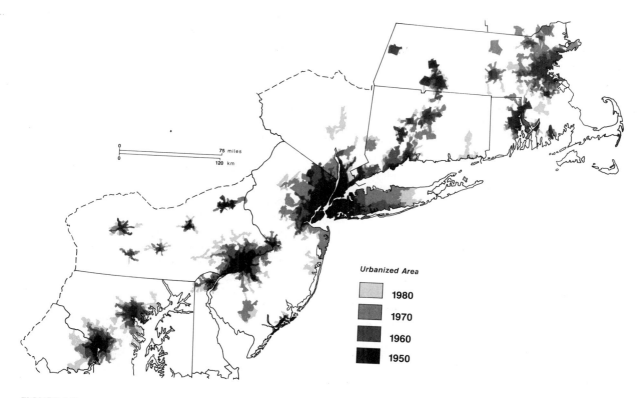

FIGURE 4-5

SPREAD OF URBAN POPULATION IN MEAGALOPOLIS, 1950–80 (after C.
E. Browning). The sprawling metropolitan regions in Megalopolis have grown at
different rates, but the tendrils of urban densities from each now reach far from
the original cores.

such influence. First, as city populations grew, a
greater number of people not directly engaged
in food production had to be fed by foodstuffs
shipped in from rural areas. The tens of millions of people in urban Megalopolis consume
agricultural products from all across America and
beyond. Many of those farming the agricultural
land close to the cities, however, chose to specialize in higher priced foods and in products
with high perishability because the distance to
market was short and the price of the final product did not need to include high transport costs.
Dairy products, tomatoes, lettuce, apples, and a
wide variety of other intensively produced "ta-

ble crops" came to dominate farm production in
sections of rural Megalopolis.

Accompanying this shift in agricultural emphasis was the second influence of expanding
urbanization. As land on the margins of urban
areas was approached by dense settlement and
intense economic activities, the price of land was
driven up. Good farmland purchased at a few
hundred dollars per acre in the 1920s or 1930s
might be sought for residential subdivision at
many times the price. For example, a 100-acre
farm purchased decades ago for $20,000 as ag-
ricultural land might eventually be sold to a real
estate developer for $1 million. The developer,

in turn, would subdivide the land into 250 home lots of 0.4 acres each and then, after putting in utilities and streets, sell each lot for $15,000, or a total $3.75 million.

Even if a farm family were able to resist the profitability of such a sale, the taxes on the land would rise sharply toward urban levels as nearby areas began to be used for urban activities. The only way to remain in farming was to pursue intensive agricultural production devoted to high-value products. The consequence is a pattern of high-value farm-product operations clustered in between the expanding urban regions of Megalopolis (Figure 4-6).

The spread of urban centers and the corresponding shifts in agricultural practices in Megalopolis did not occur only because of pressures exerted from the interior of these cities toward their peripheries. Also important has been the pull between the major urban nodes along the main lines of interurban access. Strong flows

FIGURE 4-6
AVERAGE VALUE PER FARM OF FARM PRODUCTS SOLD, 1978. The intense and generally high-value agriculture in Megalopolis may be best represented in the counties of Maryland's Eastern Shore, southern Delaware, southern New Jersey, and eastern Long Island.

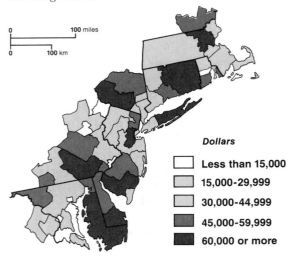

	Dollars
☐	Less than 15,000
☐	15,000-29,999
☐	30,000-44,999
☐	45,000-59,999
■	60,000 or more

of traffic developed early between the cities of Megalopolis. Each center stimulated growth in the others through complementarity as well as competition; transportation and communication lines were needed to carry the interaction. When people who continued to work within the major cities relocated their residences, it was only natural that a high proportion chose sites that allowed them easy access to the main cluster of workplaces. The arterial highways and, to a lesser extent, the interurban railroad and main feeder roads became the seams along which metropolitan populations spread first and farthest. As a consequence, the urbanized areas merged first along these main interurban connections. And the demand for increased accessibility generated the construction of even better interurban movement facilities.

As the populations of these urban areas grew, the composition of the populations also changed. Prior to 1910, the cities of Megalopolis absorbed large numbers of immigrants from Europe. During the 1840s and 1850s, 4.3 million immigrants arrived in the new world from northern and western Europe, with 70 percent of these coming from Ireland and Germany (Chapter 3). Large numbers of Italians, Poles, Russians, and Austro-Hungarians as well as other small national groups arrived from Europe throughout the latter half of the nineteenth century. These migrants passed through one or another of the large Megalopolitan ports. Those who did not continue westward into the farming areas or urban centers of the Midwest and Great Plains settled in dense clusters within Megalopolis's cities, usually forming communities of each nationality. These communities within the very large urban areas offered cultural and linguistic support for the new arrivals. Many of these communities have had enough social cohesion to remain spatially identifiable today (see pp. 118–119).

When World War I broke out in Europe, the flow of emigrants stopped suddenly and a new set of migration flows began to filter into Megalopolis. The new stream of migrants was, in fact, an expansion of an already established population movement. After 1910, the slow trickle of black migration from the southeastern states began to grow. Black migrants repeated the patterns of settlement used by migrant groups from Europe. Most blacks settled within the cities in areas that were already occupied by small black populations.

As the black migration continued through midcentury, population densities increased and generated outward expansion from the original core areas. Often after many decades of population increase within a city, black densities began to increase in outlying regions of black settlement as well. In New York, for example, the black settlement area in the borough of Queens on Long Island did not show a census tract over 90 percent black until 1930, long after much of Harlem on the island of Manhattan had become an entirely black area.

This fragmented expansion of black settlement also illustrates the increased social and economic differentiation within ethnic and racial groups in these cities. As portions of each group found better paying jobs and developed "mainstream" attitudes that helped broaden these opportunities, families moved toward the city's edge or beyond. Faced with individual and institutional discrimination, blacks and some other non-European minorities did not generally move as early or as far or in as large numbers as other urban groups, but they too fit the trend (Figures 4-7, 4-8). In this way, the urban pattern of intense core and diffuse periphery is reinforced again. The most intense ethnic and racial separation near the city center gradually changes to loss of ethnic identity as one travels away from the center.

Central-city populations have been declining since the mid-1950s at least, but much of the population leaving the large core cities relocated to one of the nearby suburbs. Thus, the total metropolitan population continued to grow. Much has been written about this flight to the suburbs on the part of those who could afford to move out and would not be barred suburban entry by racial discrimination.

Although central-city populations continue to decrease and, in general, the suburban rings continue to increase, the 1980 population census indicated that suburban populations are no longer growing as rapidly as earlier. This is an entirely new aspect of urban change that appeared during the 1970s in the metropolitan regions of the United States. There are some indications that this change is national in scope, but it is the largest and oldest cities, such as Megalopolis's ports, that demonstrate the change most dramatically.

Sometime during the late 1960s, for the first time in U.S. history, people began to leave the largest metropolitan regions—when both central city and suburbs are considered together—in greater numbers than others moved in. Suburbs were still growing, but their growth no longer exceeded central-city decreases. Smaller cities and towns and nonmetropolitan areas in general have been the primary recipients of these population shifts.

Because Megalopolis comprises the greatest single collection of old large metropolitan areas but is also a region of tremendous economic energy and potential, it is difficult to foresee the impact of the national population redistribution on this region as a whole. The core-city populations have shared in the trend (Table 4-1) and many of the suburbs showed declines when the 1980 census data were released, but the geographic advantages that stimulated commercial, financial, industrial, and cultural activities during two centuries of urban growth are still present. Much of Megalopolis is composed of medium-size cities and nearby rural areas that are located between the largest urban cores. In spite of the significant shift in national population away from the northeast, many of the smaller places within Megalopolis continue to grow. One trend in population redistribution is reducing the number who live in the region as a whole while another is rearranging the pattern of population within the region.

It should be clear by now that urban areas are landscapes of change and that the changes in

FIGURE 4-7

ETHNIC DIVERSITY IN BOSTON, 1910. Areas with greater than 10 percent of their population native to the ethnicity or race are indicated.

·FIGURE 4-8

ETHNIC DIVERSITY IN BOSTON, 1970. Areas with greater than 10 percent of their population native to the ethnicity or race are indicated. After 1910, the city's black neighborhoods spread south between areas that were primarily Irish.

Megalopolis suit the unusual character of the region. The changes here have been continuous; they have been drastic; and they have occurred on a scale unmatched anywhere else in the world.

Of course, there are clear reasons for the region's dynamic nature. Each aspect of change is a response to demands and pressures that could not be resolved without some change. Some changes came about because of conditions hundreds or thousands of miles from Megalopolis, conditions that led to immigration, for example, or conditions that altered the cost of oceanic shipping. Some changes occurred within Megalopolis in response to developments inside the region. In each case, new problems were created as previous pressures were resolved. The new problems often appeared more complex and more difficult than those faced earlier, although this may be largely because they were new. Even as officials attempt to find solutions, conditions continue to change. In a general sense, however, there are categories of urban problems of a geographic nature suggested by the experiences of Megalopolis.

85

TABLE 4–1
Megalopolis Central City Population (1980)

City	1980 Population	Population Change 1970–1980	Percentage Change
New York	7,071,639	−823,924	−10.4
Philadelphia	1,688,210	−261,786	−13.4
Baltimore	786,775	−119,012	−13.1
Washington, D.C.	638,333	−118,177	−15.6
Boston	562,994	−78,007	−12.2

SOURCE: *General Population Characteristics* (Washington, D.C.: U.S. Department of Commerce, Bureau of the Census, 1982).

PROBLEMS IN MEGALOPOLIS

It should be made clear that the problems faced by the residents of Megalopolis are more than the problems of any large city multiplied four or five times. Megalopolis is *not* a super city with super problems. Many of the difficulties of organization and growth are those found in most large urban areas. But this region also contains many large metropolitan areas competing economically while merging their territories in a complex, uncoordinated manner. And problems have arisen that are different from those of large cities—in kind as well as degree. In addition, the cities of Megalopolis are among the oldest of the presently large cities in the United States.

Most of the problems of Megalopolis can be treated in the context of three characteristics: density, accessibility, and spread. The first two include problems that are fundamentally urban, occurring in the cities of Megalopolis on a somewhat larger scale than elsewhere. The third characteristic also includes a set of problems that are megalopolitan (multiurban) in nature.

Disadvantages of Density

High population density and spatially intense economic activity are typically urban patterns. In and of themselves, neither can necessarily be expected to create problems in urban areas. The undesirable consequences of high density and intense activity are generated when either becomes greater than can be handled by social and political organizations. Problems do not arise until the results of these features can no longer be dissipated by means that are considered feasible.

Problems of high population density in Megalopolitan centers have come about through a more rapid influx of migrants than could be accommodated by housing openings. The core cities of Megalopolis have contained overcrowded residential areas for a great many years, perhaps ever since they existed as cities. As discussed earlier, migrants from Europe poured into these cities faster than they could be accommodated. Not only were new housing structures constructed more slowly than the increase in population, but also immigrants usually settled in locations far away from the new building sites. The magnet of social or cultural cohesion and the constrictions of low migrant incomes and restrictive property-tax structures resulted in extreme overcrowding in those sections of the city that had some of its oldest buildings and, thus, where owners had less incentive to maintain those buildings.

Intense concentrations of economic activity also produced problem conditions that are exaggerated by the high urban-population densities in most Megalopolitan cities. Economist Kenneth Boulding has argued that most attention in the economic system has focused on the production of goods and that too little attention has been paid to the accompanying production of "bads." He is pointing out that waste is a necessary by-

Abandoned buildings, evidence of inadequate public services, and limited local recreational opportunities are representative landscape characteristics in many sections of the older Megalopolitan core cities. (Courtesy H.U.D.)

product of human activity. Sufficient volumes of these wastes are now being produced that they can no longer be ignored. This is especially true in areas where human activity is highly concentrated—that is, in urban areas. If maps were constructed showing waste density, that is, volume of waste per unit area, a familiar general pattern would be exhibited: high values near the city core with values declining irregularly outward to the urban periphery. The problem, then, is not that waste is produced in large quantities but that it is produced faster than its negative aspects can be neutralized. When this happens, the excess spills over into the environment, the environmental balance deteriorates, and the result is recognized as pollution.

Urban areas are subjected to several kinds of pollution. Air and water pollution are among the most noticed forms. Industrial activity frequently is singled out is the primary polluting activity in urban areas; in many cities, industries do contribute greatly to environmental deterioration. It would be a mistake, however, to lay the chief blame for urban pollution on industry. The main cities in Megalopolis possess much lower proportions of heavy industrial activity than do many cities along the Great Lakes, yet their air- and water-pollution problems are severe. Washington, D.C., for example, has almost no heavy industry, but its population will testify to frequent smoggy days every summer.

Even more important than industry are what

87

are euphemistically called municipal sources. These sources are actually the urban dwellers who generate increasing individual levels of waste in such high densities that it cannot be locally absorbed. Hundreds of thousands of vehicles are driven in and out of each core city every day. Millions of homes are heated every winter and cooled every summer with public-utility produced electricity or public-utility supplied natural gas. Extremely concentrated areas of office space in the city center are artifically heated or cooled. Millions of gallons of waste water are expended from washing machines, dishwashers, garbage disposals, and flush toilets every day. And all of this is concentrated within a few hundred square miles. In the New York metropolitan region, for example, 5 million cubic yards of treated sewage residue (sludge) is produced each year. The accumulation of this sludge spread evenly over Manhattan would bury the entire island to a depth greater than 3 feet in just 12 years. And in Megalopolis, the pattern of urban concentration repeats the problem again and again within a small portion of the continent.

There are two approaches offered to the problems of pollution when the problems are addressed at all. One is the technological approach—develop and apply better ways of neutralizing the negative aspects of waste. The other is the ecological approach—find a better balance with nature's capacity to absorb waste by producing and consuming less, thereby reducing the output of waste. A third choice, the spatial approach—reduce waste density by spreading population and economic activity over a greater area—might be more often discussed, but there are already other spatial problems associated with an increasingly dispersed population.

Accessibility and Density

High population densities are strongly associated with difficulties in getting from one place to another. Where activities are intensely concentrated, there is a high demand for movement facilities. Congestion occurs when the activity sites become so densely packed and the level

of interaction between them becomes so intense that the routes for interaction become overcrowded. Real access is reduced although demand for access may continue to increase. The phenomenon is familiar as a daily cycle to all who have lived in, or visited, the core areas of major cities. Morning and evening rush hours find movement so difficult that "rush" is no longer an appropriate description of these periods of the workday.

The problems of intracity accessibility are exaggerated in Megalopolis by two features. One is the obviously large number of densely populated urban areas, including the continent's extreme case in New York City. Manhattan south of Sixty-first Street, for example, absorbs more than 1.5 million additional people between 7:00 A.M. and 10:00 A.M. each average workday, and it loses them again each evening. The other factor is related to these cities' sites. All were founded and grew because of their easy access to ocean shipping and the excellent harbor facilities provided by these sites. The coastal configurations, however, create severe problems for land movement. Boston and Baltimore have no land approach to their cores from the east and southeast, respectively. New York and Philadelphia cannot be reached directly by land from the southwest and the south. The rivers within these cities and in the District of Columbia congest movement by funneling it across a limited number of bridges and through tunnels. When they are all open to traffic, there are 14 bridges (including 3 for railroads) across two rivers in the District of Columbia, 22 bridges into or within Philadelphia, and 20 major routes of access for land movement to Manhattan Island in New York. Baltimore, Boston, and many smaller cities in Megalopolis—New York's New Jersey suburbs; Wilmington, Delaware; Providence, Rhode Island; and Connecticut's Bridgeport, Hartford, and New Haven—also possess transportation networks that are constricted and limited by the expensive need to bridge or tunnel under water barriers.

The cities of Megalopolis have approached this

All large cities in Megalopolis have responded to the demand for interaction by building numerous limited access highways. By eliminating intersections where such routes cross each other, the capacity of both is increased. At times, as in this case on Baltimore's 29th Street Expressway, the result appears more complex than the problem. (M. E. Warren/Photo Researchers.)

problem of congestion in several ways. The first method used when existing facilities become overcrowded is almost always to increase the capacity of the facilities. Thus, 4-lane roads are widened to 8 lanes, and 8 lanes are widened to 12 lanes—or even 16 lanes. Limited-access expressways are constructed parallel to other limited-access expressways, and additional telephone, water, sewage, and electricity lines are installed. Sooner or later—often shortly after construction—these expanded routes are carrying traffic up to, or in excess of, their designed capacities. It has been clear from this repeated experience that traffic congestion will not be alleviated solely by building more and wider roads. Some argue that traffic actually is encouraged to increase when new roads are built.

A second method for alleviating congestion is proposed frequently but, for a variety of reasons, is seldom used successfully. As an alternative to increasing movement capacity by building parallel routes, each existing line can be used more efficiently. An average automobile at rest occupies almost 20 linear feet of road and can contain up to five or six adults in moderate comfort, although only through carpooling are such high-passenger levels reached. A bus, on the other hand, may occupy 50 linear feet of road, but it can carry eight times as many persons as a car. This ability of mass transit to move much larger numbers of passengers per unit road space than the private car appears to offer a clear opportunity to increase movement capacity by making more efficient use of existing lines of access.

In fact, however, there are two even clearer indications that this apparently greater efficiency is not so obvious a solution. First, mass transit already carries a large proportion of commuter traffic into some Megalopolitan cores but has been in serious financial difficulties for decades. More than 73 percent of all commuters entering Manhattan in 1970 traveled by bus, ferry, railroad, or subway. Nevertheless, several commuter rail lines went bankrupt, subway fares continued to rise, and bus facilities struggled to keep pace with the demand.

The second indication, strongly related to the first as both partial cause and partial effect, is a long-term decline of mass transit ridership. Movement by private automobile gives at least the impression of greater convenience and flexibility of travel and greater personal safety to the passengers as well as the illusion of lower costs per trip. Only when fuel is less available (as briefly in 1973 and again in mid-1979) has public transit ridership increased significantly. Employer and municipal attempts to encourage carpooling have met with modest success.

Federal, state, and local governments have re-

Almost everyone wants to go somewhere else in cities, at least several times each day. Because there are so many alternative origins and destinations and so many people moving within a small area, each trip may be considerably slower than might be expected simply from the distance traveled. This is especially true when different modes of interaction must share the same movement space, as shown in this photo. (Richard Laird.)

sponded to private transit demands by constructing hundreds of miles of new expressways, building new bridges or increasing the capacity of existing ones, and in other ways expanding capacity for private movement in and around Megalopolitan centers. Meanwhile, public transportation has, with few exceptions, declined in quality and capability. If fuel prices rise rapidly again, more intraurban travelers may choose mass transit rather than private automobiles. However, the reasons for individual travel decisions are complex and not all are based on an economic rationale.

One aspect of urban landscape change that is well underway in Megalopolis is operating to alleviate the congestion resulting from centralized activities. As suburban populations have grown, many stores and services dependent on proximity to large numbers of potential customers have relocated out of the central city. Urban workplaces have also shifted toward suburban and exurban locations rather than remain in their former metropolitan core locations. With more activities available outside the central city, fewer trips are taken into what had formerly been the main focus of metropolitan congestion: the central city's central business district (CBD). A 1978 *New York Times* survey[4] of suburban residents, for example, found that 25 percent made fewer than five nonbusiness trips per year into New York City and another 25 percent said that, except for business, they never went into the city at all.

For many years, it was believed that the trend toward urban decentralization would have less effect on total traffic congestion in this larger region than in other urban areas. Megalopolis's cities contained a distinctive employment mix with a disproportionately large labor force in managerial and office work, that is, in white-collar occupations. These positions require daily decision-making or are activities that support those making the decisions. Such a concentration of white-collar jobs was supported until recently by the advantages of face-to-face interaction in the decision-making process. As a result, Megalopolis possesses a higher proportion of the continent's libraries, universities, and other sources of information and ideas than can be found in an equivalent area elsewhere. Finance and insurance (activities controlling investment); publishing firms, major magazines, and newspapers (indicating the flow of information); and cultural leadership for much of the nation (through theater, opera, symphonic orchestras, and such activities as clothing fashions) are all much more important components of Megalopolis than comparable areas elsewhere in the United States and Canada. These activities employ, or are a response to demand from, white-collar workers.

However, even Megalopolis's cities have recently experienced an outflow of decision-making jobs in unprecedented numbers. Among the nation's top 1000 corporations, for example, there are now more corporate headquarters in New York's suburbs than in the city. To some degree, this relocation may have been stimulated by an accumulation of the disadvantages associated with central-city location. But improvements in computer technology and tele-communications also have facilitated the rapid exchange of information without requiring the former level of centralization—and such improvements promise to continue.

The overall occupational structure of the central cities' employment has changed dramatically during the last 25 years, especially since the early 1970s. For example, New York City's decline in manufacturing employment has been great (Table 4-2). Losses in some sectors through 1983 were not balanced by a comparable growth in finance, service, and governmental jobs. The result was a total decline in employment of more than 30,000 jobs during the previous 15 years, whereas national employment grew by more than 50 percent. The continued decline in manufacturing lies at the root of the city's overall decrease in jobs. In fact, computer applications reduced the labor requirements of data management and information retrieval; these tasks had been expected to more than make up for the loss of manufacturing jobs. Finally, the locus of national decision making was shifting from the private sector (and New York) to the federal government (and Washington, D.C.). Just as this region's cities led North America during a period of centralization, they may be leading as strongly the early stages of a long trend toward decentralization.

Some Repercussions of Sprawl

As urbanized areas expand, many of the problems found in the metropolitan cores are exaggerated. Public services are provided at near-urban intensities farther and farther from the central cities, and an increasing area is exposed to the dumping of waste products from urban activities. Instead of polluting short stretches of a few large rivers within the cities, nearly all rivers and streams in the metropolitan region suffer from deteriorating quality. More than one-third of the suburbanites interviewed in the *New York Times* survey[5] believed their communities were becoming "more like a big city" with all its attendant problems.

[4] *New York Times*, November 12, 1978, p. B1.

[5] Ibid.

TABLE 4-2
Distribution of Jobs by Industry, New York City, 1968–83 (in thousands)

Industry	1968	1978	1983	Net Change 1968–83
Manufacturing	840.0	538.9	444.7	−395.3
Wholesale and Retail	745.7	620.2	702.7	−43.0
Transportation and Public Utilities	320.3	259.8	268.7	−51.6
Construction	102.5	69.2	87.4	−15.1
Finance, Insurance, and Real Estate	437.1	418.6	522.1	+85.1
Services	747.9	818.1	1081.4	+333.5
Government	526.0	500.7	581.7	+55.7
TOTAL	3721.7	3226.9	3688.7	−33.0

SOURCE: *Employment and Earnings, States and Areas, 1939–1974,* Bulletin 1370-11, U.S. Bureau of Labor Statistics. Washington, D.C.: 1975; and *Employment and Earnings,* Monthly Reports, U.S. Bureau of Labor Statistics. Washington, D.C.: November 1979; December 1983.

There are two somewhat contradictory trends that follow from the slow aging of the interior suburbs. One is further urban sprawl as inner suburban residents relocate farther out toward the metropolitan periphery. The opposite trend is reflected in the still-modest inflow of population as central-city residences are rehabilitated in ways that will attract middle- and upper-income families and individuals. In spite of disadvantages, the large cities of Megalopolis possess great economic and social possibilities. These "new" inner city locations are attractive to a small but significant number of America's metropolitan population.

An equally serious consequence of Spread City—as the sprawling metropolitan regions have been inaccurately called—is that they contain not one city but a great many independent municipal areas. Everyone familiar with the great cities of the United States has an image of a large city surrounded by a multiplicity of smaller suburbs. In fact, the process of suburbanization is now so advanced that many of the smaller cities near the major centers are hardly *sub*urban. The urban character that typifies the original core city now extends uninterrupted far beyond its legal boundaries and includes hundreds of independent urban territories. The extreme case, undoubtedly, is the New York Metropolitan region, which contained some 1556 distinct political units in 1982 (Figure 4-9), a situation one observer described more than two decades ago as "a governmental arrangement perhaps more complicated than any other that mankind has yet contrived or allowed to happen."[6] This pattern is repeated at varying scales throughout Megalopolis and much of the rest of urban North America.

The implications of this fragmented pattern of jurisdictions are staggering. The economic geography of the metropolitan region does not coincide with its political geography. Most people live in (and pay the bulk of their local taxes to) a city different from the one in which they work. Neighboring cities compete for the "right kind" of business or industry. The desirable industry is nonpolluting, pays a high average wage, and can contribute significantly to the municipality's tax base. The desirable resident is mid-

[6]Robert C. Wood, *1400 Governments* (Cambridge: Harvard University Press, 1961), p. 1.

FIGURE 4-9

NUMBER OF JURISDICTIONAL UNITS IN NEW YORK METROPOLITAN AREA, 1982. Governmental operations are made somewhat chaotic in a metropolitan region with so many separate jurisdictions when the urban problems to be faced are rarely contained within formal administrative units.

dle or upper income. These characteristics are desired because they permit the city to pay for the services it must provide and yet continue to remain independent of the "less desirable" central city. The portions of the population and industry most able to support these public services through their taxes are also the ones most able to move to a new location if local conditions deteriorate. As the movement out of the core city continues, the burden increases for those who remain. New York's brush with bankruptcy in 1976 and Cleveland's in 1979 are only a partial reflection of these trends, as both cities' improved solvency indicates.

The central cities' problems are not isolated difficulties peculiar to the older sections of large metropolitan regions. These cities are economically and socially woven into the spatial fabric of their entire urbanized areas. It is coordination and cooperation between the hundreds of politically independent municipalities in each region that has been lacking. Attempts have been made to develop regionwide coordination by the establishment of metropolitan study and planning regions that cross county and state boundaries. Boston, New York, the Delaware Valley area (Philadelphia), the Chesapeake Bay area (Baltimore), and the National Capital region (Washington) all established the beginnings of planning commissions whose jurisdictions far exceed the geographic limits of each large central city. The real impact of each commission, however, is variable and certainly has never been strong. Furthermore, the trend within Megalopolis and in other large metropolitan areas in the United States lately has been toward a decrease of influence by these multimunicipal agencies in regional coordination and planning.

A logical question at this point is: Has there ever been any equivalent regional planning agency for all of Megalopolis? No, there has not, and it is unlikely that one would be effective even if in existence. Megalopolis is a massive region of coalescing metropolitan territories with each territory containing a distinct urban core. But Megalopolis, as we said, is *not* a super city. The volume of interaction between the metropolitan regions is still much less than within urban areas. The opportunity for interaction between the several regions is not much greater than the opportunity each has with cities elsewhere on the continent.

Although Megalopolis is not a super city and its preeminence has declined slightly during the last two decades, it clearly remains a complex urbanized region of tremendous importance to the rest of the United States and Canada.

Additional Readings

Alexander, Lewis M. *The Northeastern United States,* 2nd ed. New York: Van Nostrand, 1976.

Bergman, Edward F., and Thomas W. Pohl. *A Geography of the New York Metropolitan Region.* Dubuque, Iowa: Kendall/Hunt, 1975.

Browning, Clyde E., (ed.). *Population and Urbanized Area Growth in Megalopolis, 1950–1970.* (Studies in Geography No. 7.) Chapel Hill: University of North Carolina, Department of Geography, 1974.

Clay, Grady. *Close-up: How to Read the American City.* New York: Praeger, 1974.

Conzen, Michael P., and George K. Lewis. *Boston: A Georgraphical Portrait.* Cambridge: Ballinger, 1976.

Meyer, David R. *From Farm to Factory to Urban Pastoralism: Urban Change in Central Connecticut.* Cambridge: Ballinger, 1976.

Muller, Peter O. *Contemporary Suburban America.* Englewood Cliffs, N.J.: Prentice-Hall, 1981.

Muller, Peter O., Kenneth E. Meyer, and Roman A. Cybriwsky. *Metropolitan Philadelphia: A study of Conflicts and Cleavages.* Cambridge: Ballinger, 1976.

O'Brien, Raymond J. *American Sublime: Landscape and Scenery of the Lower Hudson Valley.* New York: Columbia University Press, 1981.

Olson, Sherry H. *Baltimore.* Cambridge: Ballinger, 1976.

Stansfield, Charles A., Jr. *New Jersey: A Georgraphy.* Boulder, Colo.: Westview Press, 1983.

Thomas, Jean-Claude Marceau. "Washington, D.C.," in *Contemporary Metropolitan America,* vol. 4. Cambridge: Ballinger, 1976.

CHAPTER FIVE

THE NORTH AMERICAN MANUFACTURING CORE

The United States and Canada are countries in which manufacturing is a most important economic activity. The evidence of this is everywhere, in articles of clothing, items of preserved food, residential structures, means of transport and communication, and in many other ways. In spite of the clear imprint of items manufactured in Europe or Asia, domestic industry remains paramount in both countries. Similarly, manufacturing plants are numerous, especially in the United States. It is rare for any town in these countries larger than 15,000 population to be without at least some local employment in manufacturing.

Manufacturing may be found throughout the more populated regions of North America, but it is not distributed evenly across the continent. The northeastern United States together with southern Ontario—but excluding northern New England—comprises the single most significant region of manufacturing on the continent (Figure 5-1). Although the region straddles the international boundary, it has little obvious impact

on the region's shape. This territory of heavier manufacturing concentration is loosely defined on three sides by the Ohio River Valley, Megalopolis, and the southern Great Lakes. The western margin of the region is less clear; it blends gradually with agriculture-dominant landscapes across southern Indiana, Illinois, and beyond. Although manufacturing may be found west of the Mississippi River in many important clusters, this activity's impact is definitely less concentrated than within the region defined as the continent's manufacturing core.

In spite of this region's moderate extent, it is of tremendous economic significance in North American geography. The manufacturing region indicated in Figure 5-1 occupies about 3 percent of the total land area of the United States and Canada. However, this region contains factories that produce 80 percent of these countries' steel and nearly the same proportion of their motor vehicles and motor-vehicle parts. Ranked in terms of 1981 sales volume, 8 of the top 10 and 22 of the top 30 industrial corporations in

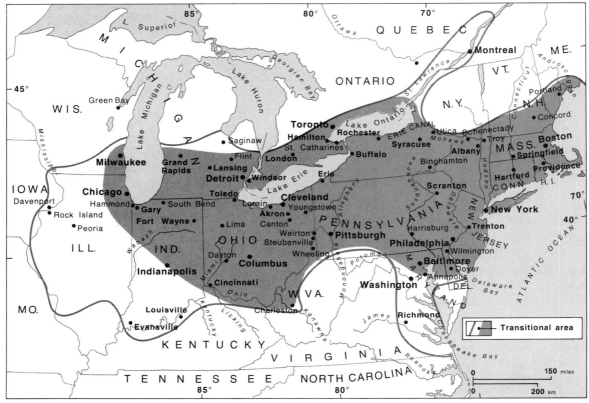

FIGURE 5-1
NORTH AMERICAN MANUFACTURING CORE REGION.

North America had their headquarters located within the region. In addition, 8 of the 25 largest U.S. Standard Metropolitan Statistical Areas (SMSAs) in 1980 and 20 of Canada's 52 cities over 50,000 population in 1981 were located in this region.

Because of its clear significance, the region frequently is described by geographers as the North American *core area*. A core is generally thought of as the center, as containing the heart, as comprising the essential characteristics of a thing or place. The geographer's approach to the idea of a core area is not easy to define precisely because it contains different elements of human activity. The political core of a country may not extend beyond the metropolitan region

of the nation's capital city, or it may be the region from which the national government receives its major support. The economic core might contain many cities, extensive agricultural lands, and numerous industrial complexes all linked together by a dense network of transportation and communications. A cultural core may be defined only arbitrarily because important components of a nation's culture may originate in separate locations. The key to the difficulty lies in identification of locations or areas that contain the bulk of the most essential components of each aspect of a country's human organization. By outlining the region of greatest overlap, the core area is defined.

This region of North America contains much

of what would allow it to be considered the continent's core area. Most of both countries' industrial capacity is within the region. The most important ports, the main centers of communication, and the primary financial centers for both countries are within, or very near, the region. Both countries' political capitals, although outside the boundaries drawn here, are on the immediate margins of the region. The intensity of urbanization is much higher in the region than in any other territory of comparable size on the continent, and the region includes the two larger clusters of coalescing metropolitan areas. Much of the region contains land yielding great quantities of essential agricultural products. Manufacturing, however, is one of the two underlying features that make this territory distinctive and that have provided the economic support for much of what gives the core its character. To be more accurate, then, the region described in this chapter should be called the North American manufacturing core.

This core area is given its highly complex, multiple, but distinctive flavor by a set of landscape features. This is a region of many large cities and even more cities of medium size. It is a region that experienced growth of numerous industrial concentrations and the transport networks demanded by them. It is a region of smokestacks and factories, some of which may appear massively drab on the outside but can be vibrant and energetically alive inside. It is a region of great ethnic diversity, for the industrial cities of the manufacturing core grew just as waves of migrants left central, eastern, and southern Europe for new lives in the United States and Canada.

As complex as this may be, a comprehensive understanding of the continental core is made even more difficult by its strongly dual character. Much of the territory occupied by the *manufacturing core* is also occupied by a major portion of the continent's *agricultural core*. In many respects, it was the vitality and productivity of the territory's farm population that created the resources and the demand for industrial pro-

duction. Success in agriculture supported the region's early market centers, but it was the gradual mechanization of agriculture that demanded diversified manufacturing support. Mechanical reapers, winnowing machines, and cultivating implements by the tens of thousands were required during the later 1800s. Tractors, hay balers, pumps, other supporting equipment, and increasingly specialized farm machinery continued to be important local sources of industrial demand during the first half of the twentieth century. And transportation lines were improved and expanded to carry the tremendous volume of agricultural products produced on the region's farms. Thus, agriculture stimulated early industrial and urban growth, and successful manufacturing and rapidly growing urban centers supported continued intensification and improvement in agriculture.

Therefore, we encounter here a single portion of North America that must somehow be treated as two thematic regions without losing sight of the interdependence between the two aspects of the same territory. One theme, the urban and industrial nature of the region's manufacturing centers, is discussed in this chapter. The other theme, the rural and agricultural character of the territory's small towns and countryside, is presented later (Chapter 11), where it offers a contrast with another strong and agrarian culture, that of the Deep South (Chapter 9).

The territories occupied by the two thematic regions of the continental core are not identical; the manufacturing core extends farther east to include Megalopolis, whereas the agricultural core reaches westward at least to the edge of the Great Plains. The regional label, the Midwest, is often applied loosely to the territory of overlap located in the United States, but a portion of the continent's core region is located in Canada. Furthermore, the landscapes that have evolved as reflections of each theme—the urban-industrial and the rural-agricultural—are too distinctive and too important to be presented without specific and separate consideration of each.

MAJOR METROPOLITAN AREA POPULATIONS IN THE MANUFACTURING CORE, 1980			
New York, N.Y.-N.J.	9,120,346	Akron, Ohio	660,321
Chicago, Ill.	7,103,624	Syracuse, N.Y.	642,971
Philadelphia, Pa.-N.J.	4,716,818	Gary-Hammond-	
Detroit, Mich.	4,353,413	East Chicago, Ind.	642,781
Toronto, Ont.	2,998,947[a]	Northeast Pennsylvania	640,396
Boston, Mass.	2,763,357	Allentown-Bethlehem-	
Nassau-Suffolk, N.Y.	2,605,813	Easton, Pa.-N.J.	635,481
Pittsburgh, Pa.	2,263,894	Grand Rapids, Mich.	601,680
Baltimore, Md.	2,174,023	New Brunswick-Perth Amboy-	
Newark, N.J.	1,965,969	Sayreville, N.J.	595,893
Cleveland, Ohio	1,898,825	Jersey City, N.J.	556,972
Cincinnati, Ohio	1,401,491	Hamilton, Ont.	542,095[a]
Milwaukee, Wis.	1,397,143	Youngstown-Warren, Ohio	531,350
Buffalo, N.Y.	1,242,826	Springfield-Chicopee-	
Indianapolis, Ind.	1,166,575	Holyoke, Mass.-Conn.	530,668
Columbus, Ohio	1,093,316	Wilmington, Del.-N.J.-Md.	523,221
Rochester, N.Y.	971,230	Flint, Mich.	521,589
Providence-Warwick-		Long Branch-Asbury Park, N.J.	503,173
Pawtucket, R.I.-Mass.	919,216	Lansing-East Lansing, Mich.	471,565
Dayton, Ohio	830,070	Paterson-Clifton-Passaic, N.J.	447,585
Albany-Schenectady-		Harrisburg, Pa.	446,576
Troy, N.Y.	795,019	New Haven-West Haven,	
Toledo, Ohio-Mich.	791,599	Conn.	417,592
Hartford, Conn.	726,114	Canton, Ohio	404,421

NOTE: [a] Indicates 1981 population.

It should be kept clearly in mind that manufacturing is carried on in many centers beyond the immediate boundaries used here to define the North American manufacturing core region. No major populated portion of the continent is without some light manufacturing. Even the more intense pattern of heavy industrial activity based on steel and metals fabrication, typical of the core region itself, declines only gradually, diminishing westward from Chicago across northern Illinois and from Cincinnati along the Ohio River. Should the region be extended to include Peoria, Illinois, Davenport, Iowa, Rock Island, Illinois, and even St. Louis, Missouri? Should the region reach downriver from Cincinnati, Ohio, to include Louisville, Kentucky, Evansville, Indiana, and the lower Tennessee River industries? A case could be made for, and

just as easily against, such extensions. Any argument of this type, however, only serves to reinforce the concept of regional boundaries as transition zones (see Chapter 1). It also serves to raise a set of questions concerning the permanence of the regional outline and, indeed, about the impact of changing patterns of economic activity on the permanence of a manufacturing core region.

Turning first to the manufacturing theme, we know where the core is, but why is it located there? What set of conditions or circumstances led to the development of so complex a mix of economic interrelationships on this portion of the continent? What is it about this region that encouraged the growth of *heavy manufacturing industries* and all of the related human activities that have come to dominate here? General answers can be suggested by reviewing the features of North America that are essential to manufacturing: raw materials, transportation, and the population characteristics of labor and market. More specific answers can be approached by considering the interplay between the several manufacturing components and the variations within the manufacturing core.

MINERAL RESOURCES AND RELATIVE LOCATION

As discussed earlier (Chapter 2), the metallic mineral resources of the United States and Canada are found in three broad zones of metamorphic rock: the Appalachians, the western mountains, and the Canadian Shield. The mineral fuels, on the other hand, are found under much of the continent, especially beneath the sedimentary lowlands west of the Appalachians and between the Gulf of Mexico and the Arctic Ocean. The quality, quantity, and variety of minerals are extremely good. In many cases, the deposits could be mined with no more than average difficulty. The United States and Canada, in other words, are large countries, and they have been well blessed with industrial resources. It is not surprising that manufacturing in the United States and Canada came to be significant.

Two additional geographic factors are critical in defining the size, character, and location of manufacturing clusters in North America—relative location and accessibility. Consider first the patterns of metallic mineral concentration in North America and their placement on the landmass: a long northerly arc covering the northeastern quarter of the continent (the Shield) and two linear areas, one extending northeast-southwest (the Appalachians) and one extending northwest-southeast (the Rockies). Next, consider these patterns with respect to each other. Clearly, the eastern and western mineral zones form a broad V cut off at the base, and the Canadian Shield is almost fitted into the northern open end of the V. As a consequence, the broad interior plains are nearly enclosed by these zones

of metallic minerals. Furthermore, much of these same interior plains are underlain by large deposits of high-quality mineral fuel, especially in the eastern section where the Shield and the Appalachians are only narrowly separated by the St. Lawrence lowland. In terms of the mineral requirements of heavy industry, then, the triangular region bounded by the Appalachians, the Canadian Shield, and the Mississippi River contains much of what is needed within a relatively small area.

These three factors, resource quality, resource quantity, and close relative location undoubtedly would have led to the development of a concentrated manufacturing core in this general region even without additional advantages. But these other advantages do exist in the form of what might be called natural accessibility resources. Accessibility is a critical concept in human geography because it defines the ability to interact between locations and the relative ease and costs of movement. Accessibility can be created or improved through human effort. Roads can be built or widened. Railroads can be constructed and maintained. Airports, canals, and pipelines are all means by which the accessibility of a place might be improved. Investment in transportation facilities, however, involves the clear risk of financial loss unless there is an overwhelmingly obvious need for these facilities. Navigable waterways, on the other hand, are natural routes of access that do not require equivalent investment or contain the same possibility of financial loss. The presence of such waterways, therefore, is a very real resource when movement facilities are needed.

The interior manufacturing core of North America possesses great accessibility resources. Connecting the mineral-rich Canadian Shield and the fuel-rich interior plains, the Great Lakes represent an internal waterway unlike any other in the world. The five Great Lakes are interconnected with only two significant changes of elevation. A small drop of about 6.7 meters (22 feet) between Lake Superior and Lakes Huron and Michigan is overcome by locks at Sault Sainte

River barges on the Ohio are usually pushed by a single boat in groups two or three abreast and as many as six units long. This group is moving slowly past one of the coal-powered electric generating plants along the river. Note the variety of freight carried in this one barge train. (Charles E. Rotkin/PFI.)

Marie. The much greater change of elevation between Lakes Erie and Ontario might have been a more serious barrier to water transport, but the Welland Canal (first opened in 1829) was built in Ontario to skirt Niagara Falls and the Erie Canal was constructed (by 1825) in New York to avoid Lake Ontario altogether. With these exceptions, the lakes offered hundreds of miles of inexpensive transportation to early European developers of North America. Well over 800 kilometers (500 miles) of cheap transport is available to those moving the Shield's iron ore to the coal in Illinois, Ohio, or Pennsylvania. Some of the basis for industrial capacity presently located along the margins of the Great Lakes—in cities such as Chicago, Gary, Detroit, Cleveland, Hamilton, Toronto, and many others—can be attributed to this natural accessibility resource.

Although the Great Lakes are closed to traffic during the winter months, this has not eliminated their transportation importance.

Within the interior core, additional means of natural accessibility are available. Flowing westward from deep within the coal-rich Appalachian region, the Ohio River crosses the interior plains for hundreds of miles, dividing the Eastern Interior Coalfield before joining the Mississippi River. Dozens of tributaries supply the Ohio with its water and also provide further accessibility, either directly because they are navigable or less directly because they offer easier routes for land movement through their valleys. Along the western margin of the core region, the Mississippi River and its tributaries provide access from the south and west. It is also a means of natural access between the industrial capacity of the core and the tremendous agricultural potential of the interior plains.

So unique is this combination of spatial and mineral resources between the Appalachians and the Mississippi that the manufacturing core in the United States is often thought of as this territory alone. References to the industrial Midwest or more dramatically to America's industrial heartland may seek to fire the imagination, but they are incomplete. The interior manufacturing core is a direct western extension of the large ports of Megalopolis. The Appalachians form a barrier to easy transportation, except along the Hudson-Mohawk valley that connect New York to the interior, but land connections were constructed to at least neutralize this barrier. A consequence is that the North American manufacturing core includes both the interior core *and* Megalopolis, the urban region through which the interior is linked to international commerce. The essence of Megalopolis is urbanism and exchange, but Megalopolis also contains

The valley of the Mohawk River in central New York is an excellent passage through the Appalachians. In addition to the river itself, a railway, two highways, and the limited access New York State Thruway provide direct surface transport connections between Megalopolis and the inland cities of the continent's manufacturing core region. (New York State Thruway Authority.)

101

massive industrial development and is fundamentally a part of the continental manufacturing core region.

Growth of the Region

The continental core region is a development of the nineteenth century. This portion of North America was not recognized as possessing distinctly greater potential for manufacturing than other parts of the continent during the early years of European expansion and settlement. The initial paths of settlement penetration were along the natural lines of access, the waterways. Families moved into this region from the Ohio River, along its many tributaries, and from the margins of the southern Great Lakes. Average population densities of 18 persons per square kilometer (46 per square mile) were reached almost everywhere by 1880, in what was to become the manufacturing core region, but the overwhelmingly dominant economic activity was agriculture. In fact, it was not until the 1870s that manufacturing industries began to appear in intensities that encourages extension of the tentative label of continental core region to the area west of the Appalachians.

The foundations for growth of the region are reflected in the gradual shift of transportation concentration as railroad lines began to be spread across the interior plains (Figure 5-2). Prior to 1830, urban and industrial development of the region was limited almost entirely to the Atlantic Coast in the port cities' immediate *hinterlands*. European settlement of the trans-Appalachian area was still tentative, limited to scattered subsistence agriculture and a few urban outposts.

Between 1830 and the outbreak of the Civil War, population density in the interior increased, agriculture intensified and began to produce a regular surplus, and demand grew for functionally efficient centers of exchange. This was the period of rapid increase in water-freight transport on the Ohio, on the Mississippi, and on the Great Lakes. The Erie Canal was constructed to provide a direct link between the interior and the rapidly growing port of New York. The other great ports of Megalopolis could not immediately match the support capabilities of the Erie Canal, but they could attempt to gain their portion of the productive interior by land transportation. Baltimore, with the Baltimore and Ohio Railroad, and Philadelphia, with the Pennsylvania Railroad, drove for the interior plains of the Midwest through gaps in the Appalachians. Not to be outdone, New York was also connected via the Mohawk Valley with the construction of the New York Central.[1] Boston, north and east of the other ports, found its main lines of access to the interior preempted by the other cities' connections. Therefore, Boston intensified the rail net in its more immediate hinterland much earlier than other cities. Canadian interest spurred construction and consolidation of rail lines between Montreal and Toronto and from there westward across Ontario to Michigan and southeastward to Portland, Maine. The last decade of this period, 1850 to 1860, foreshadowed the tremendous intensification of interaction that was to follow the Civil War interruption, as an interwoven web of rail lines was flung across Ohio, Indiana, and Illinois.

After a decade of slowed growth, the 1870s produced several changes that allowed the core region's full range of advantages to be realized. The development of steel allowed replacement of the original iron rails. This, in turn, allowed heavier railroad equipment, higher speeds, and longer hauls than had been possible before the 1870s. Similarly, the *rail gauges* and rolling stock were standardized to allow interline exchange and more efficient movement. The rapidly increasing demand for steel and the new ability to move economically the desired bituminous coal in large volume over longer distances and with more speed than earlier combined to increase exploitation of the accessible Appalachian coalfields.

---

[1] Students who might puzzle over the name, New York Central, should note the location of the Mohawk Valley within the state.

FIGURE 5-2

UNITED STATES AND CANADIAN RAILROADS: 1850, 1860, and 1880 (with permission, Association of American Railroads). The tremendous increase in developed land mobility across the northern half of the United States is in sharp contrast to the more limited rail construction in the South and in Canada.

Looking eastward across downtown Pittsburgh, its strategic location at the conflu-ence of the Allegheny River (to the north) and the Monongahela River (to the south) can be easily understood. The outline of Fort Pitt is visible in the park be-low the tall office buildings. Note also the conflict between the demands of water movement and land movement as demonstrated by the many highway and rail-road bridges. (Courtesy Gulf Oil Corporation.)

Cities in the coal region grew rapidly with the expansion of steel production. Pittsburgh is a prime example. Located well within the coal-producing region, it also enjoyed natural acces-sibility to the west by occupying a site at the head of the Ohio River and commanded good land connections to the east and west because it had been an early center of growth during the im-mediate postcolonial period. Other cities less closely tied to coal or steel production began to grow as coal and steam provided alternative sources of industrial energy to water power. Manufacturing was no longer tied to waterfront

location, although if the waterway was navigable all of the nonpower advantages remained.

These changes and the further intensification of agriculture are reflected in the rapid multi-plication of rail lines by 1880 (Figure 5-2). With the spread of coal-powered electric generators during the 1880s and a continuation of im-provements in transportation, a self-reinforcing cycle of growth was well underway before the end of the century, and the manufacturing core of North America was clearly outlined.

The technological changes that directly af-fected the manufacturing geography of the

United States have been grouped by one geographer into a framework that is useful here. Professor John Borchert suggested that the pattern of metropolitan growth could be related to conditions that existed during four periods, or historical epochs as he called them.[2] Five categories of city population size were established, and the relative growth or decline of a city during each "epoch" was indicated by its change from one size category to another.

The earliest period, 1790–1830, was called the *Sail-Wagon Epoch* by Professor Borchert. During

this 40-year period, almost all cities and towns were associated with water transportation (Figure 5-3). The Atlantic ports and towns that had their beginnings along some of the coastal rivers were the major urban centers. The greatest inland urban growth during this period occurred along the main inland waterways—the Mohawk River, the Great Lakes, and the Ohio River. Land transportation was still too rudimentary to support equivalent city growth.

The second period, 1830–70, was triggered by development of the railway, a radical innovation in land movement. The *Iron Horse Epoch* at first stimulated further growth in the already established port locations, both coastal and on the inland waterways. The new railway net-

[2]John R. Borchert, "American Metropolitan Evolution," *Geographical Review* 57, 3 (July 1967), pp. 301–332.

FIGURE 5-3

CHANGE IN RANK OF SELECTED URBAN PLACES: EPOCHS OF METROPOLITAN GROWTH (after Borchert). The relative growth experiences of many cities in the U.S. manufacturing core illustrate the shifts in the importance of each city's site and situation advantages as the country's economy and technological base developed after 1790.

Large steel mills dot the southern margins of the Great Lakes where rail-carried northbound coal meets southbound iron ore transported by Great Lakes ore carriers. This Inland Steel mill on the shore of Lake Michigan near Chicago is a good example. (G. R. Roberts.)

works were constructed to focus on the port cities. Aside from additional growth of the larger port cities in what was soon to become Megalopolis, the greatest growth occurred in such cities as Pittsburgh, Cincinnati, and Louisville (all on the Ohio River); Buffalo, Erie, Cleveland, Detroit, Chicago, and Milwaukee (all on the lower Great Lakes); and St. Louis, Memphis, and New Orleans (all on the Mississippi River). Many other cities grew and a few declined in population, but the key to growth during this period was expansion of the hinterlands of already existing port cities plus expansion behind the frontier of settlement. Because of the importance of the Ohio

and the Mississippi rivers and of the Great Lakes, most of the rapid urban growth during this period was within what came to be the manufacturing core region.

The *Steel-Rail Epoch,* 1870–1920, was stimulated by the development of steel, the replacement of iron rails with stronger and heavier steel rails, increased demand for bituminous coal, and the spread of electric-power generation. The greatest growth in national urban areas occurred in cities only peripheral to the manufacturing core, but there were several notable exceptions. The exceptions were the numerous smaller cities near the coalfields, near the Great

Lakes, or on one of the major rail connections between larger cities. These were the cities able to establish themselves because of the interconnecting rail network that so densely crisscrossed the region between the Ohio and the Great Lakes. Akron, Canton, and Youngstown, Ohio, are clear examples because they are located between the coal-and-steel city of Pittsburgh and the iron-ore port and steel city of Cleveland. The greatest decline during this period occurred among those cities most tightly tied to the navigable waterways. Within the core region, the Erie Canal cities (Syracuse, Utica-Rome, and Albany-Schenectedy-Troy), Louisville on the Ohio River, and St. Louis on the Mississippi were the most important losers west of Megaolopolis.

Perhaps even more important for the formation of the manufacturing core region, this period saw two distinct factors shift within the generalized manufacturing process. Before 1870, the greatest costs borne by manufacturers were the shipment costs and the power costs. In considering shipment of raw materials and the finished product, raw material shipment was more expensive because a portion of the raw material would be used up or discarded during the process of manufacture. Therefore, industry tended to be located near the raw materials or where they might be shipped cheaply. That is, industry tended to locate in the port cities on navigable waterways. This locational tendency was reinforced by the continuing dominance of water power as an energy source. Toward the end of the century, however, cheap and efficient railroad transportation had removed some of the advantages of waterway location for industry. The spread of electric power removed additional dependence on port cities. Although advantages remained, for some port cities or for some industries, the general effect was to make it more worthwhile for many manufacturers to locate near their markets. In the Canadian section of the manufacturing core, southern Ontario developed more rapidly than did southern Quebec. Ontario shared the important Great Lakes waterway and it also lay much closer to the growing markets of the interior core in the United States. Toronto, Hamilton, and Windsor all have the advantages of both accessibility and market proximity.

Professor Borchert indentifies a fourth epoch, the period since 1920, as the *Auto-Air-Amenity Epoch*. The main effects of transport innovations, such as the automobile and the airplane, have been to increase individual mobility and to minimize the impact of shipment costs in the production process. Thus, migration within the United States has been directed at those places perceived as having amenities (California, Florida, Arizona). However, because shipment and communication costs declined and this strengthened further the market and labor orientation of industry begun in the previous epoch, industry has been drawn to areas of greatest population growth; these are primarily the amenity areas.

The best description for the epoch might be stable growth until the 1960s with clear prospects of future decline within the manufacturing core region. Large markets and labor pools exist; manufacturing and the linkages between industries are well established. The main exceptions to the general stability have been slower than average growth in the steel city of Pittsburgh and more rapid than average growth (until the 1960s) in the automobile-production centers of Detroit, Flint, and Lansing, Michigan.

It is possible that the two countries are entering still another period, one that was not yet apparent when Professor Borchert wrote, one which might be called the *Information Technology Epoch*. As the economies of the United States and Canada become more dependent on the production and exchange of information, the means of processing and transmitting this information will encourage the growth of industries that do not need cheap bulk transportation or even large population clusters. This suggests that those factors supporting growth in the large manufacturing core cities will no longer provide those cities with special development advantages during future decades.

THE ECONOMIC CHARACTER OF CITIES IN THE REGION

In discussing the patterns of growth within the North American manufacturing core, we have dealt almost entirely with metropolitan centers. In fact, until recent decades, little manufacturing in this part of North America was to be found located beyond the immediate vicinity of urban areas. More than a coincidence of location, there is a striking parallel in much of the United States and Canada between growth patterns of urbanization and manufacturing industrialization. However, growth stimuli in one city may be quite different from forces stimulating the growth of another city. In this way, places that accumulate urban intensities of people and economic activities can possess a distinctly different economic character. As a result, the visible landscape of the cities of the North American core, as defined through the components of the many urban economies, also varies.

The economic character of North America's manufacturing core cities falls into two broad regional categories: (1) that associated with the Atlantic coastal cities and the narrow region east of the Appalachian Highlands and (2) that associated with the cities of the interior core between the Ohio River and the Great Lakes. Aspects of the former group of metropolitan areas have been described, for this is Megalopolis and its periphery. With Boston, New York, Philadelphia, and Baltimore based early and firmly on commerce and its financial underpinnings, these ports and their satellites began to accumulate population long before manufacturing became dominant in the national U.S. economy. Although manufacturing was drawn to the eastern coast by the promise of matchless local markets and tremendous labor supplies, the economies of most of Megalopolis's cities maintained a distinctly professional character.

One section of the eastern core developed manufacturing at the same time that its ports were growing. In this, New England was exceptional. So early was New England's industrial development that this region can be called North America's manufacturing *hearth*. More densely populated as a region than other portions of British colonial America, agricultural opportunities were relatively poor and nonagricultural labor became available earlier here than elsewhere. Shipbuilding industries thrived along the coast and generated countless subsidiary manufacturing operations needed to supply such a complex industrial undertaking. When factory industry began to develop in North America, New England found itself with several advantages. Experience and continued immigration from Europe were important, but the ready availability of power in New England's small but abundant rivers was primary. Although many of the textile industries that laid the foundations of the region's manufacturing development eventually shifted their sites to the South Atlantic states (Chapter 9), the highly skilled residual population drew in enough replacement industry to maintain a slow regional population growth and a high proportion of that population in manufacturing (Table 5-1).

Boston, as the regional capital of New England, characterizes many of the changes in this portion of the continental core. Boston's industry contains remnants of an earlier period in apparel and leather manufacture, but most growth since World War II has occurred in electronic components and machinery (currently called high tech industry). Many of these newer industries are located along Route 128, a limited access circumferential expressway 16 to 19 kilometers (10 to 12 miles) from the city center. There they can take advantage of the high accessibility such a location provides. The growth of new industry on the periphery of the established city also emphasizes the decline of Boston's port. The harbor and the facilities found there remain excellent, but industry in New England (and Boston) now ships most of its products by land, either to markets in the rest of the United States or south to New York for export through Megalopolis's primary port.

New York's primacy among North American

TABLE 5-1

Percentage of 1980 Labor Force [a] **in Manufacturing by State in Core Region**

Region and State	Percentage by Region	Percentage by State
New England	26.0	
Maine		25.7
New Hampshire		26.7
Vermont		22.0
Massachusetts		24.6
Rhode Island		29.8
Connecticut		29.0
Middle Atlantic	21.6	
New York		19.6
New Jersey		23.5
Pennsylvania		26.8
Maryland		11.9
Delaware		27.5
Interior	26.1	
Ohio		27.2
Michigan		26.8
Indiana		27.8
Illinois		24.5
Wisconsin		25.1

SOURCE: Calculated from *Statistical Abstract of the United States* (Washington, D.C.: U.S. Government Printing Office, 1981).
Note: [a] U.S. national average: 20.8.

harbors has been discussed. As might be expected, manufacturing industries found proximity to this node of international commerce and the population cluster around it to be most advantageous. So strong was this pull, that New York's industrial mix became extremely diversified. Industries of many types were located on Manhattan until after the beginning of the twentieth century. The increasing demand for space by an even more intensive business community—reflected in the rapid construction of skyscrapers—gradually pushed the heavier industry to the very margins of lower Manhattan or beyond the island's confines into the New Jersey tidal marshes across the Hudson River. More recently (see Chapter 4), a great many manufacturing jobs have been pulled away from the New York region altogether and only barely replaced by nonmanufacturing employment.

As important as the New York metropolitan economy may still be in terms of its proportion of the national whole, the urban region's character is dominated by office industries. These are the headquarters for activities of dozens of companies and corporations, the banking and insurance cluster, the publishing houses, and all the other service and control decision centers that require a worldwide information network and the facilities to transmit their responses rapidly. So, New York is industrial by virtue of the large number of people employed; even with some recent losses, it probably remains the economic decision-making capital of the United States.

Philadelphia and Baltimore, so different in industrial inheritance and character, have shown indications during recent years that they may be becoming more alike. Philadelphia's manufacturing base is almost as diversified as New York's, although there is a greater emphasis than in New York on food-processing industries (from the truck farms of southern New Jersey, the lower Delaware River Valley, and even Eastern Shore Maryland) and on shipbuilding and ship repair. The growth of Philadelphia's industrial base suffered somewhat from the presence of New York's better harbor and superior access to the interior only 120 kilometers (75 miles) to the north; but Philadelphia's better access to the coal and steel regions of western Pennsylvania, its respectable port facilities, and its heritage as the first political and cultural center in the United States have maintained the Philadelphia metropolitan region's growth within Megalopolis. Baltimore, on the other hand, has always been on the periphery of the manufacturing core region. Like Philadelphia, its port possessed good rail connections with the coal and steel regions of the interior, and Baltimore's industrial mix reflects this. The manufacture of transport machinery is also important in Baltimore.

Two additional industrial sectors are well represented in Philadelphia and Baltimore. Metals fabrication and chemical industries are significant in both metropolitan economies. Copper refineries and steelworks at Sparrows Point east of Baltimore, similar steelworks near Trenton north of Philadelphia, and the chemical industries in Philadelphia and around Wilmington, Delaware, between the two larger cities, emphasize the coastal connections of these regions with the heavy industrial interior. It is because of this situational factor of location that the two are likely to become even more similar in the future. Increasing national dependence on imports of iron ore from Labrador and Venezuela and continuing importation of petroleum from Venezuela, Nigeria, and the Persian Gulf states may increase the importance of this region's coastal access to the inland demand centers.

The major cities of the other, larger portion of North America's manufacturing core region, the industrial Midwest, have derived their primary character from their location relative to the rich mineral and agricultural resources of the continent's interior. If the distribution of large cities in the western portion of the manufacturing region is examined (Figure 5-1), it can be seen that almost all of them are located along the Ohio River (or one of its tributaries) or along the shores of one of the Great Lakes. These comprise two of the three zones of manufacturing in the interior core. The third zone of generally smaller centers is that between the Great Lakes and the Ohio River Valley.

The most important factor in the development of urban centers in this portion of the manufacturing core has been the movement of metallic mineral ores from the margins of the Canadian Shield to the coalfields of western Pennsylvania and West Virginia and the smaller movement of coal in the reverse direction (Figure 5-4). Iron ore is mined at the Mesabi range of northern Minnesota, at the Steep Rock deposit in western Ontario, and at the Gogebic, Marquette, and Menominee ranges in northern Michigan and Wisconsin. This ore is carried to the southern shores of Lakes Michigan and Erie in large ships designed specially for Great Lakes

travel. Some ore is carried to the southern shore of Lake Michigan, to Hammond and Gary, Indiana, where it is met by coal transported north by rail from the large Illinois coalfields. Most of the ore, however, is shipped to Lake Erie ports and there either carried south, primarily to the steel cities of the Ohio River or converted to steel in the lakeside cities using coal carried north on the return rail trip from the Appalachian fields. Ore is also shipped beyond Lake Erie to the Canadian port of Hamilton on Lake Ontario, supplementing iron ore brought up the St. Lawrence from Labrador deposits.

Of the cities of the interior core, Pittsburgh is the one whose name became synonymous with steel. Located at the junction of the Allegheny and Monongahela Rivers, both of which drain the coal-rich margins of the Appalachians, and also at the head of the Ohio River, Pittsburgh was in an excellent position to take advantage of access to both raw materials and downriver markets. The city's development as a steel-producing center was aided by the activities of Andrew Carnegie, the power behind what eventually became the U.S. Steel Corporation. As Pittsburgh grew, industries dependent on steel crowded onto the narrow river plains to take advantage of proximity to their supplies and low-cost water transportation. Metal-fabricating industries, machine parts, and other industrial consumers of large quantities of steel located their plants in and around Pittsburgh. Nearby cities also benefited from the powerful pull of steel at Pittsburgh. Youngstown, Canton, and Steubenville in Ohio, Wheeling and Weirton in West Virginia, and New Castle and Johnstown in Pennsylvania shared to some degree in the industrial growth of this region and obtained steel and steel-products industries (Figure 5-5).

Urban-industrial growth did not occur solely at the source region of coal. The iron ore shipped across the lake system had to be transferred to railroads at points along the Lake Erie shore for final movement to the Pittsburgh region. Metal-products industries and other industry wishing to take advantage of the transportation focus

FIGURE 5-4

MAJOR COAL AND IRON ORE TRAFFIC FLOWS. Shipments of ore across the
Great Lakes and both coal and iron ore between the Ohio River and the southern
shores of the lakes have been fundamental to the development of urban industrial
concentrations in the interior section of North America's manufacturing core region.

along the lake shore and the *break-in-bulk* of the
materials transfer process chose to locate at each
of the port cities where iron ore was unloaded.

Cleveland was the largest of these Lake Erie
port cities. Cleveland's initial growth was stimulated by a canal connecting the narrow and
winding Cuyahoga River with a tributary of the
Ohio. Although the city quickly outgrew this small
initial advantage, it was enough to give Cleveland a head start on its nearby urban competitors. The diverse industrial base that resulted took
advantage of the accessibility offered by the lakes
and by the major east-west railroads in the United
States connecting New York with Chicago and
the agricultural core to the west. Cleveland's
growth also spilled over into adjacent ports, such
as Lorain, Ashtabula, and Conneaut, Ohio, and
perhaps as far east as Erie, Pennsylvania, as far
west as Toledo, Ohio, and to the complementary growth centers inland, such as the rubber-
producing city of Akron.

Buffalo is located at the eastern end of Lake
Erie, and this location has had three major effects on its industrial character. Somewhat ac-

FIGURE 5-5

RAW STEELMAKING CAPACITY IN THE U.S. MANUFACTURING CORE, 1979, RELATIVE TO 1960. With several notable exceptions, most U.S. steelmaking capacity remains along the southern Great Lakes and the Ohio River-Pittsburgh area. Furthermore, most plants had increased their capacity since 1960.

cessible to the same factors that generated steel and metals manufacturing elsewhere along the lakeshore, a significant portion of Buffalo's manufacturing is involved in this type of industry. In addition, until recently, Buffalo was the premier flour-milling center on the continent. Although there has been some decline in the volume milled in Buffalo, wheat from the Plains states is still brought to the western Great Lakes and then carried in bulk to Buffalo for refining before sale in the large eastern markets of Megalopolis. With the opening of the St. Lawrence

Seaway, however, some of Buffalo's locational advantages as a transfer site have been weakened. A more recent advantage was the harnessing of Niagara Falls for hydroelectric power. This drew chemical and aluminum industries, among others, to the Buffalo region because of these industries' high electric-power requirements.

Of the two major Ontario port cities in this section of the manufacturing core, it is Hamilton that shows the strong mix of steel-making and metals-fabricating industries typical of the eastern Great Lakes region. Much larger than

Manufacturing core cities along the Great Lakes contain industrial clusters, but this land along the Cuyahoga River in Cleveland possesses one of the most intense concentrations of heavy industry in the region. (Charles E. Rotkin/PFI.)

Hamilton is Toronto, with a more diverse industrial base. The reasons for Toronto's size, a city having little obvious locational advantage over Hamilton, are complex (see Chapter 6), and the result is a full mix of *light industry* rather than specialization in one or two heavy industrial sectors.

The cities on the narrow water passage between Lakes Huron and Erie, Detroit and Windsor, would seem to be the wealthiest in terms of locational advantages owing to water transport. Detroit, however, did not grow rapidly during the early period of manufacturing in North America, possibly because it is located more than 80 kilometers (50 miles) north of the primary New York-Chicago rail connection. The more direct, but politically more difficult, rail route across southern Ontario via Buffalo helped

stimulate some growth, but it was not until the rise of railroads' chief land transport competitor, motor vehicles, that Detroit developed the urban character it still possesses.

The motor vehicle industry is complex, in the sense that an extremely large number of components are required to make up a single automobile or a single truck or bus. As the most successful automobile manufacturers concentrated in Detroit and nearby cities and the demand for automobiles skyrocketed, a wide variety of suppliers were drawn to southeastern Michigan. Detroit became the fifth largest city in the United States, a rank it held until the 1970s. Although in some ways Detroit is as specialized in its industrial mix as Pittsburgh and as susceptible to similar fluctuations and trends in demand and competition, the range of complementary industries is greater. This has lent Detroit a somewhat greater buffer against decline in employment than Pittsburgh has had, even though Pittsburgh's economy began to diversify. For example, although metropolitan Detroit's non-farm payroll employment increased at only 57 percent of the national rate between 1970 and 1980, metropolitan Pittsburgh's equivalent employment increase was even smaller, less than 35 percent of the national average.

So strong is the pull of complementarity between the Superior iron ranges and Appalachian coalfields that, with only two exceptions, the largest cities of the interior manufacturing core are located along the southern Great Lakes, within or on the margins of the coalfield itself, or else they possess politically distinctive markets to serve. The smaller of the two remaining metropolitan centers is Milwaukee. Milwaukee's industrial character is typical of Great Lakes port cities but with an addition of agriculture-related industry. In addition to the usual industrial mix of heavy industry and motor vehicle manufacturing, Milwaukee is one of the leading centers of the continent's brewing industry, largely a result of the large number of German immigrants who settled in Wisconsin during the late nineteenth century. A significant food-processing industry is also present in Milwaukee, as it is a major focus for the middle Dairy Belt of the state.

Chicago is easily the dominant city of the interior manufacturing core. So important did this city become that for many years it has been called the Second City, recognizing Chicago's millions (3,005,072 in 1980) as second only to those of New York in number. Although Los Angeles grew rapidly, was virtually the same size as Chicago in 1980 (2,966,850 in the city and with a much larger urbanized area than Chicago), and surpassed Chicago's population soon after the census was taken, the informal "capital" of the Midwest persists as the strongest focus of the U.S. interior.

A look at the map of the United States should indicate several strong situational advantages for Chicago's location, but the city had to overcome a series of early disadvantages to grow to its present size and economic importance. First, the city occupied an undesirable swampy lake margin. Although ideal for mosquitoes and similar pests, it was viewed with little favor by early human inhabitants. The groundwater condition also provided Chicago's drinking water with a distinctive flavor that was often unappreciated. Finally, after the city had begun to establish itself as a regional center, nearly the entire city was consumed in the great 1871 Chicago fire.

It is testimony to the array of situational advantages that the city exists. Located along the southwestern shore of Lake Michigan, Chicago occupies the optimum location for people and goods transfer between lake transportation and the rich agricultural region to the west and southwest. The Illinois and Michigan Canal, located in part through the heart of the city, was completed in 1848 to link the Great Lakes and the Mississippi River system, the two greatest waterways of the continent. This focus of water transportation was supplemented four years later as Chicago was connected to New York by rail.

Whether these two transport connections or the many succeeding rail links toward the interior were the critical ones, Chicago drew on the support such as a focus of transportation provides and grew rapidly, absorbing many thousands of European immigrants throughout the century and spawning a radial network of rail lines into Illinois, Wisconsin, and the agricultural states beyond. The meatpacking industry developed around the city's large stockyards. Other industries, such as furniture and clothing manufacturers, located in Chicago to take advantage of the growing local market and the good access to markets farther west. After the turn of the century, the steel industry was introduced to the Chicago region, south of the city itself but still along the lakeshore at Gary and Hammond, Indiana, and easily accessible to the city's unparalleled railroad network. Already a city of 1 million by 1890, Chicago doubled in size before 1910 and exceeded 3 million by the mid-1920s. The self-reinforcing cycle of population increase (la-

This small portion of Chicago's downtown illustrates the intense demand for space and the accommodation made for water traffic on the Chicago River and its access to Lake Michigan at the top of the photo. (Courtesy Kee Chang/Chicago Association of Commerce and Industry.)

bor and local markets), economic growth, more population growth, and so on, clearly was operating before the turn of the century. The volume of Chicago's manufacturing activities (more than 840,000 employed in the SMSA in 1979) is matched only by the immense diversity of products manufactured. In this volume and diver-

sity, Chicago is at least partly an effective regional counterbalance to the set of intense economic nodes in Megalopolis. It is also appropriate that Chicago's industrial base formed on both the typical Great Lakes port pattern of steel and metal-products manufacturing and the agricultural orientation provided by the lands further west. Changes in meat-production techniques and shipping closed the city's largest stockyards, but Chicago's vitality and its situational consequences were caught effectively in Carl Sandburg's description of the city as "Hog Butcher, Tool Maker, Stacker of Wheat, Player with Railroads and Freight Handler to the Nation."

The remaining manufacturing clusters and major urban centers within the continent's manufacturing core region fall into one of the other two general zones. One is the Ohio River complex that includes Pittsburgh and its immediate neighbors, the industrialized Kanawha River valley, the older port city of Cincinnati, and many smaller centers in between. Almost all of the Ohio River clusters are related to the accessibility the river provides in moving coal and other Appalachian raw materials at low cost or to the return flow of materials from the Mississippi and beyond. Small, but representative, cities with names like Ironton and Ashland are sprinkled along the eastern half of the Ohio River, most with small steel plants or metals-fabricating factories. When the plants close in these small one-industry towns, whether temporarily or permanently, the effects are devastating. The Kanawha Valley industries in West Virginia, on the other hand, have used coal as a raw material as well as a fuel and specialized in chemicals, synthetic textiles and artificial rubber, plastics, and glass. Cincinnati also has gone through a change in industrial mix based on a shift in transport emphasis from a port city for agricultural-product transfer to a more balanced manufacturing city, but one that bears the typical Ohio River stamp of metals industries.

Defining the other zone of manufacturing in the region are the diverse and scattered centers located between the Great Lakes and the Ohio River and between the Appalachian coalfields and the gradually dominant agricultural economy toward the Mississippi River. These cities, Indianapolis, Fort Wayne, South Bend, Indiana; Dayton, Lima, Columbus, Ohio; Battle Creek, Jackson, Flint, Michigan; and many, many others, contain manufacturing activities that are partly a spillover from the larger centers surrounding the core's interior, partly a response to the intensely successful agriculture of the region (see Chapter 11), and partly a consequence of the network of interaction set up across the land area between these larger centers.

THE DYNAMICS OF NORTH AMERICAN MANUFACTURING

Long-term trends in the United States and Canadian economies have been consistent with general patterns elsewhere. Agriculture's proportion of the total national economic effort has declined gradually during the last century, whereas manufacturing maintained its proportion until the last several decades. The drop in the labor force engaged in agriculture from nearly 50 percent in 1880 to less than 3 percent in 1980 has been matched by an increase in what are called service or *tertiary* activities. The tertiary sectors of both economies, comprised of activities that could loosely be described as "serving" the population (retail and wholesale, transport, recreation, finance, and many others), have grown as income became more generally distributed among the countries' inhabitants. With this, benefits from the *primary*, or extractive, sector (agriculture, mining, lumbering) and the *secondary*, or manufacturing, sector (steel, chemicals, consumer products, etc.) could also be more broadly shared without requiring the bulk of the labor force for their production.[3]

---

[3] Some geographers are also beginning to study the distributions of what are called the *quaternary* and *quinary* sectors of the economy. Activities in the quaternary sector are basically those involved in information creation and transfer, whereas control functions make up the quinary sector. Examples of the former are education, portions of the communications industry, and research and development components of governmental and private industry. Quinary activities can be represented by decision makers, such as those in business headquarters or governmental policy forums. There are indications that these important sectors, long included with tertiary activities, are growing in relation to the other three much as the tertiary sector grew after the late nineteenth century.

While the balance among the several economic sectors was gradually shifting in both North American countries, the geography of these sectors was changing as well. The application of new technology to agriculture permitted increases in total production even as the farm population declined. The tremendous growth in manufacturing in the continental core region created a demand for labor that attracted migrants from Europe and from rural America, especially the South. The sprawling industrial agglomerations of the region grew to city size and filled with multiethnic populations heavily dependent on the secondary sector. Alterations in this pattern during the last several decades are a result of the balance between forces encouraging industrial relocation and those resisting such a change.

The stability of a manufacturer's location is primarily an extension of the factors that led the firm to establish a plant at its specific site in the first place. Where are the firm's raw materials or its suppliers of component parts? Where are the most important markets for the item produced? How good (i.e., how reliable and how expensive) are the transportation facilities on which the firm depends? Where is the labor supply? How important are local and regional amenities for maintenance of satisfactory life-styles by the firm's employees? How important are each of these (and similar locational factors) in relation to each other? Because the patterns of these factors are subject to changes beyond the immediate control of any single firm, the set of conditions that made a manufacturing site desirable at one time could make that same site undesirable some years later. Because a firm is unable to enjoy for long the best location for its activity, they must pay higher costs owing to shifts in the factors affecting their production location. The costs incurred because of these shifts have been called *inertia costs of location.*[4]

The costs of locational inertia, then, are the costs each firm must bear if it is no longer situated in the optimum location. They are called inertia costs because it is the stability of a plant that creates the costs. The stability is necessary because a fixed investment in some location is required. The inefficiencies of a location may be slight initially, but changes in technology, population distribution, the locations of suppliers or competitors, and many other factors could raise this inefficiency considerably. If the costs of remaining in a location despite its locational inefficiency (i.e., the inertia costs) are small, the firm will not move. If, however, these costs grow until they exceed the costs of abandoning the existing facilities, moving to a new site, and constructing a new production center, then the firm is likely to move.

Consider the implications of this for individual firms within the North American manufacturing core and for the shape of the core itself. Those manufacturers engaged in production that is primarily supplied by, or consumed by, other firms within the core are not likely to suffer a rapid rise in inertia costs. So interdependent and interconnected are the suppliers, producers, and consumers of manufactured products that each supports the others' locational stability by their own. In a study that examined the changing factors affecting the location of one automobile manufacturer, it was estimated that the best site for that firm shifted only 16 kilometers (10 miles) in the 28-year period between 1935 and 1963.[5]

Much has been made recently of higher relative economic growth in the southeastern and the southwestern United States, but the total locational inertia of the manufacturing core region is immense. Changes are occurring because national population and per capita income patterns are shifting. The demand for some products is rising more rapidly than others. Energy costs have been rising rapidly, and this affects the relative importance of transportation, fuel availability, and other factors of production and

---

[4] Stephen S. Birdsall, "The Development and Cost of Location Stability: The Case of Oldsmobile in Lansing, Michigan," *Tijdschrift voor Economische en Sociale Geografie* 62 (January–February 1971), pp. 35–44.

[5] Ibid., p. 43.

TABLE 5-2

Percentage of State Employment in Manufacturing Relative to National Average, 1920–80
(U.S. Average = 100)

State	1920	1950	1970	1980
Massachusetts	167	144	104	119
Rhode Island	191	170	129	143
Connecticut	175	164	135	139
New York	127	115	90	94
New Jersey	156	146	122	113
Pennsylvania	135	137	128	129
Ohio	135	141	132	130
Indiana	110	134	141	134
Illinois	108	124	113	116
Michigan	136	158	132	128
Wisconsin	111	118	119	120

SOURCE: Calculated from *U.S. Census of Population* (1920) and *Statistical Abstracts of the U.S.*, (Washington, D.C.: U.S. Government Printing Office, 1951, 1971, and 1981).

distribution. But again, these shifts have been slow, they are long term, and they appear important primarily because the core has been predominant economically for more than a century.

The persistence of manufacturing employment in core-region states can be appreciated by examining those patterns over a long period (Table 5-2). If we consider the percentage of employment in manufacturing for the entire United States each year as our standard and compare state percentages to those standards, changes in the national emphasis on manufacturing do not affect our state-by-state comparison.

Within the manufacturing core, changes in the relative state concentration in manufacturing reflect the two broadest zones within the region. Most states east of the Appalachians were engaged more heavily in manufacturing in 1920 than states of the interior core. But it has been the eastern portion that has experienced relative decline in manufacturing employment since 1920, whereas the interior states have maintained or even increased their relative emphasis

on manufacturing. Although these figures are rather general, together with those in Table 5-1, they suggest that the enormous economic mass of the interior manufacturing core region will resist rapid change, even though specific industries within the core may experience difficulty and absolute shifts in manufacturing employment may continue to alter the economic composition of other regions in North America.

NEIGHBORHOOD ETHNICITY

The character of the major urban centers in the manufacturing core was affected also by their population composition. As discussed earlier, most of these cities grew during the period of heavy European immigration before World War I. Incoming migrants sought employment opportunities where they were most clearly available—in the growth centers along the Atlantic Coast and across the interior manufacturing core. The large number of new arrivals and their ethnic diversity created a distinctive complexion to this region's cities.

When looking for a place to live within one of the core's cities, the choice of a neighborhood by the newer immigrants was strongly influenced by the ethnic residential patterns already in existence. Large ethnic neighborhoods formed and persisted through several generations. For many who lived in these cities, attitudes and goals toward a future life in the city were partly a function of the dominant Anglo-American culture and partly a function of the culture found in their portion of the urban cultural mosaic. Although suburbanization has clouded this mosaic to some degree, many of the neighborhood influences remain and continue to affect the urban cultural geography.

In his book on former Mayor Richard Daley of Chicago,[6] Mike Royko catches the flavor and significance of the city's ethnic patterns quite well. Although he is describing Chicago in his per-

---

[6] Mike Royko, *Boss: Richard J. Daley of Chicago* (New York: New American Library, 1971), pp. 30–32.

sonal, rather wry manner, the scenario could as easily be for any of the large northeastern and midwestern cities in the United States.

Chicago, until as late as the 1950s, was a place where people stayed put for a while, creating tightly knit neighborhoods, as small-townish as any village in the wheat fields.

The neighborhood-towns were part of larger ethnic states. To the north of the Loop was Germany. To the northwest Poland. To the west were Italy and Israel. To the southwest were Bohemia and Lithuania. And to the south was Ireland.

It wasn't perfectly defined because the borders shifted as newcomers moved in on the old settlers, sending them fleeing in terror and disgust. Here and there were outlying colonies, with Poles also on the South Side, and Irish up north.

But you could always tell, even with your eyes closed, which state you were in by the odors of the food stores and the open kitchen windows, the sound of the foreign or familiar language, and by whether a stranger hit you in the head with a rock.

In every neighborhood could be found all the ingredients of the small town: the local tavern, the funeral parlor, the bakery, the vegetable store, the butcher shop, the drugstore, the neighborhood drunk, the neighborhood trollop, the neighborhood idiot, the neighborhood war hero, the neighborhood police station, the neighborhood team, the neighborhood sports star, the ball field, the barber shop, the pool hall, the clubs, and the main street.

Every neighborhood had a main street for shopping and public transportation. The city is laid out with a main street every half mile, residential streets between. But even better than in a small town, a neighborhood person didn't have to go over to the main street to get essentials, such as food and drink. On the side streets were taverns and little grocery stores. To buy new underwear, though, you had to go to Main street.

With everything right there, why go anywhere else? If you went somewhere else, you couldn't get credit, you'd have to waste a nickel on the streetcar, and when you finally got there, they might not speak the language.

Some people had to leave the neighborhood to work, but many didn't because the houses were interlaced with industry.

On Sunday, people might ride a streetcar to visit a relative, but they usually remained within the ethnic state, unless there had been an unfortunate marriage in the family.

The borders of neighborhoods were the main streets, railroad tracks, branches of the Chicago River, branches of the branches, strips of industry, parks, and anything else that could be glared across.

The ethnic states got along just about as pleasantly as did the nations of Europe. With their tote bags, the immigrants brought along all their old prejudices, and immediately picked up some new ones. An Irishman who came here hating only the Englishmen and Irish Protestants soon hated Poles, Italians, and blacks. A Pole who was free arrived hating only Jews and Russians, but soon learned to hate the Irish, the Italians, and the blacks.

That was another good reason to stay close to home and in your own neighborhood-town and ethnic state. Go that way, past the viaduct, and the wops will jump you, or chase you into Jew town. Go the other way, beyond the parks, and the Polacks would stomp on you. Cross those streetcar tracks, and the Micks will shower you with Irish confetti from the brickyards. And who can tell what the niggers might do?

But in the neighborhood, you were safe. At least if you did not cross beyond, say, to the other side of the school. While it might be part of your ethnic state, it was still the edge of another neighborhood, and their gang was just as mean as your gang.

So, for a variety of reasons, ranging from convenience to fear to economics, people stayed in their own neighborhood, loving it, enjoying the closeness, the friendliness, the familiarity, and trying to save enough money to move out.

Additional Readings

Berry, Brian J. L., et al. *Chicago: Transformation of an Urban System*. Cambridge: Ballinger, 1976.

Borchert, John R. "Major Control Points in American Economic Geography," *Annals of the Association of American Geographers*, 68 (1978), pp. 214–232.

Cutler, Irving. *Chicago: Metropolis of the Mid-Continent*, 3rd ed. Dubuque, Iowa: Kendall/Hunt, 1982.

Jones, Peter d'A., and Melvin G. Holli (eds.). *Ethnic Chicago*. Grand Rapids, Mich.: Wm. B. Eerdmans, 1981.

Lewis, Peirce. "Small Town in Pennsylvania," *Annals of the Association of American Geographers*, 62 (1972), pp. 323–351.

Proctor, Mary, and Bill Matuszeski. *Gritty Cities*. Philadelphia: Temple University Press, 1978.

Schenker, Eric, Harold Meyer, and Harry C. Brockel. *The Great Lakes Transportation System*. Madison: University of Wisconsin, Sea Grant College Program, 1976.

Sinclair, Robert, and Bryan Thompson. *Metropolitan Detroit: An Anatomy of Social Change*. Cambridge: Ballinger, 1976.

Vernon, Philip H., and Oswald Schmidt. "Metropolitan Pittsburgh: Old Trends and New Directions," in *Contemporary Metropolitan America*, vol. 3. Cambridge: Ballinger, 1976.

Warren, Kenneth. *The American Steel Industry, 1850–1970: A Geographical Interpretation*. Oxford: Clarendon Press, 1973.

THE NATIONAL CORE REGION OF CANADA

Although Canada is a larger country [9.8 + million square kilometers (3.8 million square miles)] than the United States [9.3 + million square kilometers (3.6 + million square miles)], it contains a much smaller share of the continental core region. Small as this portion may be, it is nearly coincidental with the national core of Canada. Canada's core region is entirely within the provinces of Quebec and Ontario, paralleling Lakes Erie and Ontario in a band that reaches past Lake Ontario along the banks of the St. Lawrence River to Quebec City (Figure 6-1). However, the Canadian national core also includes cities, such as Ottawa, Montreal, and Quebec, that are of strong cultural and political importance to that country but that are not really part of the continent's manufacturing core.

As small as this region may be when compared to the full national territory of Canada, it contains a large share of the nation's population and economic activity and was also the historical hearth of the country's development. More than 55 percent of the Canadian population lives within southern Ontario and southern Quebec, a regional share expected to reach almost 60 percent by the end of the century. A disproportionate amount of the national industrial capacity and urban population is also found here—industry associated with the advantages of accessibility provided by the Great Lakes and the St. Lawrence or with the mineral-resource regions nearby.

It was into the riverine lowlands that French settlers spread in the 150 years after 1608; English settlement was more extensive inland from the lakes in the late eighteenth and early nineteenth centuries. By 1867 when the new Dominion of Canada was formed, more than 90 percent of the Canadian population resided in the southern portions of Ontario and Quebec and around the margins of the Maritimes to the northeast. Thus, it has only been during the most recent century of Canada's existence that permanent settlements have been established in much of the vast interior; the core from which this extension has occurred remains dominant.

FIGURE 6-1
THE NATIONAL CORE REGION OF CANADA.

In considering the essential character of this region, three themes may be suggested. The first deals with the small size of the region. What factors have limited growth within the core to the relatively small territory it occupies? Ordinarily, one might expect a region that is the development hearth of a rapidly growing and prosperous country such as Canada to have extended its area along with the growth in its economy, but this territorial expansion of the core did not occur. A second theme is the great diversity found within this region. Although the geographic concept of a core region is subject to differing definitions, a certain uniformity of intensity and function is often an underlying assumption to the delimitation of such a region. Strikingly, aspects of the diversity that have threatened the very existence of Canada as a single political unit find their clearest expression within the core region. Finally, the character of the national core is modified because it is also part of the larger core region for the entire continent. Thus, the third theme is the relationship between the United States and Canada as seen through the integration of the national core into the continental core.

The core region of Canada is located far from the center of the country, a consequence of historical sequence and environmental patterns. The initial European entry into this portion of North America was from the northeast via the St. Lawrence River. Jacques Cartier reached Montreal

Island as early as 1535; further extension of French interest along the Great Lakes followed the turn of the seventeenth century. After the American Revolution, additional settlement by Loyalists occurred along the northern shores of Lakes Ontario and Erie as population spilled across the U.S. border. Because of the importance of the St. Lawrence as an *entry* route, the lands reached by this passage were given the administrative labels of Lower Canada, Middle Canada, and Upper Canada during the nineteenth century. Although the regional names are no longer in use, what are now the Atlantic Provinces (Chapter 7) were Lower Canada, the land in southern Quebec along the St. Lawrence River was Middle Canada, and the most southerly lands of all were Upper Canada, that area bordering Lakes Ontario and Erie.

Settlement in Canada was for a long time keyed to production of *staple products* as export goods bound for Europe, principally Great Britain. Early dependence on cod fishing off Newfoundland and fur trapping in the St. Lawrence Valley provided settlers with the means to purchase imports from Europe. As fur trappers exhausted the local supply and fur trapping shifted westward, Canadians turned to alternative staple products as the bases for their export economy: timber, metals, and agriculture. Population pressure increased during the nineteenth century within the St. Lawrence Valley and southern Ontario and this population growth continued the exploitation of primary products. The Canadian economy, however, did evolve beyond the early, nearly absolute dependence on the export of staples from the core region—by expansion of influence from the United States, by growth in domestic manufacturing industries, and by the spread of settlement beyond the core—but these factors supported the evolution of the national economy in only a partial and mixed manner, as we will see.

As a northern country, Canada's most desirable early agricultural resources were located on the more southerly Ontario peninsula and to a lesser degree along the narrow St. Lawrence lowland. With population growth in Canada prior to confederation (1867) supported by the production of staples for export and with the rapid and leading growth of the nearby U.S. core, the pressure on land resources in the Canadian core became considerable. Dense settlement, however, was not pushed much beyond the present limits of the core. By the time population expansion became truly continentwide, southern Ontario and southern Quebec had accumulated enough inertia—economic, political, cultural—and enough focus resulting from transportation developments to remain the leading growth region in Canada. But why did the developing core remain so small during its period of formation? What factors restrained expansion so severely?

LIMITS OF THE CORE

One of the most substantial factors limiting the size of the country's core region is physiographic. As discussed in Chapters 1 and 17, much of the continent north of the manufacturing core region is underlain by the Canadian Shield. The margins of this vast zone of igneous and metamorphic rock describes a long arc from the Arctic coast to the Great Lakes and the St. Lawrence Valley. Within the core, the Shield is present under a thin layer of soil from the mouth of the St. Lawrence estuary southwestward to about Canada's present capital, Ottawa, then south again almost to Lake Ontario, and finally nearly due west to Georgian Bay (Figure 6-2).

Keeping this outline in mind, it is clear that southern Quebec is very much restricted by the thin and poorly drained soils of the Shield. The agricultural and commercial populations hugged the narrow band of land along the St. Lawrence, in most cases within a river valley only a few dozen miles wide. Although the Shield is not as obviously present in southern Ontario, it helped restrict the agricultural opportunities of the expanding Canadian settlement during the nineteenth century. It was not the topography of the Shield that was the essential barrier to farming. Much of the Shield has the appearance of a gently

FIGURE 6-2

FEATURES OF THE PHYSICAL ENVIRONMENT. Each of these four indicators of the environmental support for farming reflects the importance of southern Ontario and the St. Lawrence valley in the Canadian core's agriculture.

rolling low plateau dotted with small lakes and marshes and covered with a mixed forest. Nonetheless, forests were removed in other parts of North America to meet the demand for farmland, whereas in this case, the land was too poor and the climate was too harsh for successful farming.

The land of the Shield is not suited for agriculture because it is poorly drained and the soils are too thin and acidic. As discussed earlier, the glacial ice cap crept outward during the Ice Age over the Shield and beyond. As the glaciers moved, loose material was scraped from the underlying rock surface and deposited far to the south. Then, as temperatures gradually rose and the ice melted, little cover remained on the crystalline bedrock of the Shield. In the 8000 to 10,000 years since the glaciers retreated, too little time has passed for the slow process of soil formation to have developed anything more than a relatively thin layer of soil in the shallow pockets on the Shield.

The combined forces of a scouring glacial sheet of ice and a resistant base of rock also resulted in a heritage of poor drainage for this region. Although most streams and rivers flow off the Shield to the Great Lakes, to the St. Lawrence, or north into Hudson Bay, their paths are rarely direct or uninterrupted. Most surface water on the Shield moves leisurely from lake to lake in the general direction of the Shield's margins but with little apparent local organization. The result is a *hydrography* that discourages agricultural use because of an abundance of small swampy pockets and poor drainage. This also discouraged commercial development because only a few rivers, such as the Ottawa, could be used for the movement of goods.

South of the Shield in Ontario and in the St. Lawrence Valley in Quebec, the pattern of soils also was influenced strongly by the long period of glacial activity. Most of the better agricultural soils in the Canadian core region are derived directly from the material deposited by the glaciers as they melted (*glacial till*) or from the out-

wash material carried by the tremendous volumes of water leaving the melting glacier. The low relief of much of this land masks a complex pattern of alfisols (see Chapter 2), some of which are deep and moderately fertile, and substantial pockets of clay and sandy soils that are less suited to agricultural use. A recent estimate of soil capabilities in southern Ontario places 60 percent of the land in categories of soil that could support field crops with proper management, but two additional factors must be kept in mind. First, the technology for "proper management" was not available to Canadian farmers until well into this century; second, the 60 percent of this land that is arable is only a small portion of early-settled Canada.

The other environmental feature that contributed to the early limits of Canadian core development was climate. Canada is a high-latitude country. The short growing seasons and cool summers would have made the agricultural efforts of its European settlers difficult north of the core region, even if the Shield's soil and drainage were less formidable. Lakes Huron, Erie, and Ontario bound the Ontario peninsula, and their proximity hastens the spring thaw and postpones the autumn freeze to a small degree. The valley of the St. Lawrence and the lowland between the St. Lawrence and the Ottawa rivers also benefit from an extended growing season, almost all sections averaging greater than 120 frost-free days annually (Figure 6-2). But as settlers moved inland, farther from the moderating influence of the Great Lakes and farther north, they experienced growing seasons that were appreciably shorter. More important, they found that summer days did not accumulate enough heat energy for good crop growth. Measured as the number of degrees over 5.5°C (over 42° F) for the number of days this temperature was exceeded, the better agricultural areas in Canada are those exceeding 3000 annual degree-days. The boundary of the national core in Quebec coincides closely with the line indicating this level of mean annual heating, and all of the Ontario portion of the core receives an average of more than 3000 degree-days.

MAJOR METROPOLITAN AREA POPULATIONS IN THE CANADIAN CORE, 1981	
Toronto, Ont.	2,998,947
Montreal, Que.	2,828,349
Ottawa-Hull, Ont.-Que.	717,978
Quebec, Que.	576,075
Hamilton, Ont.	542,095
St. Catharines-Niagara Falls, Ont.	304,353
Kitchener, Ont.	287,801
London, Ont.	283,668
Windsor, Ont.	246,110
Oshawa, Ont.	154,217

Mositure availability is generally not a problem in the Canadian core. Although mean annual precipitation is below 100 centimeters (40 inches) in most areas, the summer temperatures are cool enough that *evapotranspiration* rates are not high. Only along the Lake Erie and Lake Ontario margins are temperatures warm enough to produce a small water deficiency, but it is not sufficiently severe to hamper agricultural production.

The location of the core in Canada, therefore, is partly a result of major routes of continental entry and partly a result of the inertia provided by strong transport ties to Europe and later to western portions of the country. With limited technology, individual farmers could not establish settlement north and west of the Great Lakes across the Shield. With a diffused political organization, arguments for connecting the national core region with the western three-fourths of the country for any purpose beyond the export of staple products were also unconvincing until late in the nineteenth century. By this time, the agricultural base was strong and southern Quebec and southern Ontario were well settled with two cities rapidly approaching 100,000 (Montreal and Toronto) and another five exceeding 10,000 (Quebec, Ottawa, Kingston, Hamilton, and London). Patterns of transport and communication were well established. The economic and political foci for the country had already developed here. The inertia of national dominance was clearly present in this region.

DIVERSITY WITHIN THE CORE

The southern margins of Quebec and Ontario share many characteristics. Both are located along the main route of access between Europe and the Canadian interior. Both developed urban commercial centers that came to dominate the economic organization of each portion of the core and to compete with the other center for ultimate dominance of all of Canada. Both experienced the gradual population shift from rural to urban residences common to most economies undergoing industrialization. Both possess the moderately restrictive environmental base discussed above. And the economic organization of both areas is tied through investments and subsidiary ownerships to the economic structure of the United States. In spite of similarities such as these, striking diversity exists within the Canadian core.

Perhaps the most pervasive contrasts within the core are those related to Canada's cultural dualism. A large majority of the population of Quebec can trace its ancestry to French origins, whereas most of those living in southern Ontario are of British heritage. Although Canada has received large numbers of migrants from other countries of Europe and some population movement has occurred between these two provinces, the cultural boundary zone continues to approximate the provincial boundary (Figure 6-3). The French Canadian cultural base is reflected in linguistic distinctiveness, attitudes, settlement patterns, and frequently in a political posture defensive of French Canadian culture. This last feature is due to the minority position of French culture within Canada. About one-fourth of the total Canadian population lives in Quebec. In addition, over 200 million more English-speaking people live immediately to the south in the United States.

It could be argued that French Canada has remained distinct within British North America largely because it has maintained just such a defensive posture with regard to any natural or planned assimilation trends. By the time the British conquered the French in this portion of North America (1763), most of the land in the St. Lawrence lowland was occupied by French-speaking farmers. Quebec and Montreal were rapidly growing commercial centers, and nominal French influence had spread beyond the Great Lakes to the Ohio and the Mississippi rivers along routes frequently traveled by traders and fur trappers. Seizure of governmental control from the French by the British did effectively end additional cultural infusions and most migration from France, but it did not mean that

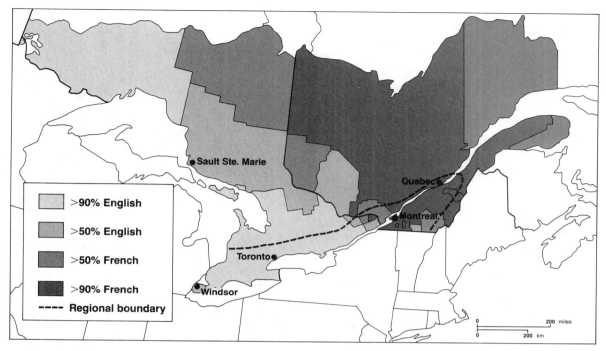

FIGURE 6-3

LINGUISTIC DOMINANCE IN ONTARIO AND QUEBEC. The political boundary between these provinces does not match the linguistic boundary perfectly, but there are a relatively few in southern Ontario who use French as their first language and relatively few in Quebec who consider English their mother tongue.

the cultural base in Quebec was destroyed. Quebec's economy simply became redirected toward British goals. Following the American Revolution, British Loyalists moved into areas less well settled and possessing greater opportunity, so that there was little immediate dilution of the French Canadian culture within Quebec's lowlands. Even during the nineteenth century, Montreal maintained a French dominance in this portion of the core region as urban commercial and industrial growth stimulated the growth of that city. The cultural distinctiveness persisted even though the city attracted many non-French Canadians as well as many from rural Quebec. The rural French Canadian population, more isolated from Anglo cultural influences, also grew rapidly. A highly

urban province with 80 percent of its population in urban places in 1981, the nonurban population of Quebec is also larger today than it was a century ago.

To accommodate the cultural and regional diversity of this large country, efforts toward unification—formally completed in 1867—required the governmental organization to be in the form of a *federation*. A federation is a form of government in which a group of states or provinces are united under a single, mutually acceptable, sovereign central authority, even while they retain other governmental powers. Ideally, a federation (or a confederation as it was called in Canada) permits many of the advantages of a larger unified organization while leaving room

127

The French Canadian cultural presence in this area of Quebec is visible in both language and architecture. (Courtesy Gouvernement du Québec Ministère de l'Industrie, du Commerce et du Tourisme.)

to accommodate local regional diversity. If the differences are persistent and deep enough, however, there may be times when the required accommodations conflict with each other or with the demands of the central authority. In the United States—also formally a federation—the conflict led to a civil war.

In Quebec, the central/regional tension has revolved around issues of French Canadian cultural independence and equality within a largely non-French country. The country is officially *bilingual*. This ostensibly means that French and English are of equal importance in the governmental operation of the country. In theory, a member of the federal civil service in British Columbia, for example, could address others in the province in French and expect to be understood. In fact, with extremely few speaking

French as their first language, the western provinces do not participate in this bilingualism, in spite of its official status. In recent decades, questions of cultural dualism, cultural integrity, and equality have focused on the validity of bilingualism. Outside French-speaking portions of the country, this element of Canadian federal governmental policy has accumulated considerable opposition. Either as cause or effect, there has been a correspondent heightening of Quebecois nationalism within *La Belle Province.*

Much more is involved than language differences, but contention over linguistic parity between English and French has long been used as a political issue in Canada. For example, when a study by the Canadian census in the late 1960s projected that a majority of Montreal's population would be non-French-speaking by the year

2000, French Canadian nationalism was aroused further. A provincial law was subsequently passed requiring that all signs in Quebec be in French and primarily in French alone. Although this was ruled unconstitutional by the Canadian Supreme Court, a new law was passed that clearly affects potential English-language migrants to the province. It requires that all children permanently residing in Quebec attend French-language schools unless the parents themselves were educated in Quebec in English-language schools.

The long-term significance of the larger issue of separatism, or independence for French Canada, lies in the location of Canada's population and productivity (about 25 percent of both are in Quebec), and the location of the French-speaking population within the provinces (about 85 percent are in southern Quebec). As cultural differences are translated into strong political feelings, the Canadian core becomes threatened with division.

The British impact on southern Quebec was not trivial, although it was southern Ontario that received the most broadly based cultural impression. Aside from a mild French imprint, the European sequence of entry into, and taming of, the Ontario wilderness really began with the influx of Loyalists forced out of the new United States. They were followed by large numbers of Americans and British. Settlement proceeded slowly enough that the land was surveyed and *platted* before the population arrived. Roads were planned and settlement was directed inland from the main urban centers. The population that migrated to southern Ontario was drawn by governmental policy from the British Isles, partly to offset American invasions but also to balance the large and growing French Canadian population that had been farming along the lower St. Lawrence for more than a century. Many of the British settlers were Scots and Irish, although most were probably English. By the time of confederation, southern Ontario provided a definitive Anglo-Canadian culture hearth for new population diffusion around the Great Lakes and onto the Canadian prairies.

RURAL LANDSCAPES

A measure of agricultural land suitability is actually visible in broad terms in southern Quebec. The imprint of each culture on the landscape has created differences in its appearance. Land was assigned to early French settlers within seigneuries, and, in these land grants, the *rang* system of settlement was used. Under the *rang* system, each farmstead was located on the river, on the lake shore, or on a road constructed parallel to the shore and at the head of a long narrow lot that extended inland perpendicular to the shore or road. The advantages of this arrangement for settlers were equal access to the main transport lines and proximity to their land. As demand for land increased, other *rang* were demarcated by the construction of new roads, farther inland but also parallel to the shore or the first road. The best farmland in southern Quebec (and even in a few areas of southern Ontario) had been laid out in this manner before the British takeover. What is popularly known as the French long-lot pattern, therefore, can still be used as a general indication of the better farmlands in southern Quebec (Figure 6-4).

The rural landscape in Ontario and in the less well-endowed or less accessible lowlands of southern Quebec was laid out by the British in a square or rectangular pattern. Individual homesteads tend to be more dispersed under such an arrangement. Although French Canadians gradually moved into many of these areas during the nineteenth and early twentieth centuries, the pattern of landholdings was rarely changed from the initial British form.

The sequence of agricultural emphasis is a consequence of both the environmental conditions and the interplay of cultural histories. For many decades, grain and animal production dominated the farmers' efforts. In the less productive interior, near-subsistence farming was often supplemented by maple sugar manufacture and lumbering during the long nonfarming season. As the rural Quebec population grew, migration to the United States or to nearby Ca-

This scene illustrates the typical form of French Canadian agriculture in the St. Lawrence lowland. When this land was distributed by the French crown, it was parceled out in narrow, parallel strips extending inland from the river. Each "long lot," as the fields are called, has access to the river, or to a road built parallel to the river at some distance to the interior. This small, linear village (about 65 miles from the city of Quebec) lies along the first road inland from the river, visible at the bottom of the photograph. The rather irregular topography and the presence of numerous rocks in virtually every field indicate that farming is not easy here, nor is it especially remunerative. (Lawrence R. Lowry.)

nadian cities drew off some of the increase, but many French Canadians were encouraged to remain true to their rural traditions and expand the settlement frontier onto the Shield or into the equally marginal land of the northern Appalachians. During the nineteenth century, other parts of Canada, more suited to wheat production, began to exert a comparative advantage over even the better land in southern Quebec. The emphasis of production shifted, and dairying, served by hay and oats, became the most impor-

tant farm pursuit. Many specialty crops, such as vegetables, sugar beets, fruits, and poultry, were also significant. As jobs became available in resource industries and factories, much of the marginal land was abandoned and allowed to revert to forest.

The pattern of agricultural change is similar in southern Ontario, but the more southerly location and the more extensive lowland generated a somewhat different result. Increasing demand for food in the early Canadian core and

FIGURE 6-4

LONG-LOT FARMS IN QUEBEC (from *A Regional Geography of North America*, second edition, by G. S. Tomkins, T. L. Hills and T. R. Weir, copyright 1970, Gage Publishing Limited. Reprinted by permission of Gage Publishing Ltd).

in the eastern United States led most new farmers in southern Ontario to grow wheat during the early 1800s as their main cash crop. By mid-century, however, wheat yields declined sharply as the alfisols lost their initial fertility under constant wheat production. Although farming became diversified, wheat remained prominent in Ontario agriculture until the Prairie Provinces were settled. Mixed farming—both feed crops and meat animals are produced—became the primary farm type until about World War II. Innovations in storage, transport, and daily operation during the 1920s and 1930s gradually encouraged a substantial proportion of farms in the region to undertake dairying. Interestingly, there appears to be little geographic pattern to

dairy farming vis-à-vis livestock farming in southern Ontario. Dairying is dominant near the urban areas of Windsor, Hamilton, and the Niagara peninsula and in the nonurban eastern counties, but livestock farming exceeds dairying in value of production near Toronto, London, and Kitchener. In recent decades, farmers in the core region of Ontario have experienced the same pressures as those in southern Quebec to increase their capital investment, with consequent mechanization and increased specialization of production (see Chapter 11). In both sections of the Canadian core, this may lead to further land abandonment along the margins of the region and to further intensification of production within the most favored areas of production in the core itself.

There are also a number of significant specialty crops grown in southern Ontario. Along the middle Lake Erie shore and to the northeast, sandy soils support production of large quantities of high-value, flue-cured tobacco. The high annual returns for production of this crop are apparently sufficient to outweigh any anxiety about its long-run demand. Ontario's tobacco production appears secure.

Less sure of the future are those who depend on the fruit produced within the region. Located between Lakes Ontario and Erie and at a relatively southern latitude (for Canada), the Niagara peninsula possesses a set of climate and soil conditions approaching the optimum for some fruits. Winters are moderated, and spring blossoming is retarded because of the peninsula's position between the lakes. Grapes, peaches, and sweet cherries are the most important of the fruits grown in the region. Future fruit production does not appear secure, however, as urban areas on the peninsula expand outward to accommodate their growing populations.

A recent study of land-use change in the Niagara Fruit Belt concluded that "most of the urbanization has been occurring in the most intensive fruit-growing areas"[1] (Figure 6-5). More

[1]Ralph R. Krueger, "Urbanization of the Niagara Fruit Belt," *Canadian Geographer* 22 (Fall 1978), p. 189.

The Niagara Peninsula, especially the land along the southwest shore of Lake Ontario, contains many fruit orchards and grape vineyards. Although this region continues to be an important source of apples, peaches, and grapes in Canada, urban activities—such as the subdivision development shown in this photo—are encroaching rapidly on the agricultural uses. (Wayne Farras/The Regional Municipality of Niagara, Niagara Planning and Development Department.)

serious for the long-run agricultural productivity of the region, a great deal of the urban development has occurred on the best fruit soils. There have been calls to preserve what remains, but this is one of only a few recent attempts in either the United States or Canada to control rural-to-urban land-use change in a large region (see Chapter 15). The success of such attempts is still unclear.

URBAN AND INDUSTRIAL DOMINANCE

Even though the core is important as a region of agricultural production, this portion of Canada is defined as the national core because of its complex pattern of urban and industrial development. As the country's population began to spread beyond the Great Lakes to the prairies and the Pacific Coast, this region remained the

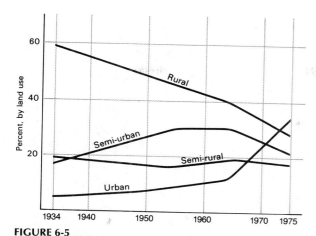

FIGURE 6-5

CHANGE IN LAND USE IN THE NIAGARA
FRUIT BELT, 1934–1975 (after Krueger). The
rapid conversion of valuable fruit-producing land to
urban uses has become of great concern in this section of Canada.

source of much development capital, most of the
migrating population, and the primary benefi-
ciary of the industrial growth that accompanied
this economic stimulation. The manner in which
this growth occurred, however, was variable and
reflects differences within the core. Differences
in the responses of each section of the core to
stimuli arising in the United States have also been
important.

Both portions of the core have urban-industrial centers that far exceed the size of all others
in their province: Montreal in southern Quebec
and Toronto in southern Ontario. Greater
Montreal, with a population in 1981 of 2,828,349,
benefited from important military and commercial site advantages during the colonial period.
The city's original setting lies at one end of an
island in the St. Lawrence River and adjacent to
both the confluence of the Ottawa River with the
St. Lawrence and the first major rapids on the
St. Lawrence. Thus, a strong defensive position
was combined with control of two major water
routes to the interior. Montreal's even greater
advantages of situation along with a municipal
administration that exploited these advantages

became more important stimulators of growth
during the nineteenth and twentieth centuries.
Montreal surpassed the city of Quebec in population between 1825 and 1830. Its location on
the river at the point where shipments were required to circumvent the rapids was important
to early commercial development. True dominance in southern Quebec was not assured until
the decade of the 1850s when the river was
bridged and Montreal was connected by railroad to Portland, Maine, in the east and to Toronto in the southwest. Access to the southeast
enhanced Montreal's position at the northern end
of the Lake Champlain-Hudson River valley and
increased its access to New York City.

As a commercial, financial, and transportation center, Montreal, by confederation in 1867,
was well into a self-reinforcing growth spiral of
industry attracting labor and a larger market that,
in turn, drew more industry that drew more labor, and so forth. A distinctive result was the
development of a large non-French population
in the city. Most of the financiers and industri-
alists in the city were of British origin, whereas
most of the labor force was French Canadian.
The French-speaking population exceeded the
English-speaking population in number by 1871,
but Montreal retains a much stronger bicultural
character than does Quebec. This latter city at
the head of the St. Lawrence estuary had fewer
of the advantages that stimulated Montreal's industrial growth. Although Quebec's metropolitan area population had grown to 576,075 by
1981, it still retains the strong French Canadian
flavor expected of a cultural, religious, and provincial administrative center.

Urban growth in southern Quebec outside of
the immediate Quebec and Montreal spheres is
a relatively recent phenomenon. In every case,
growth of the new centers can be traced to re-
source-based industrial development. Some, like
Trois-Rivières (50,466 in 1981), combine access
to Shield resources—pulp timber and, more recently, hydroelectric power—with imported materials. Bauxite is processed at nearby Shawini-

The city of Quebec was founded at the mouth of the St. Lawrence atop a high
river bluff to control the river traffic. During the centuries that followed, the city's
commercial, cultural, administrative, and industrial activities gradually over-
whelmed its original limited purpose. These activities now support more than one
half million people in the metropolitan area. (N.F.B. Phototèque O.N.F.)

gan for example. Other centers, such as Thetford
Mines (19,965) and Sherbrooke (74,075), have
grown through utilization of local minerals (as-
bestos at Thetford) or abundant local labor and
diversified industrial investment. In all cases,
Montreal dominates through transport and financial ties.

Toronto, with a metropolitan population of
2,998,947 in 1981, is southern Ontario's equivalent and competitor to Montreal. At first glance,
Toronto appears to have few of the natural site

or situation advantages that would support the
growth of such a metropolis. It is located on Lake
Ontario, possesses a good harbor, and lies at the
head of the Toronto Passage, a lowland route to
Georgian Bay in the north. But other sites along
the lakeshore, perhaps at either end of the lake,
would seem to provide equivalent advantages to
urban growth. Toronto was not located at a focus of natural access as was Montreal. And Toronto did not possess Montreal's early momen-
tum of urban growth.

Toronto has been able to approach, and in some ways exceed, Montreal's economic growth because it has several advantages of function and location that are not present in Montreal and that can be added to its natural harbor resource and lake accessibility. It was decided shortly after the beginning of British rule in southern Ontario that the region's capital would be located at York (now absorbed by the urban area of Toronto) rather than at Kingston, Niagara, or London. It was clear that to administer properly, the capital had to be well connected by road, later by rail, to the rest of the territory. Toronto's political function, therefore, provided the city with a network of land transportation that made it the natural focus of all Ontario prior to confederation. Of equal importance, this land transport network, which focused on Toronto, drained a larger and more productive agricultural region than was accessible to Montreal. Toronto's superior hinterland provided it with an alternative economic base to Montreal's strong commercial growth potential.

These early stimuli for urban growth were

Toronto's skyline, showing clear evidence of the city's continued growth as an industrial, commerical, and communications focus, is enhanced by Ontario Place, the lakefront recreational complex in the foreground. (Courtesy Toronto Convention and Visitors Association.)

further increased by developments in the United States. As long as the bulk of both the commercial and industrial activity in the United States was concentrated along the northeastern seaboard, Montreal's greater proximity was to that city's advantage in spite of Toronto's good access via the Erie Canal route through the Mohawk and Hudson river valleys. During the late nineteenth century, however, as the centers along the southern Great Lakes began to grow in industrial capacity and the U.S. manufacturing core extended into that country's Midwest, Toronto's location on the Great Lakes and easy accessibility by rail through Windsor-Detroit and Niagara-Buffalo assured it of continued urban and industrial expansion.

The relative strength of Toronto's hinterland and the region's proximity to major U.S. centers of manufacturing are also shown in the variety and intensity of urban-industrial concentrations (1) around the western end of Lake Ontario in what has come to be called The Golden Horseshoe, (2) in cities adjacent to the southern Michigan industrial concentration, and (3) in cities between these two areas that were well connected to both by railroads. At the extreme western end of Lake Ontario, with a 1981 metropolitan population of 542,095, Hamilton developed a large iron and steel complex by utilizing iron ore brought from the Lake Superior region through the Welland Canal and, more recently, ore brought from the Shield in Labrador via the St. Lawrence. Coal was also imported inexpensively from the Appalachian region by direct rail connections through Buffalo. The complex of foundries and rolling mills undoubtedly encouraged development of further industry within Hamilton as well as transport and farm machinery in nearby St. Catharines. Hydroelectric power from Niagara Falls, good water access to both the St. Lawrence and the interior Great Lakes, and abundant agricultural production have attracted such industries as chemical, textile, metals refining, and food processing to this region, one that many Americans picture merely as a tourist attraction.

Around the western end of Lake Ontario lies a small crescent-shaped region of especially intense economic activity and spatial interaction. From the automobile center of Oshawa, through Toronto and Hamilton, to the uncomfortable blend of high-value agriculture and chemical and metal-fabricating industries on the Niagara peninsula, the land is occupied intensely and productively. Sometimes called by the Amerindian name Mississauga, the same name as one of Toronto's larger suburbs, the region has been widely recognized by the more memorable appellation of The Golden Horseshoe. As attractive as this place-name may be, there is some danger of misunderstanding by the unwary. The horseshoe shape is apparent, but the "golden" label refers only to the money-generating potential of the region's activities. As Maurice Yeates put it: "The word 'golden' can hardly refer to the environment, for much of it is cement, steel, and macadam, and although areas of beauty exist, there has been little regard for preservation."[2] This portion of the Canadian core also has the highest per capita income levels in the country. Thus, the advantages of cheap transportation; good access to raw materials, energy, and markets; proximity to concentrations of labor and supportive industry; and the early growth of Toronto and Hamilton have combined to stimulate urban and manufacturing growth throughout the Golden Horseshoe. In many respects, this region possesses much of the promise and most of the problems of Megalopolis (Chapter 4), although they are found here on a far smaller scale.

The southwestern Ontario industrial clusters also take advantage of good accessibility resources, proximity to other manufacturing activity, and, to some degree, local mineral resources. Petroleum deposits with commercial value were discovered in Oil Springs, near Sar-

---

[2]Maurice Yeates, *Main Street* (Ottawa: Macmillan of Canada, 1975), p. 165.

nia, shortly after the middle of the nineteenth century. A refinery at Sarnia by the beginning of the twentieth century and subsequent connection of the city by pipeline with the large petroleum-producing regions in Alberta made this city of 50,892 (in 1981) a center of petrochemical industries. The complex interindustry linkages of production and supply have spilled the southeastern Michigan automobile-producing region across the Detroit River into Windsor (metropolitan population of 246,110 in 1981), where it is supplemented by food and beverage industries. In both cases, location on the continent's major industrial waterway supported growth within the manufacturing sector of the Canadian economy.

Several additional industrial centers have developed between the Toronto-Hamilton area in the east and the Lake St. Clair area in the west. Such cities as Guelph, Kitchener-Waterloo, Brantford, and London are all located on major east-west railroad connections across the southern Ontario peninsula. Limited more to local resource advantages than the other manufacturing clusters, the industries in these cities are involved in food processing, in some activity dependent on large and accessible supplies of wood, or in a manufacturing process, such as electronic-machinery production, drawn to the labor supplies of established urban centers.

With the preceding discussion in mind, the urban and industrial differences between the Ontario and Quebec portions of the Canadian core should be clear. Although both possess a single dominant urban center and smaller industrial complexes located near good transportation routes and accessible to local resources, the industrialization process is more widely spread across southern Ontario. In a similar manner, Ontario also contains more large cities, after Toronto, than does Quebec, after Montreal. One consequence of these differences has relegated Quebec to a relatively low ranking of all provinces in terms of average family income (see Figure 7-5). The spatial variation of income in southern Quebec is undoubtedly great. Many

sections of Montreal and Quebec City as well as other urban centers contain small regions of population with high annual income. In other parts of these cities and across wide sections of rural Quebec, other people reside and labor each year to earn incomes that can be called no more than marginal. In this pattern of income distribution, southern Quebec is no different from other major regions of the continent. It should be noted here, however, that Quebec possesses only a moderate overall average income, yet it occupies a substantial portion of the national core of Canada where incomes might be expected to be higher.

One of the essential differences between the Ontario and the Quebec portions of the national core is the access that the former has via the Great Lakes to a large region of intense economic activity: the U.S. portion of the interior manufacturing core. The St. Lawrence Seaway was completed in 1959, opening the Great Lakes to direct oceanic shipping. The Seaway also permitted access between southern Quebec and the industrial interior of the continent, thus reducing the apparent advantage enjoyed by southern Ontario as well as generating large amounts of much-needed electricity. After 25 years of operation, however, the impact of the Seaway on the Canadian region most directly involved with the St. Lawrence has not been extraordinary. Considerable industry is located along or near the St. Lawrence, but much of this was present before the Seaway was opened. Most of the benefits from the Seaway have been enjoyed by ports around the Great Lakes, for it greatly increased their sources of supply and potential marekts. These benefits flowed to both the U.S. and the Canadian portions of the continental manufacturing core, emphasizing its international character. But a major consequence for Montreal has been the loss of many transshipment activities and potential industry associated with break-in-bulk freight handling. Clearly, the increased amount of traffic that now moves up and down the St. Lawrence can have no real benefit for southern Quebec as long as it does not stop.

REGIONAL ORGANIZATION AND CONTINENTAL INTEGRATION

Two additional characteristics of the Canadian core region should be considered. One involves the spatial organization of the region—how and why places in southern Ontario and Quebec are tied to each other through their interaction. The other is related to the general economic integration of the Canadian core with the U.S. manufacturing core. Some of the integration is visible in transportation lines, the flows of people and goods across the political border, and so forth. But what may be even more important are the less obvious financial exchanges that represent corporate investment and control, primarily by U.S. interests over Canadian companies.

Regional Organization

Implicit in much of the earlier discussion of the content and diversity of the Canadian core is the idea that places within the core are important to the Canadian economy at least partly because they are located where they are. Quebec was located at the head of sea navigation but later on poorly connected by rail to other places. Montreal was located at a site on the St. Lawrence River where water transport was broken by rapids, and from which New York City was directly accessible. Windsor was located next to the larger center of Detroit but separated from it by the river. London was located about halfway between Toronto and Windsor in the middle of a good agricultural area and where two mainland transport lines crossed. Similar notes could be made for all major urban centers. Some, like Guelph, were small settlements before the stimuli for their growth began to appear. Others, such as Kingston, have gone through a growth-decline-growth sequence as changing external factors altered the importance of each city's location and, thereby, its development support. Regardless of such differences, knowledge of the location of economic activity is often critical to an understanding of the reason for a city's existence.

The Canadian core, as defined, is partly a culture hearth and partly a political focus; it is the diversity of culture and the way these differences have been translated politically that often threatens to divide the national core. But as part of a larger continental core, it is primarily the economic and urban aspects of this region that have tended to hold it together. The components of this region do not exist in isolation, possessing specific characteristics and linked to other places only by their relative location. Instead, each place is tied to other places by the movement of people, goods, or ideas. This pattern of interrelationships is fundamental to the location of economic activities and to the general manner in which regions develop. Because of its explicitly spatial nature, when the interrelations are economic, the pattern they form is said to be an expression of the *space economy* of the region.

Examination of a region's space economy is useful because it allows us to deal with the full set of interrelated parts with emphasis on the relations rather than on the place characteristics alone. One way in which the notion of a region's space economy can be approached is through the principles suggested by A. K. Philbrick.[3] He called the theme underlying his suggested principles areal functional organization instead of space economy because his principles pertain to noneconomic functions as well, but they can be used to analyze a region's economic interconnections and in this way be an approach to understanding the space economy. Philbrick's principles deal with the manner in which economic, social, and political functions are organized by people within the areas in which they operate. Our concern here would be with the application of these principles to the organization of urban-economic functions in the area of southern Ontario-southern Quebec.

Four of Philbrick's original principles are useful for us. First, human activity has focus. The spatial pattern described by nearly all human activity is one that has a place of concentration or

[3] Allen K. Philbrick, "Principles of Areal Functional Organization in Regional Human Geography," *Economic Geography* 33 (1957), pp. 296–336.

emphasis (focus) connecting otherwise unrelated places. For an individual, the focus may be a residence; for a business, the focus may be company headquarters; a city is the focus for its hinterland; and so forth. This principle states the way in which human activity is organized. Through this organization, the places where this activity occurs can be said to possess *focality*. The second principle states that the sites on which human activity focuses are localized. This simply recognizes that each focus of activity has a unique and discrete position on the earth's surface and that position can be viewed as relative to other places where activity occurs. Third, these nodes of activity are interconnected, and the interconnections define units of organization that are more complex than individual activity. Through this principle, Philbrick observed that lines of transportation and communication are the means by which an area is organized. The fourth principle deals with the fact that different activities may have their focus in the same location. They may do so because of the complexity or quality of their respective hinterlands or they may be sharing mutually beneficial advantages of proximity to each other.

In an oversimplified view, it is for these general reasons that urban places form. The fourth principle also recognizes that some places are the focus for a greater variety of activities than are other places. Large urban centers usually have more functions than small centers. Those that have a greater number of functions are said to be of a higher order than those that have fewer functions. Philbrick's fourth principle, therefore, states that there is a hierarchy of functional organization in which lower-order places are connected with (or focused on) higher-order places as well as nearby places of the same order. In other words, each city tends to have a set of smaller places tied to it by transport and communication links and, in turn, to be one of another group of cities focusing on an even larger metropolis. Note that nothing is said here about *where* these cities are or should be located. The concern throughout is with identification of the

way in which areas are organized and tied together by means of interconnected functions.

The functional organization of the Canadian core area illustrates many of these patterns (Figure 6-6). First-order individual residences and second-order villages are scattered throughout the settled area. Higher-order places emphasize the linear shape of the core region extending between Quebec and Windsor and the urban-economic differences between the two core provinces. The core's functional organization (and its space economy) is dominated by the two nodes of Toronto and Montreal. Each fifth-order city is the focus of a pattern of fourth- or lower-order cities (smaller and economically less complex than either Montreal or Toronto). The functional connections emphasize the cultural and provincial differences between Montreal and Toronto. The only exceptions are Cornwall's link with much closer Montreal and Ottawa's interconnections (as Canada's political capital) with both major cities. In addition, these large fifth-order centers are strongly connected with each other by road, rail, air, and telecommunications (thereby defining the main axis of core organization) and with fifth-, sixth-, and seventh-order foci in the United States and around the world.

The region's pattern of functional organization also illustrates dramatically the relative paucity of urban-economic development in southern Quebec when compared with southern Ontario. Quebec City is the only fourth-order city in the province, whereas four such cities have developed in Ontario between Toronto and southern Michigan. This difference in economic base is shown again with third-order nodes. Aside from Hull, Ottawa's twin city across the river, Quebec has only two of the larger third-order centers, although Cornwall has some northeast orientation. Southern Ontario, on the other hand, can support nine such nodes, focusing on Hamilton, London, Kitchener, or Toronto. It is only when considering the small third-order centers (less than 30,000 population) that the Quebec portion of the core equals that of Ontario. These

FIGURE 6-6

THE FUNCTIONAL ORGANIZATION OF THE CANADIAN CORE. Toronto and Montreal are the focuses for their regions, but there is clearly more urban development in Toronto's immediate hinterland of southern Ontario than in Montreal's equivalent hinterland of southern Quebec.

towns are agricultural market centers, contain small-scale industrial supply activities, or may have received strong economic stimulation too recently to have grown large.

Regional Integration

A significant facet of the Canadian core's space economy that does not receive appropriate emphasis by this discussion of functional organization is the region's position within the continental core. Most of Canada's national core is also part of North America's great manufacturing core region. The Canadian portion is joined to the larger U.S. part by the Great Lakes and the upper St. Lawrence as much as these waters separate the two parts. Canada's core is an integrated section of the continental core through road, rail, and air connections focusing on Toronto and Montreal and on the lower-order cen-

ters in the Niagara peninsula and in extreme southwestern Ontario.

A more subtle form of continental economic integration has also been taking place. Since the turn of the century—but more rapidly after World War I—American business and industrial firms have invested heavily in the Canadian economy. Only 14 percent of the foreign capital invested in Canada in 1900 originated in the United States. This proportion had increased to 53 percent by 1926 and surpassed 80 percent (and $25 billion) by 1966. This invasion of U.S. capital was largely an attempt to circumvent tarrifs imposed on U.S. manufactures by Canada. By establishing branch plants in Canada, the American firms gained access to Canadian markets *and* British markets without paying the import penalty of tarrifs. In some cases, this meant rapid development of Shield resources or construction of new manufacturing plants to em-

TABLE 6-1

Percentage of Taxable Income from Manufacturing, 1972, by Region and Control of Company

Company Control	Atlantic Provinces	Quebec	Ontario	Prairies	British Columbia	Total
United States	2.7	18.9	62.2	9.6	6.7	100
Canada	3.6	28.8	43.2	9.8	14.4	100

SOURCE: W. Clement, *Continental Corporate Power* (Toronto: McClelland & Stewart, 1977), Table 63, p. 298.

ploy Canadian workers and supply the Canadian market.

Most American investment has been in the mining and the manufacturing industries. More than 50 percent of Canada's mining industry, especially mineral fuel mining, is owned or operated by parent U.S. firms; fully 40 percent of Canada's manufacturing industry is controlled from the United States. Because Canadian manufacturing is concentrated in the southern Ontario-Quebec core region, these two provinces have drawn attention from U.S. investment capital. As reflected in taxable income from manufacturing, nearly 20 percent of the total generated by U.S.-controlled companies occurred in Quebec and more than 60 percent in Ontario (Table 6-1).

This level of U.S. capital investment has been viewed with considerable alarm by some Canadians. Much as Quebecois nationalism reflects French Canada's strong concern and resistance to cultural assimilation by the larger Anglo culture, a broader Canadian nationalism has been expressed that reflects fears of economic assimilation by the larger American interests. Although U.S. economic involvement is widespread, a considerable share of its involvement is concentrated within the heart of Canada's national economy—within its national core region.

The political boundary between the two countries is, in fact, a selective and semipermeable barrier. For example, it is quite easy for citizens of one country to enter the other for recreation or business. The boundary may be indicated by no more than a strip of paint across the road or a few brief questions from a customs agent. A

truly integrated continental economic core region is one result of such a transparent border. But another consequence has been intense cultural and economic influence in Canada from the much more populous United States. Canada's population is politically and culturally distinct from its large southern neighbor, but it is only 10 percent as large as the American population. In spite of its internal divisions, the Canadian core region is strongly influenced by, and integrated within, a much larger continental culture and continental economy.

Additional Readings

Bradbury, John H. "State Corporations and Resource Based Development in Quebec, Canada: 1960–1980." *Economic Geography*, 58 (1982), pp. 45–61.

Cartwright, Don. "Language Policy and the Political Organization of Territory: A Canadian Dilemma." *Canadian Geographer*, 25 (1981), pp. 205–224.

Easterbrook, W. T., and M. H. Watkins. *Approaches to Canadian Economic History*. Toronto: McClelland & Stewart, 1967.

Foley, James (ed.). *The Search for Identity*. Toronto: Macmillan of Canada, 1976.

Innis, Harold A. *The Fur Trade in Canada*, rev. ed. Toronto: University of Toronto Press, 1956.

Marseau, Jean Claude. *Montreal in Evolution*. Montreal: McGill-Queens University Press, 1981.

Skidmore, Darrel R. *Canadian American Relations*. Toronto: John Wiley & Sons Canada, 1979.

Sussman, Gennifer. *The St. Lawrence Seaway: History and Analysis of a Joint Water Highway*. Montreal: C. D. Howe Research Institute, 1978.

Watson, J. Wreford. "Regions of Canada: Central Settled Canada," in *Canada: Its Problems and Prospects*. Don Mills, Ont.: Longman Canada, 1968, pp. 130–177.

West, Bruce. *Toronto,* rev. ed. Toronto: Doubleday Canada, 1979.

Whebell, C. F. J. "Corridors: A Theory of Urban Systems," *Annals of the Association of American Geographers,* 59 (1969), pp. 1–26.

CHAPTER SEVEN

THE BYPASSED EAST

If you look at a map of the eastern seaboard of North America, you might notice a lack of large cities along the coast north of Boston. Few major overland routes extend inland from this coast, and interior cities are generally even smaller than those along the coast. This area, comprising the Atlantic Provinces of Canada plus northern New England and the Adirondacks of New York, can be referred to as the Bypassed East (Figure 7-1).

The Bypassed East is near, even astride, major routeways, but not on them. Ocean transportation can easily bypass the region. Fine harbors farther south along the coast have easier and more direct access to the interior. There is little advantage for oceanic trade to pass through the harbors of the Bypassed East. To the north, the coastline of the Atlantic Provinces is hollow, indented by the wide, deep estuary of the St. Lawrence River. Cartier sailed up the St. Lawrence hoping that he had found the westward sea route to Asia from Europe. At the present site of Montreal, he found rapids blocking his path, and he abandoned his search. What he found was an

easy passage deep into the heart of what was to become Canada. It was to be Montreal, not one of the many fine harbors of the Atlantic Provinces, that became the major eastern port for Canada. Like northern New England, the Atlantic Provinces were settled early, but as Canadian settlement pushed westward, these provinces were increasingly isolated from the regions of growth and development.

The Bypassed East, then, is in a transportation shadow (defined as an area of limited transportation development located near an area of much greater facility availability). This has led to slow regional economic growth and even stagnation, for geographic accessibility is a key element in the growth prospects of a place (see Chapters 4 and 5).

As discussed in Chapter 1, we have placed southern and northern New England in two separate regions. The strong image of New England as a unified region is held almost universally outside the area and often accepted within it, but this is a historical carryover. Certainly, it was once a section of the United States

FIGURE 7-1
THE BYPASSED EAST.

with considerable uniformity in economy and institutions. Today, however, southern New England is a part of metropolitan America. Northern New England, for the most part, is not. Northern New England is much more like Canada's Atlantic Provinces in their characteristic problems and prospects, even though these two parts of the Bypassed East are separated by an international boundary. There is perhaps no better example of the fact that regional boundaries can and do shift over time, although that shift may not be recognized by many observers if the previous region was clearly identified.

THE PHYSICAL ENVIRONMENT—THE JOY AND THE HINDRANCE

> The summer folk call it paradise mountain,
> While we call it poverty hill.

These lines from a well-known folk song are not about any part of the Bypassed East, but they could be. Much of this part of the continent is, simply stated, beautiful. The Presidential Range in the White Mountains of New Hampshire contains some of the most rugged landscape in the eastern United States. The extensive shoreline thrusts out into the Atlantic and meets the ocean's waves with a heavily indented coast. Dramatic headlands are mixed with many small coves bordered by rocky beaches. Large empty areas, almost totally lacking in settlement, are only hours away from some of the largest cities on the continent.

Topography

Most of the Bypassed East is a part of the northeastern extension of the Appalachian uplands. However, the structure of the area bears little surface resemblance to the clearly delineated Blue Ridge—Ridge and Valley—Appalachian Plateau sequence of the southern Appalachians (see Chapter 8).

The Adirondacks, in northern New York, are a southern extension of the Canadian Shield (see Chapter 17). This broad upland was severely eroded by continental glaciation, so that the surface features are generally more rounded than angular. Although elevations in the Adirondacks are not great, the areal extent of this highland is considerable.

A large upland plateau covers most of New England. This upland is old geologically and has also been heavily eroded by the actions of moving water and ice. One result of this long history of erosion is that elevations throughout the region seldom top 1500 meters (4600 feet). Widespread scouring by continental glaciers during the Pleistocene rounded most of the hills and mountains across the plateau. Only where elevations were high enough to remain above the moving ice can one find more rugged mountains.

The two major mountain areas of northern New England are the Green Mountains of Vermont and the White Mountains of New Hampshire. The Green Mountains are lower in elevation, less than 1500 meters (4600 feet) at their highest, and were completely covered by ice during the Pleistocene, with the result that their tops are well rounded. The White Mountains, by comparison, are nearly 600 meters (1975 feet) higher, and their peaks rose like islands above the ice. Thus, their upper slopes are rugged and steep, with an appearance far more like some of the younger mountains in the western part of the United States. Farther south, where the upland plateau has been heavily eroded by flowing water, several isolated peaks stand well apart from the major mountain areas to the north. The largest of them is Mount Monadnock in southern New Hampshire. Its name is given to all such areas of hard rock that have become low rounded mountains as the surrounding rocks were removed by water erosion.

Although northern New England (with New York) is a land that draws character from its mountains, its people find their homes and their livelihoods in the valleys and lowlands. The largest such areas are the Connecticut River valley between New Hampshire and Vermont, the Lake Champlain Lowland along the northern Vermont-New York border (basically an extension of the ridge and valley area of the southern Appalachians), and the Aroostook Valley in northern Maine. The latter valley resulted from the erosion of an area of soft rock by the St. John River. A number of smaller lowlands border the seacoast, and innumerable streams have sliced the plateau throughout the area. A detailed map of the distribution of population would look like an inverted map of elevation, with most people in the lowlands (Figure 7-2).

The mountains of the Atlantic Provinces are

FIGURE 7-2

EMPTY AREAS OF NEW ENGLAND (adapted from *The Geographical Review,* **44,** (1954), Lester E. Klimm). This map of areas distant from any paved road or dwelling unit is based on data from the early 1950s. A 1980s map would show little change.

generally even lower and more rounded than those of northern New England. Elevations do not rise much above 700 meters (2200 feet). The surface patterning of the southern Appalachians is missing almost totally here, although the basic grain of the topography does follow the same northeast-southwest trend. In the southern Atlantic Provinces, uplands in peninsular Nova Scotia and interior New Brunswick are separated by a wide lowland, most of it ocean covered, but it also includes Prince Edward Island (PEI to Canadians).

As in northern New England, the lowlands of the Atlantic Provinces are the home of most of their people. The central lowland, which includes Prince Edward Island, is underlain with soft rocks, limestones and sandstones. The Annapolis Valley, in southern Nova Scotia, is a part of the same structure that underlies the Connecticut River valley. These Triassic lowlands reappear frequently along the interior east coast, even as far south as central North Carolina. The larger cities throughout the provinces, and nearly all of the population of Newfoundland, are clustered around the area's harbors.

Climate

This region is no place for those who desire a dry, warm climate. It is a place where polar, continental, and maritime weather systems meet, and the result is a climate that is seldom hot, often cold, and usually damp. Because of its location on the eastern side of the continent, the wind systems tend to push continental conditions into the area and to limit the maritime impact on the location, especially away from the coast. The

MAJOR METROPOLITAN AREA POPULATIONS IN THE BYPASSED EAST, 1981	
Halifax, N.S.	277,727
St. John's, N.F.	154,827
Burlington, Vt.	114,070[a]
Saint John, N.B.	112,974
Lewiston-Auburn, Maine	73,778[a]

[a] Indicates 1980 population.

substantial climatic difference between the coast and interior is further increased by higher inland elevations.

The Labrador Current that flows southward along the Bypassed East is cold. Even in late summer, only the most intrepid swimmers are willing to dip themselves into its waters for more than a short time. Still, the climatic conditions along the coast are moderated substantially by proximity to the water. The growing season near the coast is as much as 70 days longer than the interior average of 120 days. Average midwinter temperatures at coastal sites are often 3° to 6° C (5° to 10° F) higher than at nearby interior locations. Midsummer temperatures, in contrast, are slightly higher in the interior than along the coast.

The mild maritime influence brings frequent cloud cover and fog, particularly along the southern coastline. Southeast winds blowing across the warmer waters of the Gulf Stream and then over the colder waters of the Labrador Current closer to shore create a dense fog by the time they reach land. This presents a substantial climatic problem, for the fog and cloud cover serve to cool temperatures further during the summer. It is consequently difficult to grow crops that require more summer heat and sunlight than is usually available.

Almost all parts of the region receive substantial precipitation, usually between 100 and 150 centimeters (40 to 60 inches) annually. Precipitation is usually scattered evenly throughout the year. Snowfall is generally substantial, with most places receiving between 25 and 50 percent of their total precipitation in the form of snow. Many interior locations average at least 250 centimeters (100 inches) of snow annually.

POPULATION AND INDUSTRY IN A RUGGED LAND

The Bypassed East is not an easy place in which to live and work. Its harsh climate and thin, rocky soils limit agriculture, except in a few particularly blessed locations. Few mineral resource deposits of substantial size had been found until recently. Coupled with a small local market and relative isolation, this has limited the development of manufacturing. The advantages that the area does offer thus become relatively more important.

Early Settlement

This was not always the Bypassed East. Its foreland location, jutting far out into the Atlantic Ocean, meant that its shores were among the first land in North America encountered by European explorers and settlers moving into the New World. The first European settlement in Canada was opened by the French in 1605 at Port Royal on the Bay of Fundy. Less than a decade later, the English settled in Newfoundland, where they quickly established a series of fishing villages. By midcentury, many of the small harbors of central and southern Maine housed British villages. Settlement was kept out of the interior by the local American Indian population until the middle of the eighteenth century. The Indians in the lower St. Lawrence Valley were aided by the French in hopes that the British would be kept as far from French settlements as possible.

These early European settlers utilized two prime local resources. The rich fishing banks off the Atlantic Provinces—possibly visited by European fishermen well before Columbus made his first voyage to the New World—were immediately important. The banks, shallow areas in

Peggy's Cove, Nova Scotia, is typical of the many small fishing villages of the Bypassed East. The villages do much to help shape our perception of the region's character, but their limited size and small-scale fishing fleets contribute to the low regional income. (Nova Scotia Communication and Information Center.)

147

the ocean at the outer margins of the continental shelf, underlie waters that are rich in fish. Their shallowness allows the sun's rays to penetrate easily through much of the water's depth. This encourages the growth of plankton, a basic food for many fish. Cold-water fish, such as cod and haddock, are abundant. Using this nearby resource, the early European settlers soon began a substantial export of salted cod.

The other prime resource of the region was its trees. England had already lost much of its domestic forest cover to agriculture. As England began to develop an important navy and merchant marine fleet, it was forced to turn to the forests of Scandinavia for its ship masts and for lumber. This was unsatisfactory, for it made England dependent on a foreign power for one of its vital military and economic resources. Consequently, the English sought an alternative source within their territories.

The white pine was dominant in the forests of New England. A magnificent tree, it reached heights in excess of 60 meters (190 feet) and stood straight and proud. Its wood was clear, light yet strong, and easily cut. Almost all of the virgin forests are gone now, and the second- and third-growth forests that remain are short and insignificant in comparison. A few lone giants remain; in some isolated areas, small groves remind us of the forests that once covered the region. Forest resources allowed Maine to become a center for ship construction. The best of the white pines were reserved for the ship masts of the Royal Navy. Still, there was more than enough for all.

Agriculture was the third major occupation of the early settlers, but farms tended to be small and production limited. Early farming was primarily a subsistence activity. By the end of the 1700s, however, most of the lowlands south of Newfoundland, and even much of the slope land, was in farms.

The peak of agricultural development in northern New England probably came just after the start of the nineteenth century. But two developments elsewhere in North America soon began to pull people off their farms, first in a trickle, then in droves. One development was the opening of the West. In the United States, settlement moved beyond the Appalachians onto the rich farmlands south of the Great Lakes early in the century. Then, in the 1820s, the construction of the Erie Canal, and later other canals farther west, made the markets of the east coast more accessible to Western farmers. The poor farms of upper New England rapidly lost what little market they had to crops imported from places like Ohio and Indiana. New Englanders left their farms and joined the migration westward, exchanging bad land for good. Although an accurate count is not possible, it is estimated that about 1 million New Englanders moved off the region's farms and headed west during the nineteenth century. It was said that the only crop those New England fields could produce was rocks and that their only export was people.

The pattern was much the same in the Atlantic Provinces. The settlement of the middle St. Lawrence River valley in southern Quebec and more especially the rapid expansion of settlement in southern Ontario in the decades after the American Revolution provided superior agricultural land (see Chapter 6). Construction of the Welland Canal to bypass Niagara Falls in 1829 increased the accessibility of southern Ontario.

A second blow to the region's agricultural fortunes also occurred during the late 1700s and early 1800s with the development of manufacturing in southern New England, where the Industrial Revolution began in the United States. Industrial growth created a great demand for labor. That demand was first met with New England farmers seeking the higher wages and steady income offered by manufacturing employment. An increase in child and female labor, particularly in the textile mills, further enhanced the value of manufacturing work over farming.

Agriculture Today

Agricultural decline has continued across most of the Bypassed East for the last 150 years. To-

FIGURE 7-3

MAJOR AGRICULTURAL AREAS OF THE BY-PASSED EAST. Most of the region's farming is concentrated in its few lowlands of substantial size.

day less than 10 percent of the land in the three states of northern New England is in farms; 100 years ago, the amount was closer to 50 percent. Until the last decade or two, many northern New England towns had patterns of population decline that had lasted for a century or more. Farming retreated off the slopes, allowing them to return gradually to forest. Even in the valleys, soils were often too infertile, the climate too cold, and the farms too small for successful agricultural production.

The pattern of decline is perhaps not quite as great in the Atlantic Provinces, but it is still substantial. From a peak late in the nineteenth century, total land in farms has declined by about two-thirds and improved land has fallen by about the same proportion. Productivity remains poor compared to the rest of Canada, and regional declines in acreage and in production remain far larger than for other parts of the country.

Where farming in the Bypassed East remains important, it tends to specialize in single-crop production and to be concentrated in a few favorable locations. There are four significant areas of agricultural production in the region (Figure 7–3).

Aroostook Valley. The Aroostook Valley in northeastern Maine is the Bypassed East's newest major agricultural area. Commercial farming did not become important until late in the nineteenth century. It was soon one of the leading potato-producing areas in the United States.

The silty loam soils are ideal for potato growth, and the short growing season encourages a superior crop that is used widely elsewhere as seed potatoes. Large-scale, mechanized farming predominates; French-Canadian farm workers are employed heavily during the harvest season.

The Aroostook Valley potato growers have gone through a difficult period during the past several decades. Changing American eating habits have led to a declining market demand for potatoes. A result has been frequent overproduction with consequent depressed market prices. Potato growers, primarily in Idaho and Oregon, created well-organized merchandising operations, developed many new kinds of manufactured potato products, created effective advertising, and used coordinated marketing. Aroostook Valley farmers, proudly independent, lost many of their traditional markets to these other areas. Today the valley's agricultural economy is depressed and production is down. Many farmers have gone out of business. Others have tried to diversify their crops, but so far they have met with limited commercial success.

The Lake Champlain Lowland. Megalopolis represents by far the largest concentration of demand for agricultural produce in the United States. Many of these needs can be supplied by crops grown anywhere. Such products may be relatively low-bulk, high-value crops, like tea, which can be shipped great distances from overseas growing areas with only a small percentage increase in price. Or they may be crops, like wheat or corn, which can be handled roughly without damage or spoilage, so that they can be shipped great distances in bulk containers. Or they may be processed near the area of production, so that they can be shipped and stored easily, such as frozen and canned vegetables.

Fresh milk meets none of these requirements. It is a relatively high-bulk, low-cost product. It spoils easily and cannot be stored for any extended period. Consequently, urban areas usually rely on nearby agricultural areas for fresh milk. Most parts of the United States and Can-

149

ada are subdivided into *milksheds,* or nodal regions that supply a city with its milk needs.

The Champlain Lowland's proximity to Megalopolis gives it a substantial market advantage over more distant areas for milk sales. It is in the milkshed of both New York City and Boston. In addition, its summers are mild and moist, a climatic condition that encourages the growth of fodder crops. These cool summers are also well suited to dairy cows. Vermont has long led the United States in the per capita production of dairy products, and most of that comes from the Champlain Lowland.

Prince Edward Island (PEI). An overwhelmingly agricultural economy is found on PEI. The local economy provides little alternative employment, and most of the workers are engaged in agriculture. It may also be that no more beautiful agricultural landscape can be found in North America than this rolling, sloping land of small, green farms. Although agriculture is somewhat diversified, the major commercial crop is seed potatoes.

One of the delightful aspects of PEI for tourists is the faintly worn appearance of many of the farmsteads. They have been here for a long time. Declining demand for its principal crop, increased competition from other sources, and too many farms that are too small to be profitable in a modern farm economy are all part of farming on PEI. The comfortable, used, lived-

Prince Edward Island is a land of gently rolling hills and small farms. Its agricultural economy suffers somewhat because of its isolation from the major markets of eastern Canada. (Information Service, Prince Edward Island.)

in appearance of the landscape reflects in part an agricultural economy that has stagnated and no longer has sufficient sales for capital to invest in improvements.

The Annapolis Valley. Nestled into southwestern Nova Scotia, protected from northwest winds by a low mountain ridge, the Annapolis Valley is another attractive agricultural area. The valley, about 130 kilometers by 15 kilometers (80 miles by 9 miles), has long been one of the main apple-producing areas of Canada. Once again, the principal crop has suffered in recent years by increased competition from sources closer to its traditional markets, in this case Great Britain and the urban centers of Quebec and Ontario. Crop diversification has been more successful in recent years, especially in the production of fresh produce and dairy products for nearby Halifax. With more than a quarter of a million people, Halifax is easily the largest urban center in the Atlantic Provinces.

Forestry

Parts of northern New England offer a good example of the destructive potential of uncontrolled logging. Limited organized reforestation followed the cutting of trees, so that much of the forest that replaced the original trees is of poor quality for both lumber and pulp production. Neither is it as handsome as the forest it replaced. None of these three New England states is an important lumber supplier today, and production is not increasing significantly. This is an area where the great majority of the land is in trees, so the lack of a larger wood-products industry may be somewhat surprising. The slow growth rate for trees in the cool climate plus the continued lack of large-scale corporate operations in much of the area are other contributing factors.

An exception to this pattern of limited output is the pulpwood production of northern Maine. Here, on some of the most inaccessible large tracts of land in the eastern United States—where a small number of private owners control most of the land—forest industries remain important.

Many of the roads and recreation facilities in these Maine Woods are also privately owned. The area contains some of the finest white-water canoeing streams in the country as well as large numbers of glacial lakes and marshes. Disagreements over their utilization versus their preservation has been a source of conflict between the pulpwood producers on the one hand and canoeists and conservationists on the other hand.

Forestry plays a much greater role in the economy of the Atlantic Provinces and is the leading supplier of exports from that area. Only a small percentage of the labor force actually works in the woods, but the manipulation of the cut wood accounts for nearly half of the entire manufacturing employment of the Atlantic Provinces. Pulp and paper are the key products, with the most important areas of logging located in northern New Brunswick and on Newfoundland. The largest paper mills are on the north shore of New Brunswick close to the river mouths. The logs can be floated to the plants for manufacture, and the finished products can be loaded directly on large ships for transport to market.

Fishing

Fishing remains an important, if troubled, part of the economy of the Atlantic Provinces. Canada is among the world's leaders in the export of fish products. Nova Scotia leads all of the provinces in total fish catch each year, followed in order by British Columbia, Newfoundland, New Brunswick, and PEI. Most of Canada's fish exports come from the Atlantic Provinces.

There are two kinds of ocean fishing in eastern Canada. Inshore fishing, the most important, uses small boats and requires a relatively small capital investment, with lobsters and cod the most valuable catch. Maine and Nova Scotia lobstermen each provide about half the total catch annually in their respective countries. Deep-sea fishing on the banks off the coast requires far larger boats and more capital investment.

The inshore fishing industry faces several substantial problems. Pollution is a growing

The huddle of small fishing boats on Cape Cod are in sharp contrast to the large Soviet factory ship preparing to receive a catch from a small catcher boat. (Top, FAO Photo; bottom, Cape Cod Chamber of Commerce.)

problem in some areas, forcing fishermen to move to offshore locations. The inshore fishermen, with their small investments and small incomes—often living in one of the many small fishing villages scattered along the coastline of Nova Scotia and Newfoundland—cannot easily obtain the money needed to buy and outfit large, modern oceangoing vessels. Whey they move onto the banks with their smaller vessels, they are in direct competition with large fleets of modern ships, including factory ships that process the fish at sea.

The provincial government of Newfoundland, concerned about the uncompetitive nature of its inshore cod-fishing industry, initiated a program of centralization some years ago by encouraging fishermen and their families to move, at government expense, to larger communities. There they could form larger, more effective cooperative fishing operations and, at the same time, help cut the high cost of providing a variety of government services to the many small fishing communities. The plan has had substantial success; out of a total of 1200, about 500 of these small communities (called outports in Newfoundland) have been abandoned. A confounding problem, however, is the already high urban unemployment in the province.

Newfoundland was for many years not a part of Canada. Only in 1949, after several years of disastrously low fish prices in Europe had wrecked havoc with Newfoundland's economy, did residents of the province vote by a narrow margin to join Canada. The former name for the eastern provinces, the Maritimes, was changed to the Atlantic Provinces to reflect the incorporation of Newfoundland.

Offshore fishermen in both countries face heavy competition from European and Japanese ships. Americans and Canadians have lagged far behind their foreign competitors in turning to larger fishing vessels. Still, the offshore industry is expanding, especially in Newfoundland. In 1979 Canada established a 200-mile (320-kilometer) offshore management zone closed to unrestricted fishing by foreign vessels. The number of registered Canadian fishing boats increased sharply in the following several years.

Minerals Production

Offshore fishing has recently been threatened by the high demand for domestic petroleum in the two countries. Fears of pollution from offshore oil drilling in the rich fishing banks in the United States were overruled in 1979 when the Department of the Interior granted exploration leases to several oil companies. Major resources of petroleum and natural gas have been discovered in the Hibernia field near the Grand Banks off Labrador, and natural gas has been found near Sable Island 300 kilometers (180 miles) off Nova Scotia. It is hoped that the Hibernia field will supply Canada with its first substantial eastern petroleum production by the early 1990s.

The possible environmental impact is not the only problem with developing these offshore fields. Development costs will be many billions of dollars. Ownership conflicts between the provinces and the central government are difficult to solve. Nova Scotia and the Canadian federal government recently agreed that the province would get all governmental revenues from its offshore natural gas until the province's per capita income is 16 percent above the national average, opening the way for expanded exploration.

Mining other than for petroleum and natural gas in the Bypassed East is not currently of great importance. This was not always the case. Iron ore has been mined in the Adirondacks for more than 100 years, and the reserves there are still substantial; but total output from the mines is relatively small. Although New York is usually among the top five states in iron ore production, only the top two, Minnesota and Michigan, can realistically be called major producing states.

Nova Scotia has some of the larger coal reserves in Canada, but production fell steadily to about 2 million tons annually before reviving somewhat in the late 1970s. There is simply not a large demand for the coal in the Atlantic Provinces; American coal undersells it in the more distant markets of Ontario. The coal has been used as coking coal at a small steel plant in Sydney. Unfortunately, the plant is too small and outmoded to be efficient, and it is kept open with governmental financing. Steel from the plant is limited to the local market. Steel produced elsewhere in larger quantity, with much better economies of scale, can be shipped to the Atlantic Provinces and still undersell the local product. Again, the isolation of the region from the major markets of the two countries has hurt the local economy.

The igneous and metamorphic rocks of northern New England have made the area an important producer of building stone for years. Many granite quarries operate in central Vermont and along the central coast of Maine. Vermont is also the leading marble-producing state in the country. The value of all of these rocks is small compared to the minerals industries found in other parts of the continent, but it is still an important element in the economy of the two states, especially Vermont.

The largest mining operation in the Atlantic Provinces is based on the development of the iron ore deposits in Labrador near the Quebec border, but it has little impact on the region (Chapter 17). Although these deposits are located in Newfoundland, the ore travels by rail southward to a railhead at Sept Îsles, Quebec, where it is loaded on lake vessels and sent up the St. Lawrence River. The only benefit of that mining for the Atlantic Provinces is the tax that Newfoundland collects. Otherwise, its impact is significant for southern Quebec, not the coast.

Cities and Urban Activities

By a slight majority, most of the residents of the Bypassed East are urbanites. However, the region contains few substantial urban areas. The two largest cities of northern New England are Burlington, Vermont, and Bangor, Maine. Each has only about 50,000 residents in its urban area. In Canada, Halifax, Nova Scotia, had 277,000 residents in its urban area in 1981, and St. John's, Newfoundland, and Saint John, New Bruns-

wick, had 150,000 and 110,000 metropolitan residents. Although substantial at the regional level, these are relatively small places nonetheless.

The small size of the major regional centers is a good indication of what may be the greatest single reason for the relatively low per capita income levels of the region. Most higher-income occupations in the two countries are urban based, and this area lacks urban occupations. Although they are less than half of the total, a high percentage of the work force is engaged in primary occupations, which are traditionally among the lowest paying on the continent. Because of the absence of a large local market and because of poor access to major urban areas, primary industries have not served as the foundation for the development of a more broadly based manufacturing economy, as they have elsewhere in Canada and the United States. As we have said earlier, a majority of the manufacturing jobs in the Atlantic Provinces relate directly to the manufacture of the raw materials produced in the region.

Several cities in the region do have some locational advantages, and they are growing. Halifax, for example, is eastern Canada's second leading ocean port following Montreal. During winter, use of the St. Lawrence is limited by ice. Although a channel is cleared through the ice, the volume of traffic using the river is greatly reduced. It is during winter that the harbor at Halifax is at its busiest. The city has developed a diversified economic base and is the most prosperous center in the Atlantic Provinces. St. John's is too far east to serve as a major port of entry for Canada, but its excellent harbor promises city growth as the principal port for eastern Canada's deep-sea fishing fleet. Usually ice free, its harbor is an important winter facility for fishing fleets from many countries as they fish the offshore banks.

THE FUTURE—PERPLEXING

The future of the Bypassed East is not easy to

Halifax, Nova Scotia, is the largest city in the region. Its port function is enhanced during the winter, when the St. Lawrence freezes over, blocking access to Montreal. (Nova Scotia Communication and Information Center.)

predict. We can make some generalizations about the present, and these can help in trying to anticipate the future. First, let us reiterate a few points. This is an area with a strong continuing economic emphasis on the collection and manufacture of raw materials, especially fish and wood products in eastern Canada. Although a majority of the region's population is urban, it is substantially more rural than Canada and the United States as a whole. The entire region is in a transportation shadow, near some of the busiest routeways in the world, bypassed in favor of alternative paths that provide more effective penetration into the continent.

An important result of all of this is regional poverty. In 1982 the average per capita income in the Atlantic Provinces ranged from $7528 (Canadian) in Newfoundland, easily the lowest in the country, to Nova Scotia's $9041. In all other provinces the statistic was over $10,000. In the United States in 1981, average per capita income in northern New England ranged from a low of $8655 in Maine to $10,073 in New Hampshire (Table 7-1). The national average was $10,514. Only Mississippi had a lower per capita income than Maine after adjustment for differences in cost of living (living costs were substantially higher in New England than in much of the South). Average income levels in New Hampshire were boosted by the higher incomes in the urban areas along the southern edge of the state. Per capita income averages for the states of southern New England ranged from Rhode Island's $10,723 to Connecticut's $13,748, the highest for any state, except Alaska.

Northern New England

There does seem to be reason to anticipate that economic change will occur in northern New England. The 1980 national census indicated that the long-term demographic stagnation had ended. Maine, New Hampshire, and Vermont were the only states outside the South and West to grow at a rate above the national average, and it is expected that the trend will continue (see Figure 1-2).

There seem to be several reasons for this shift in regional population fortunes. One is the gradual northward growth in Megalopolis. As the cities of the urban region expand, as new peripheral areas urbanize and become a part of urban America, and as people search farther outward to find residences away from the large cities, the Megalopolitan periphery has been pushed steadily northward in New England (Figure 7-4).

Northern New England is also attracting a number of new manufacturing facilities, which are locating in many areas and tend to be light industry with a medium-sized work force. In part,

they are locating in the region because the employers and their workers find the small-town and rural environments typical of the area to be good places to live. Also, the construction of several interstate highways into the region during the 1960s has provided greater accessibility to the south and west.

Recreation, second-home development, and retirement. The tourist industry has been northern New England's boom industry since the end of World War II. Magnificent scenery; rugged coasts; fishing, skiing, and canoeing; and just plain driving around looking at the beauty of the place—all of these are a part of this tourist growth. The region is fortunate enough to have several tourist seasons. First there is the summer, the traditional time to vacation for most American families. Next comes the "Fall Foliage Festival," when thousands of automobiles fill the roads, so that their occupants can get a glimpse of the oranges, yellows, and reds before they are covered with road dust. In winter comes skiing— "Christmas in Vermont," "White Christmas," and all the rest. Only the spring, the season of mud and flies, has not been fitted into the travel habits of Americans.

The economy of the Adirondacks area is also heavily dependent on tourism. Lake Placid, home of the 1932 and 1980 Winter Olympics, is but

TABLE 7-1
Average per Capita Income, 1960–81, for Selected States

	1960		1970		1981	
State	Income	Percentage of U.S. Average	Income	Percentage of U.S. Average	Income	Percentage of U.S. Average
Maine	$1,825	83	$4,766	83	$ 8,655	82
Vermont	1,864	85	4,924	89	8,654	82
New Hampshire	2,172	99	5,417	96	10,073	96
South Carolina	1,396	63	4,665	80	8,050	77
Mississippi	1,196	54	2,547	65	7,256	66
United States	2,201		5,893		10,517	

SOURCE: *U.S. Statistical Abstract*, 1979, 1983.

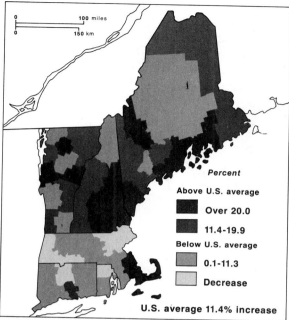

Percent

Above U.S. average

■ Over 20.0

■ 11.4–19.9

Below U.S. average

0.1–11.3

Decrease

U.S. average 11.4% increase

FIGURE 7-4

POPULATION CHANGE, 1970–1980. Population growth rates are above the national average across most of northern New England. This is in sharp contrast to the slow growth or decline that characterizes much of the rest of the "Snow Belt."

purchase and upkeep costs. In a number of the counties of northern New England, there are more of these part-time dwellings than there are permanently occupied houses. No one is quite sure what the impact of this is or will be, but at a minimum, it has created a small flourish in the local construction industry. As with tourism, this second-home development arises from the region's proximity to Megalopolis. Nearly a quarter of America's population lives within an easy day's drive of northern New England. Residents of Quebec and Ontario also find it a nearby and important recreation area.

Finally, many of the coastal communities of Maine, the small college towns of Vermont and New Hampshire, and old villages throughout the region have become popular retirement centers. These retirees, although generally not rich, are not poor, and they do make a significant contribution to the regional economy.

Eastern Canada's Prospects

The future prospects of the Atlantic Provinces are more clouded than those of northern New England. The Canadian government has recognized the region as a problem area. It set up a regional development commission to try to identify solutions to the problem of regional poverty and to help in their implementation. The Atlantic Development Board was created not only to advise on regional economic problems but also to oversee federal infrastructure investment. In 1969 the government established the Department of Regional Economic Expansion (DREE) to encourage employment growth in the slow-growth areas of the country. While it existed (the program was eliminated in 1982), roughly half of all DREE funds were given to the Atlantic Provinces, with Quebec getting most of the rest. Included in the overall program were direct payments to private industry encouraging them to locate and helping them survive in less developed areas as well as grants to improve such things as transport facilities and water and sewer systems and funding support for manpower development programs.

one of many ski areas. The state of New York oversees much of the area through its Adirondack State Park, the largest state park in the country. Although the state encourages outdoor recreation on its lands, many local residents argue that the growth should be limited by restricting construction of new tourist facilities.

Second homes, long a part of the life-style of the well-to-do in America, have in the last 30 years become the goal of much of the rest of America as well. Strung along the seashores and around the lakes and strewn across the mountains is a growing collection of vacation homes, which are occupied by the owners for a few months or a few weeks each year and then rented for as much of the rest of the time as possible to help pay the

Nearly year-round tourism is a cornerstone of the economy of northern New England. Located near Megalopolis, its many attractions are within an easy day's drive of perhaps 50 million Americans. (Dick Smith/New Hampshire Office of Vacation Travel.)

It is difficult to assess the impact in the Atlantic Provinces of this considerable federal concern. Personal incomes began to converge with the national average after 1950. The process of convergence may have intensified somewhat in the late 1960s, although the recession of the late 1970s and early 1980s was especially hard on the Atlantic Provinces (Figure 7-5). Out-migration has slowed, and it may have turned around in

New Brunswick and PEI. Larger urban areas in the Atlantic Provinces are growing fairly rapidly. The area's relative economic condition is improving and will probably continue to do so.

The Atlantic Provinces do not have many of the apparent advantages that explain the recent population growth of northern New England. They are not near any substantial urban areas and are definitely not going to benefit from ur-

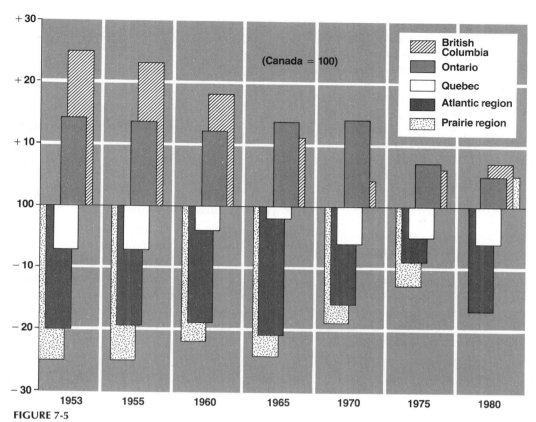

FIGURE 7-5

REGIONAL INCOME TRENDS BY PROVINCE. New resource developments in the Prairie Provinces and federal policies to encourage growth in low income areas, particularly Quebec and the Atlantic Provinces, contributed to a trend of lessening regional disparaties in the late 1960s and much of the 1970s. The recession of the late 1970s and early 1980s was particularly difficult for eastern Canada.

ban overflow. The improved accessibility that the interstate highway system provided to northern New England would be difficult to develop, partly because the distances are greater and partly because the band of Canadian territory that joins them to the rest of Canada is so narrow. High-quality, modern highways have been built, but they do not carry much traffic. Canadians from elsewhere do not find their Atlantic Provinces so attractive for recreation or retirement that they are willing to travel there in large numbers. Al-

though tourism is increasing substantially, especially to Nova Scotia and the beaches of PEI, many of the region's tourists come from the United States.

Still the Bypassed East

It is possible that the Bypassed East is a region in the process of disintegration. Northern New England is in close proximity to Megalopolis. Millions of people sitting almost at its doorstep want to move in—at least for a visit. And with a growing overland connection into southern New England and New York, northern New England is now increasing in population far more rapidly than the country as a whole. Many of the residents of the area may not want that growth, for it will change the reasonably quiet style of their lives and, later, the appearance of their landscape. The long-term pattern of outmigration from the Atlantic Provinces has slowed. There has been a substantial increase in nonprimary jobs, and that pattern too is expected to continue.

As we have seen, however, at the same time that this shift in population fortunes is occurring, northern New England's relative personal income level remains well below the national average and that of the Atlantic Provinces, although it had increased slowly with massive federal investment, may now be declining relative to the national average. These provinces remain the poorest of Canada's major regions, and they will probably continue to be so for many years, unless petroleum and natural gas development create major revenues.

There is some justification for cautious optimism about the future of the region. Certainly many analysts in eastern Canada, and most in northern New England, anticipate improvement in the region's economy. Real economic growth, however, will be slow, and the region's relative inaccessibility will be one important reason for this slow growth.

Additional Readings

Atlantic Provinces Economic Council. *The Atlantic Vision—1990.* Halifax, NS: Atlantic Provinces Economic Council, 1979.

Clark, Andrew H. *Acadia: The Geography of Early Nova Scotia.* Madison: University of Wisconsin Press, 1968.

Collie, M. *New Brunswick.* Toronto: Macmillan of Canada, 1974.

Hale, Judson. *Inside New England.* New York: Harper & Row, 1982.

Jorgensen, Neil. *A Guide to New England's Landscape.* Barre, Mass.: Barre Publishing, 1971.

Macpherson, Alan G. (ed.). *The Atlantic Provinces.* Toronto: University of Toronto Press, 1972.

Macpherson, Alan G., and J. B. Macpherson (eds.). *The Natural Environment of Newfoundland.* St. John's: Memorial University of Newfoundland, 1981.

Nutting, Wallace. *Maine Beautiful.* New York: Bonanza Books, 1924.

Pierce, Neal R. *The New England States: People, Politics, and Power in the Six States of New England.* New York: W. W. Norton, 1976.

CHAPTER EIGHT

APPALACHIA AND THE OZARKS

The Appalachian uplands, stretching from New York to Alabama, and the area of the Ozark-Ouachita mountains are separated by some 400 kilometers (250 miles) of land that includes the Mississippi Valley. They are actually two parts of a single physiographic province (see Figure 2-1). We place these two territories in the same region because of their topographic similarity and because, in both sections, there is an unusually close association between topography and human settlement.

Early European settlers, when they reached the shores of colonial North America, heard tales of a vast range of high mountains to the west. As they moved westward up and into those mountains, they discovered that the elevation of the mountains had been exaggerated. Westerners of the United States of today often hold a disparaging view of the Appalachians, suggesting that they are not worthy of the term "mountains." Their attitude is in some ways understandable. Crests in both the Rocky Mountains and the Sierra Nevada rise more than a mile higher than the loftiest Applachian peaks. Only in a few small areas do the Appalachians or Ozarks approach the dramatic mountain vistas so common in the western mountains.

Although the distinction between such features as mountains and hills is far from exact, most who concern themselves with such questions would agree that much of the Appalachian topography should be called mountainous. Local relief is greater than 500 meters (1600 feet) in many areas, and it is sometimes greater than 1000 meters (3200 feet). Slopes are often steep. Even today these mountains create major problems for transportation development.

Whatever the features are called, the rugged landscape of this region has played a major role in its settlement history. The human geography of Appalachia remains closely intertwined with its topography. Without the mountains, the area would merely be a part of several adjoining, or coexisting, areas, such as the Deep South. With them, Appalachia exists as a distinctive and identifiable American region (Figure 8-1).

161

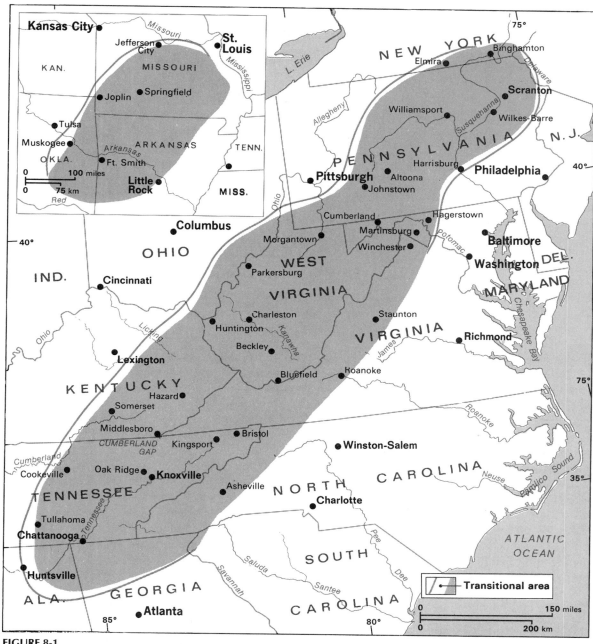

FIGURE 8-1
APPALACHIA AND THE OZARKS.

A VARIED TOPOGRAPHY

The ruggedness that characterizes Appalachia is far from uniform. Appalachia is composed of at least three physiographic provinces, each having a distinctive topography (Figure 8-2). There are also substantial differences in the pattern of human occupation in each of these three sections. These subareas are arranged in three parallel belts lying roughly northeast-southwest.

The easternmost belt is the Blue Ridge. Composed of ancient Precambrian rocks, this section has been severely eroded. Its highest elevations are currently only a fraction of their former levels. Erosive activity has been greatest along the eastern margins of the Blue Ridge. The piedmont section of the Atlantic southern lowlands bounds the Blue Ridge along Appalachia's eastern side from New York to Alabama. Although outside Appalachia, the piedmont is in fact a part of the same underlying rock structure as the Blue Ridge. On the piedmont, the former mountains have been eroded into a rolling upland plain that, in appearance, is not at all like the mountains to the west. These visual differences plus the major differences in settlement encourage most observers to treat them as parts of two separate regions.

The Blue Ridge generally increases in elevation and width from north to south. In the south, especially south of Roanoke, Virginia, it is the most mountainous part of Appalachia. The changes in elevation from the piedmont onto the Blue Ridge are usually abrupt and substantial. In Pennsylvania and Virginia the Blue Ridge is a thin ridge between the piedmont and Great Valley to the west; along the North Carolina-Tennessee border, it broadens to a width of nearly 150 kilometers (95 miles). The Potomac River, the James River at Roanoke, and the New River in northwestern North Carolina provide the best of the few low gaps through the mountains south of Pennsylvania. Movement to land beyond the Blue Ridge was funneled through these openings for many decades. Stonewall Jackson was able to surprise Union forces several times during the Civil War with his unexpected ability to charge large numbers of men across the Ridge quickly.

To the west of the Blue Ridge, one encounters the Ridge and Valley section. This is part of the great expanse of sedimentary rock beds that lie between the Blue Ridge and the Rocky Mountains. The eastern edge of these beds has been severely folded and faulted. A result is the characteristically linear topography of the Ridge and Valley.

The Ridge and Valley section averages about 80 kilometers (50 miles) in width. It is occupied by many ridges following the northeast-southwest trend of the region. They usually rise 100 to 200 meters (300 to 600 feet) above the separating valleys. Gaps in the ridges are relatively infrequent and usually have been created by rivers that cut across the grain of the area. Several miles wide, the valleys provide some of the best farmland in Appalachia. The ridges throughout this section generally are composed of relatively resistant shale and sandstone, and the valleys usually are floored by limestone, which is more easily eroded than the ridge materials. Limestone, when broken down by erosive activity, provides needed minerals for a good agricultural soil.

Between the Blue Ridge and the first of the ridges of the Ridge and Valley section is the Great Valley. Running virtually the entire length of the region, the Valley (which is hilly rather than flat in most areas) is historically one of the important routeways in the country. In Pennsylvania, it may be called the Lebanon or Cumberland Valley, in Virginia it is the Shenandoah Valley, and in Tennessee it is the Tennessee Valley. Whatever its name, the Valley has tied the people of Appalachia together more than any other physical feature, save the mountains themselves.

The westernmost part of Appalachia is the Appalachian Plateau, also known as the Cumberland Plateau in Tennessee and the Allegheny Plateau in New York and Pennsylvania. Bounded by a steep *scarp* (slope) on the east

RIDGE AND VALLEY PROVINCE

BLUE RIDGE PROVINCE

(called the Allegheny Front), the topography of this region has been created through stream erosion of the horizontal beds of the interior lowland. The Allegheny Front was the most significant barrier to western movement in the country east of the Rocky Mountains. Erosion of

164

FIGURE 8-2

TOPOGRAPHY OF APPALACHIA AND THE
OZARKS (reproduced from Raisz, *Physiographic Diagram of the United States*). The linearity of the topography of the eastern Appalachians and southern Ozarks is sharply contrasted with the jumbled surface to the west and north. This difference emphasizes the importance of underlying rock structure on topography.

the plateau created a rugged, jumbled topography with narrow stream valleys bordered by steep, sharp ridges. Except for limited areas, such as portions of the Cumberland Plateau, level land is scarce. Most communities are forced to squeeze

themselves into small level spaces in the stream valleys (Figure 8-3). The differences between the Ridge and Valley region and the plateau region are clear to anyone flying over the land. The symmetry—parallel ridges and valleys—of the former suddenly gives way to the rumpled, apparently patternless landscape of the plateau.

The Ozark-Ouachita uplands follow a topographic regionalization broadly similar to the Appalachians, with the "grain" now east-west instead of northeast-southwest. The Ouchita Mountains to the south exhibit a series of folded parallel ridges and valleys similar to the Ridge and Valley section of Appalachia. They are separated from the Ozarks section by the structural

FIGURE 8-3

HAZARD, KENTUCKY. The population of most of Appalachia is concentrated in the often narrow valleys. The shaded wooded areas correspond closely with uplands or areas of substantial local relief.

165

MAJOR METROPOLITAN AREA POPULATIONS IN APPALACHIA AND THE OZARKS, 1980	
Pittsburgh, Pa.	2,263,894
Knoxville, Tenn.	476,517
Johnson City-Kingsport- Bristol, Tenn.-Va.	433,638
Chattanooga, Tenn.-Ga.	426,540
Lexington, Ky.	317,629
Huntington-Ashland, W. Va.-Ky.-Ohio	311,350
Charleston, W. Va.	269,595
Johnstown, Pa.	264,506
Springfield, Mo.	207,704
Fort Smith, Ark.-Okla.	203,511

trough of the Arkansas River Valley, itself similar to the Great Valley. The Boston Mountains-Ozark Plateau is an irregular, hilly area of eroded plateaus much like the Appalachian Plateau section.

THE APPALACHIAN PEOPLE

Early Settlement

The mountains and the forests catch the eye of casual visitors to Appalachia, but its people have given the area much of its special character. The Appalachian "hillbilly," a person made up far more of myth than of reality in today's United States, is a necessary part of those mountains to countless Americans. Popularly characterized as independent, proud, ignorant, and poor, the hillbilly is as much a part of the popular image of Appalachia as is the rural black in the Deep South. Without the hillbilly, this is just another rather scenic, rugged part of the country. With the hillbilly, it occupies a distinctive spot in the American folk landscape.

Settlers did not push through the Blue Ridge into the Appalachian highlands until late in the colonial period, 150 years after initial occupation of the East Coast. The easiest and first-used passageway into the Great Valley and the mountains beyond was in southeast Pennsylvania,

where the Blue Ridge is little more than a range of hills. Many Pennsylvanians found the mountain lands to the north and west inhospitable. Consequently, they gradually spread their settlement down the Valley into Virginia. Most of these people were either German or *Scots-Irish*. (The latter are Scots who had earlier settled in northern Ireland as part of the Protestant British crown's program to secure its claim to Ireland.) Both groups were soon joined by others moving inland from the southern lowlands.

Then, late in the eighteenth century, people began settling the valleys and coves of the surrounding highlands. For reasons that are still in some dispute by historians, the Scots-Irish played a major role in this movement. Some argue that these people were a particularly footloose and independent breed, constantly searching for the freedom of the frontier. Others maintain that they merely happened to represent the main immigrant segment of the population at the time, and, hence, they were the most likely contributors to frontier settlement. Whatever the case, the Scots-Irish, joined by many people of English and German background, dominated early highland settlement.

The land they chose was poor in comparison with areas farther west. Its ruggedness and the cool upland climate rendered most of the region unacceptable for the plantation economy. Only in some of the broader lowlands, such as the Shenandoah and Tennessee valleys, did a few sizable plantations develop. One consequence was that few of the counties of the southern Appalachians have ever had large black populations.

When American settlers came to this area in the late eighteenth and early nineteenth centuries, the region provided adequate settlement potential for smaller farms. About 10 to 20 hectares (25 to 50 acres) of cleared land was about all a farmer could handle. Such plots were available in the stream valleys. The forests teemed with game, wood was plentiful, and animals could graze in the woods and mountain pastures. By the standards of the time, this was reasonably good land. Thus, although the more rugged areas of the region were ignored by early settlement, a large farming population soon occupied the mountains.

Much of the region gradually grew more isolated and separate from other areas. As flatter, richer agricultural land to the west was opened and grain production was mechanized, the small Appalachian farm became increasingly marginal economically. Even famous pathways through the region, such as the Cumberland Gap at the western tip of Virginia and the Wilderness Road from there to the Bluegrass Basin of Kentucky, were in fact winding and difficult, to be bypassed whenever possible. There was no easy passage at all across the southern Appalachians. Major railroad lines skirted the area; their owners preferred a longer, but easier, route. Atlanta's early growth, for example, was largely a result of its role as a rail junction at the southern edge of the mountains. Rural residents living in areas of rugged topography and far from large urban centers were particularly affected by limited transportation facilites. A study conducted in Kentucky in the 1960s found that limited accessibility, or geographic isolation, was a major element in the lack of economic opportunity for the state's Appalachian residents (Figure 8-4).

Appalachia, particularly southern Appalachia, was slow to develop any substantial urban pattern. In part, it shared with the rest of the South an emphasis on agriculture that continued well after other regions of the country had begun their rush toward manufacturing and urban living. Also, the products of Appalachia were few, and the demand for the goods and services of cities was limited. In addition to these char-

FIGURE 8-4

ISOLATION IN EASTERN KENTUCKY (adapted from Bowman and Haynes, *Resources and People in Eastern Kentucky*, Johns Hopkins University Press, 1963). The greater isolation of the Appalachian uplands is contrasted with accessibility of the Bluegrass Basin and the Ohio River valley. Appalachian highways are usually in the larger valleys.

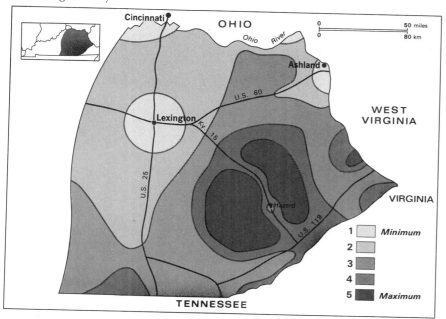

acteristic southern conditions, however, was added the paucity of transportation. The region seemed to exist largely outside a well-identified pattern of development that was common elsewhere: transportation improvements fostered city growth, and city growth encourage transportation improvements. Even by the standards of the South, Appalachia had neither.

One major result of this lack of both plantations and urban development in the southern Appalachians was that few new migrants were added to the early Scots-Irish, English, and German settlers. The area remained overwhelmingly northwest European in ethnic background and, increasingly, conservative Protestant in religion. These people tended to stay where they were; and as time passed, their attachment to their family, community, and land grew.

Southern Appalachian Cultures

This regional immobility led to the development of a cultural distinctiveness uncommon in the rest of the nation. In a nation where migration, education, and commerce were adding more and more disparate elements to the total population and yet increasing its interregional similarity, Appalachia became increasingly unusual by simply remaining the same. Many inhabitants of the region grew uncomfortable in any other place:

> The fierce loyalty of mountain people to home is mostly a loyalty to the only culture in which they feel secure and which operates in ways they know and appreciate. . . . For mountaineers, moving is a kind of death to his way of life. It cuts him off from his sustaining roots.[1]

Henry Caudill, in his sensitive look at one section of the mountains in *Night Comes to the Cumberlands*,[2] noted that in West Virginia during the

Depression, few native West Virginians left the state, whereas more than three-quarters of all foreign-born coal miners left to go to larger metropolitan areas.

It is time to take a closer look at these Appalachian people. We already know that they are overwhelmingly white, Anglo-Saxon and Protestant. They are also relatively poor. Appalachia is easily the largest predominantly white, low-income area in the country. Large swatches of the region—perhaps most notably a band from eastern Kentucky through central Tennessee—have high percentages of families living at or below the federally defined poverty level (Figure 8-5). This level, established by the U.S. Department of Agriculture, indicates how much income a family of four should have to meet what is considered a minimally acceptable standard of living. In 1981 the minimum threshold was $9287

FIGURE 8-5

POVERTY COUNTIES. Appalachian Counties with more than 15 percent of the population below the poverty level are usually in the south and are coal producers.

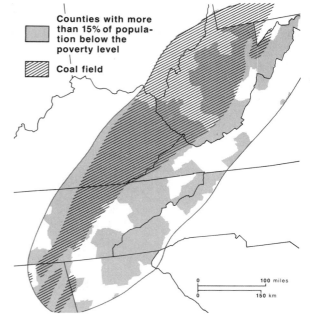

Counties with more than 15% of population below the poverty level

Coal field

0 100 miles
0 150 km

[1] Jack E. Weller, *Yesterday's People: Life in Contemporary Appalachia.* Lexington: University Press of Kentucky, 1965, p. 86.
[2] Henry M. Caudill, *Night Comes to the Cumberlands.* Boston: Little, Brown, 1962, p. 179.

for a family of four. In some areas, especially eastern Kentucky, Appalachia's major coal producing area, much of the blame for this poverty can be attributed to a great decline in the regional demand for labor. Labor demand decreased as coal mining was mechanized following the outbreak of World War II.

The region's people are also conservative in attitude. Many of the country's most conservative Protestant churches trace their roots to Appalachia. Others are found where mountain people have moved and taken their religion with them. Appalachia is certainly part of the Bible Belt, as any Sunday traveler there will attest after sampling the array of religious programs on the radio. Politically, most elected officials are decidedly conservative, although strands of rural populism are found. One interesting element of

the region's politics is the strength of the Republican Party, which has survived and even flourished there. The hill-lands of such states as North Carolina, Georgia, Tennessee, and Alabama long supplied many of the South's small trickle of successful Republican politicians. Similarly, the Missouri Ozarks have often sent Republicans to Congress. A strong belief in the central importance of the family and the individual have survived massive involvement by the federal government in the region's economic affairs. Some say this regional willingness to accept federal aid while denying the government's authority is hypocrisy. It is more likely simply a reflection of the population's fundamentally conservative nature.

Appalachia's legendary isolation, where the mountain people living deep in mountain coves

Narrow mountain valleys and coves, poorly connected to outside paved highways by an often eroded dirt road, remain the home of many of Appalachia's poorest people. (U.S.D.A.)

169

and far up steep valleys had almost no contact with the outside world, is certainly gone. The reasons for this change are of key importance to the region and will be discussed shortly. As Thomas Ford noted in *The Southern Appalachian Region: A Survey:*

> As a general rule, at least in our time, the techno-logical aspects [of a culture] are the first to change, followed more slowly by adaptations of social organization to new techniques. Most resis-tant to change are the fundamental sentiments, beliefs, and values of a people, the ways they feel their world is and should be ordered. So it is not implausible to suppose that the value heritage of the Appalachian people may still be rooted in the frontier, even though the base of their economy has shifted from subsistence agriculture to indus-try and commerce and the people themselves have increasingly concentrated in towns and cities.[3]

The economy of the area has changed greatly, but the attitudes of its people have not. Many regional leaders hope that these basic attitudes and beliefs will never change.

This comment on the continuation of pro-vincialism in the face of great economic change suggests some further words on the nature of isolation. Many have related this provincialism to residential isolation in a land of steep slopes, twisting valleys, and poor or nonexistent roads. At one time, this may have been a valid argu-ment. Today, however, the provincialism is bred of the strong bonds of family and community. Topography is only an incidental factor. These bonds tie their members together and lessen their association with others. One survey in southern Appalachia showed that 95 percent answered yes to the question, "Do you and your relatives call on each other for help?"[4]

Most areas of Appalachia are also among the lowest in the nation in terms of the percentage of the total population born in another state. Appalachia's population is basically one that has been little affected by outsiders.

REGIONAL UNITY AND DISUNITY

By now you may have noticed that most of our attention has been devoted to that part of this region that stretches from West Virginia to northern Alabama, with much less discussion of the northern Appalachians from Pennsylvania north and the Ozark-Ouachita uplands. This has been deliberate. The southern portion is the most clearly Appalachian region and the one that most Americans recognize as Appalachia. Actually, much of what has been said here about the re-gion's residents fits the Ozarks as well. The ste-reotyped Ozark hillbilly is as much a part of the legends of the Ozarks as are their brethren in Appalachia, and the legends are equally inap-propriate to both areas. Regional attitudes are similar. Even the ethnic background of the set-tlers is nearly the same.

The northern Appalachians are far less clearly associated with the broader region. Certainly they share the mountainous topography, and some of the early developmental problems created by steep slopes were also common. Beyond that the similarities grow more tenuous. First, although the northern Appalachians are by no means an area of wealth, poverty is far less evident than it is farther south (see Figure 1-3). Also, more re-cent immigrants followed the early northwest-ern European settlers into the area. This is es-pecially true in Pennsylvania and northern West Virginia where coal mining attracted many East Europeans in the late nineteenth and twentieth centuries. In New York's highlands, early set-tlers came largely from New York and New England, a different source than that for Penn-sylvania and the upland South. Many cultural patterns in the northern Appalachians, with re-ligion a notable example, are not at all the same as those of the southern highlands. Fundamen-talist churches are less common; in many coun-ties, especially in Pennsylvania, Catholics and

---

[3] Thomas R. Ford, *The Southern Appalachian Region: A Sur-vey.* Lexington: University Press of Kentucky, 1962, p. 29.
[4] Ibid, p. 52.

members of various Eastern Orthodox churches are in the majority (Figure 3-7).

Transportation within the northern Appalachians soon became far better than that in the southern Appalachians. In part, this was because the mountains were less continuous and lower and, thus, more easily breached. Also, as the upper Midwest boomed, the northern Appalachians became the center of the continent's major belt of commercial and manufacturing growth (see Chapter 5). Transport lines connecting the eastern and western portions of the manufacturing core region soon ribboned through the mountains. The economic consequence of this was far more development within the northern Appalachian area, especially in central and western Pennsylvania and New York, as compared to the southern Appalachians.

This leads naturally to the question, "Why include the northern section in Appalachia at all?" Some argue that the inclusion is recent and only developed when the federal government began pumping money into poverty areas to encourage economic growth, with the northern states asking to be included in order to obtain part of the money. A more fundamental reason is that this northern section is distinctly different from the areas that surround it, regardless of political boundaries; it is linked to southern Appalachia by the pattern of mountains and relative poverty they share. West Virginia is the only state situated entirely within Appalachia. Every other state's capital, largest urban area, and major areas of economic development are beyond the margins of the Appalachian region. Only Tennessee provides a partial exception to this general pattern, having both Knoxville and Chattanooga within the region. Both, however, have only a little more than half Nashville's 850,000 people.

The relationship that exists in Appalachia between the Appalachian sections and the rest of each state is a significant aspect of the region. In the South at the time of the Civil War, there seemed little reason for the mountain people to join the rest of their states in secession because they had little of the plantation economy and way of life. Much of the population of Virginia's mountains chose to withdraw from the state as a consequence of its withdrawal from the Union and they formed West Virginia in 1863. There was strong opposition to secession in the Appalachian sections of much of the rest of the South as well. Still, the ties within each state were strong, and in spite of considerable dissension, the Appalachian sections chose to accompany their states out of the Union.

This is one indication of the impact that the state of residence has on most Americans. The states fund public education; a part of that education is usually the development of a sense of loyalty through exposure to the history and politics of the state. State aid and state programs are important aspects of local economic conditions. Today state universities often provide success in athletic competition, a success felt and shared by people across the state. Most newspapers, and to a lesser extent radio and television stations, restrict their attention to one state primarily and their material contains heavy doses of state news and information. Because the state does provide so much, the activities of its government are important for everyone in the state. In short, the state's boundaries can be viewed as something of a container, serving to unite what may otherwise be disparate areas. For Appalachia, this means that the portions of its states that are outside the margins of the region have a substantial influence within the region. This, too, is a major element in the set of differences between northern and southern Appalachia. The states of the South share many similarities, just as do those of the Northeast, and the two areas are different from one another in many ways. These distinctions help differentiate the northern and southern Appalachians.

ECONOMIC AND SETTLEMENT PATTERNS

The national image of Appalachia is unquestionably that of a predominantly rural area. In some ways, this is valid. The urban percentage for the region is only about half the national average. Most of the population is classified by the Bureau of the Census as either rural or rural nonfarm residents (the latter are people who live in rural areas but have urban occupations). Parts of Appalachia, most notably in eastern Kentucky, have the highest rural population densities in the United States. Many counties have densities of 60 or more per square kilometer (150 or more per square mile). Clearly, Appalachia's high rural density is not supported by a large-scale, commercial agricultural system. Rather, small farms and minerals dependency (primarily coal) are the keys to this dense rural population.

Agriculture

Appalachia is the primary region for owner-operated farms in the United States, with Kentucky and West Virginia leading the country in that category. One study in an Appalachian county in North Carolina looked at all farms with more than 6 hectares (15 acres) of possible cropland and found that an astounding 99.3 percent were owner occupied. Without any important commercial crop in Appalachia, there was little early growth of farm tenancy, and that pattern has remained. This is a land of independent small farmers, however small the farm income may be.

It has been suggested that the best description of the condition of Appalachian agriculture is that it is the result of too few resources divided among too many people. The average farm in Appalachia is small, containing only about 40 hectares (100 acres). Furthermore, the rugged topography, poor soil, and short growing season in much of Appalachia have resulted in a limited amount of available cropland and a greater relative emphasis on pasturage and livestock. Because the fields are small and scattered in the valleys, the efficient use of large farm machinery is nearly impossible. The net result of all this is that farm incomes are low. A great many of the region's farmers have become part-time operators, turning to other jobs to provide supplementary income that will allow them to remain on their farms.

The growth in farm size common in most other parts of the United States results from the advantages of increased mechanization and the availability of substantial tracts of good land (Chapter 11). Land must be reasonably flat for mechanization. In Appalachia, such large tracts of level land are seldom available. In parts of the region, average farm size is actually declining as farms are split through sale or inheritance. These smaller farms are a good indication of the importance of off-farm income for many Appalachian farmers.

The type of agriculture found in most of the region is called general farming. In fact, this is the major general-farming region in the country. This term is really a catchall that covers a multitude of products and simply means that no discernible product or combination of products dominates the farm economy. Extensive animal husbandry of some form is the most common and probably best agricultural use of the steep slopes. A number of specialty crops, such as tobacco, apples, tomatoes, and cabbage are locally important in some valley areas, with small plots of tobacco being the most common cash crop in the southern Appalachians. Another important cash crop was corn, which was illegally distilled into the region's famous moonshine and rushed to lowland markets by runners who are now legendary. High production costs, alternative job opportunities, and the expansion of the legal sale of alcoholic beverages have combined to cut moonshine production drastically in recent years. There are indications that marijuana may have replaced moonshine as the region's primary contraband export. Corn remains the region's leading row crop, but it is normally used on the farm for animal fodder.

There are important exceptions to this pattern of semimarginal agriculture within Appalachia. The Shenandoah Valley of Virginia, for example, was early called the breadbasket of Virginia. Competition from wheat grown in the fertile grasslands of the Deep South and Great Plains forced the Valley out of the national market in the late nineteenth century. Although winter wheat is still grown, hay and corn for fodder and apples are now the Valley's major crops, with turkey raising also locally important; it remains a productive and beautiful farming region. Dairying and apple production are important in the many valleys of central Pennsylvania. The Tennessee Valley is a substantial agricultural district, with fodder crops and livestock most important. Still, these relatively large productive areas are exceptions.

Coal

Farming, although important, is obviously an inadequate explanation for the large rural population of Appalachia. Over much of the region, farming's chief partner is coal. Almost all of the Allegheny Plateau is underlain with a vast series of bituminous coal beds that, taken together, comprise the world's largest such coal district. Coal seams, with thicknesses of as much as 3 meters (10 feet), are interbedded in the flat-bedded sedimentary rocks of the plateau. The coal seams have been exposed by the same streams that have, through their erosive activity, created the rugged topography of the plateau. The existence of these beds along a certain elevation on a slope is clearly evident on topographic maps, where symbols for mines often extend for many miles along the same contour line.

The coal of Appalachia became important shortly after the American Civil War. It was the development of new types of coke-burning iron and steel furnaces that created this demand because coke is processed from bituminous coal. The thick coal seams of southwestern Pennsylvania and northern West Virginia provided the fuel for Pittsburgh's rise to its status of Steel City during this period. As the nation turned to electrical power in the present century, coal from Appalachia provided fuel for electric-generating facilities along much of the east coast and in the interior manufacturing core.

After the better part of a century of growth, the coal industry fell into a period of decline following World War II. Production dropped as petroleum and natural gas replaced coal as major fuel sources. This, coupled with the decision by the leadership of the United Mineworkers Union to maintain the importance of coal by minimizing its cost through mine mechanization, led to drastically high unemployment in many coal-mining areas. Out-migration resulted, and many coal counties lost a full quarter of their population between 1950 and 1960. Other workers felt themselves too old, too poorly educated, or simply too fond of where they lived to move. The resulting economic depression, blending with the poverty common to Appalachia, created areas of particularly severe problems, with eastern Kentucky again the best example.

More recently, production levels increased from around 1960 to record levels in the late 1970s and early 1980s (Table 8-1). The great decline in North America's steel industry (after electricity generation, the country's largest user of coal), initiated by the economic recession of that latter period, reduced the demand for Appalachian coal. In the mid-1970s, the steel industry used approximately 75 million tons of coal annually; in 1983 it used only half as much. Today the United States is producing more coal than ever before, but an increasing share is from the west (see Chapter 12). Nearly 100 million tons of Appalachian coal is exported annually, shipped almost entirely through the port of Norfolk, Virginia. The thick seams along with mechanization make this coal cheap on the world market, and it finds ready buyers. Within the United States, growing power demands coupled with the increasing concern over availability and cost of petroleum supplies has reemphasized the need for coal in electric power generation. The federal government discouraged the use of coal in power plants during the 1960s because of atmospheric pollution, but it now actively encourages coal use to lessen American dependence on foreign petroleum suppliers (Figure 8-6). New generating plants (with high cooling towers and smoke stacks) use huge quantities of locally mined coal to produce electricity, much of which is transmitted to areas outside the region, and they are an increasingly visible part of the landscape

TABLE 8-1
Bituminous and Lignite Coal Production, Major Mining States, 1947–1983
(in thousands of short tons)

State	1947[a]	1960	1970	1983
Eastern States				
Alabama	19,048	13,011	20,560	22,470
Illinois	67,680	45,997	65,119	59,830
Indiana	25,449	15,583	22,263	30,852
Kentucky	84,241	66,847	125,305	139,750
Ohio	37,548	33,957	55,531	32,110
Pennsylvania	147,079	65,425	80,491	67,794
Virginia	20,171	27,838	35,016	35,000
West Virginia	176,157	118,944	144,072	112,000
Western States				
Arizona	—	—	132	11,506
Colorado	6,358	3,607	6,025	16,500
Montana	3,178	313	3,477	29,447
New Mexico	1,443	295	7,361	21,195
North Dakota	2,760	2,525	5,693	18,471
Texas	—	—	—	37,779
Wyoming	8,051	2,024	7,222	109,612
Total United States	630,624	415,512	602,932	784,374

SOURCE: U.S. Bureau of Mines, *Minerals Yearbook*. Washington, D.C.: U.S. Governmental Printing Office, 1948, 1960, 1970. *Coal Age*, 89 (February 1984), p. 56.

[a] Year of maximum production prior to the 1970s.

of the central and northern Appalachian Plateau.

The proven coal reserves of the United States are great; they will last at least three centuries and, at current use levels, perhaps much longer. Declining domestic liquid petroleum supplies (estimated reserves have declined steadily for over a decade) have emphasized the importance of coal. Pilot projects to extract natural gas from coal have been successful, although at present their product is expensive. It is anticipated that gas produced in this way will become competitive as technology improves, production levels grow, and as natural gas from other sources becomes more expensive. There is also hope that petroleum will eventually be extracted from coal at competitive costs.

Appalachian coal is mined in several different ways. Underground or shaft mining was used first and is still quite important, especially in the northern parts of the region. Modern underground mining techniques—huge mobile drills and continuous mining machines rip the coal out of the seams and then deposit it on conveyor belts for the trip to the surface—now mean that tons of coal per minute can be removed from a seam. Surface or strip mining, which is far less expensive as long as the coal seams are near the surface, has increased greatly in importance. In the central region (primarily eastern Kentucky, western Virginia, and southern West Virginia) where the most important producing section is today, large machines remove the rocks along a slope above a coal seam and then simply lift off the uncovered coal. In the past, the waste rock (*overburden*) was simply dumped down the slope. Extraction along several seams on a slope by this method creates a peculiar, stepped appearance

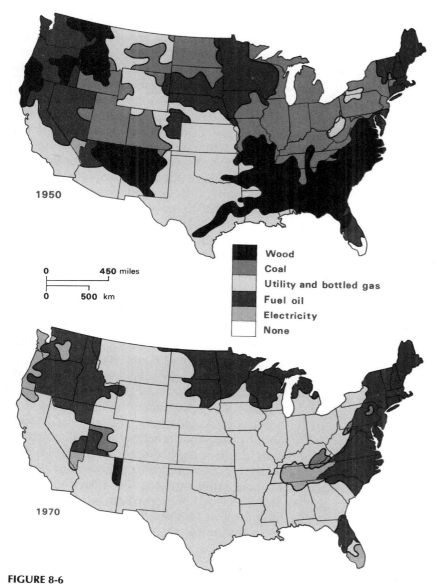

FIGURE 8-6

PRIMARY HOME HEATING FUEL, 1950 and 1970. Coal has been largely abandoned as a home heating fuel in the United States, with natural gas (where available) its usual replacement. What would a doubling or tripling of natural gas and petroleum prices do to this pattern?

175

that looks, from a distance, almost like a series of increasingly smaller boxes piled on top of one another. In areas of somewhat flatter terrain, such as southeastern Ohio, massive scoop shovels lift off the overburden and dig out the coal seam.

About half of the coal mined in Kentucky and most mined in Ohio and Alabama is from strip-mine lands. Still, most of the coal from Pennsylvania, Virginia, West Virginia, and two-thirds of that from Appalachia as a whole is from shaft mines. By comparison, nearly all of the coal from the western states of Wyoming, Texas, Montana, New Mexico, and North Dakota is from surface mines.

Today every coal mining state has passed laws requiring the reclamation of strip-mine lands. The land is now returned to something like its original contouring and replanted with vegetation. Pennsylvania has been the leader in regulating strip-mine reclamation, requiring, since the early 1970s, a detailed plan for reclamation before new mining can begin.

The first important coalfield in Appalachia was not the bituminous fields of the plateau. Their exploitation was preceded by operations in the anthracite field at the northern tip of the Ridge and Valley in Pennsylvania. Anthracite is a much harder, smokeless coal that was important in home heating. Anthracite was also a major fuel for ore smelting until techniques to produce coke from bituminous coal were developed during the Civil War. The decline in use of coal for heating, coupled with the lack of alternative uses for anthracite, led to an economic depression that has lasted since the peak period of production during World War I; although substantial anthracite reserves remain, production today is minimal. More anthracite may possibly be mined as national coal consumption increases, but because the seams are fractured and folded (this is a part of the Ridge and Valley), mining costs are high.

Coal has been very much a mixed blessing to the people of Appalachia. It has long been the economic mainstay for large parts of the region and has employed, either directly or indirectly, hundreds of thousands of workers. It is difficult to imagine the economic structure of Appalachia without coal. Still, the list of problems brought to Appalachia by coal is long. Tens of thousands have died in mine-related accidents. Black lung, the result of many years of breathing too much coal dust, has affected countless others. Both problems are mostly associated with underground mining. The recent recovery of production in response to increased market demands has been accomplished mainly by greater mechanization and unemployment remains widespread. Mining employment has increased since the early 1970s but remains at less than half the number employed in the late 1940s. Many manufacturing plants have been located in the area in response to the available labor force. However, most pay low wages and many attract a labor force that is largely made up of women. Consequently, the wife is now the employed worker in many families, whereas the husband remains at home. Most mineral rights are held by corporations who obtained them early and at low price. Their taxes remain low, and most coal profits leave the region.

The impact of coal mining on the physical environment of Appalachia has been great. Strip mining in particular creates great gashes across the landscape. It is extremely difficult, if not impossible, to return rugged topography to its former shape; until recently, few have even tried. Current reclamation legislation focuses on new mining activity, ignoring what has already happened. Acid mine drainage has killed the fish and plant life in thousands of miles of the region's creeks and streams. In many places, housing has been crowded into narrow valleys that bend and contort settlements like a pretzel, to the detriment of both the people and the landscape. Damaging floods have been a repeated problem in many of these locations. Even the rich greenness of Appalachia's summer vegetation often cannot hide the gray of vast slag heaps and coal dust. A beautiful part of America has been damaged in a severe way. Full recovery will be a matter for the centuries.

Strip mining of coal in the rugged topography of much of Appalachia can be very disruptive. This view is near Beckley, West Virginia. (Charles E. Rotkin/PFI.)

Beyond Appalachia's margins, northeastern states and Canada's eastern provinces are concerned about the consequences of acid rain resulting from the burning of high-sulfur coal (most Appalachian coal has a high sulfur content) in electric-generating facilities. The operations release sulfur into the atmosphere where it combines with water to create sulfuric acid. This acid may then fall in precipitation—either as snow or rain. The greatest concentration of such generating plants is along the Ohio River. Prevailing winds carry their emissions northeastward.

Other Mining Activities

Three other Appalachian mining activities deserve mention because of their local importance.

The Tri-State district in the Ozarks, where the borders of Oklahoma, Kansas, and Missouri meet, has long been a major area of lead mining. Southeastern Missouri, outside Appalachia, has produced lead for over 250 years, and surface mines there remain the country's most important. Missouri has supplied most of the lead ever mined in the United States and currently produces more than three-quarters of the total national production.

The first oil well in the United States was drilled in northern Pennsylvania in 1859 and that state led the country in production through most of the nineteenth century. Today the area supplies only a small part of the nation's crude oil needs, but it remains an important producer of high-quality oils and lubricants.

Finally, southeastern Tennessee is the most important remaining area of zinc production in the United States. In addition, several mines around Ducktown, near the North Carolina and Georgia borders, are the only major copper producers east of the Mississippi River.

REGIONAL DEVELOPMENT PROGRAMS

Like coal, Appalachia's rivers have also been a mixed blessing to the region. Some of the streams have been important transportation routeways, and water power was used by the earliest gristmills and sawmills. These rivers also had a darker side,

for they frequently flooded their narrow valleys during periods of heavy rain. The southern highlands are the moistest area of the continent east of the Pacific Coast. Out of a desire to control one of these rivers, the Tennessee, the largest and perhaps most successful regional development plan in American history was implemented.

The Tennessee Valley Authority (TVA)

The Tennessee River, with its headwaters in the mountains from Virginia to Georgia, was once a river that flooded as often as any in the South. It also flowed through some of the poorest parts of the inland South. President Franklin D. Roo-

Pineville, Kentucky, in the spring of 1977. Flooding is an ever-present problem on the Appalachian plateau, where the towns and streams crowd together in narrow, steep-sided valleys. (Billy Davis/The Courier Journal.)

FIGURE 8-7

THE TENNESSEE VALLEY AUTHORITY. The Authority has become a major influence on life in the Tennessee Valley, with the more than 40 dams on the area's streams being the most obvious aspect of that influence.

sevelt, early in his first term, conceived a plan to harness the river and to use it to improve the economic conditions of the entire Tennessee Valley. Many contended that the federal government had no constitutional authority to create such a regional development scheme. As a result, the Tennessee Valley Authority (TVA) was first given the charge to develop the Tennessee River for navigation, a right clearly granted to the federal government as a part of its control over interstate commerce. Today a 3-meter (9-foot) barge channel exists as far upstream as Knoxville.

In a sense, most of the other activities of the TVA can be viewed as logical extensions of the initial commitment. Navigation development included the construction or purchase of a series of dams to guarantee stream flow and reduce flooding. As long as the dams were there, it was natural to include water-power facilities with them. Today most of the more than 30 dams controlled by the TVA on the Tennessee and Kentucky rivers have power-generating facilities (Figure 8-7). A number of these dams were developed privately and TVA control was acquired subsequently. The TVA became the electricity supplier to a broad region centered on the river. As the power needs of the region grew, demand exceeded the capacity of the water-powered facilities. The TVA turned to the construction of other power sources. Today about 80 percent of the electricity it produces comes from thermal plants, including 10 that burn coal, and several nuclear-powered facilities. The TVA

179

uses 50 million tons of coal annually and is now Appalachia's largest coal user. Several additional nuclear facilities are planned or under construction.

TVA power was once cheaper than most alternatives in the country, although that is no longer the case. The inexpensive electricity attracted to the Valley a few industries that are heavy users of power. Most notable, perhaps, was the large aluminum-processing facility at Alcoa, south of Knoxville. The country's first atomic research facility was placed at Oak Ridge, west of Knoxville, partly because of the availability of large amounts of power there. Knoxville, Chattanooga, and the Tri-Cities of Bristol, Johnson City, and Kingsport in Tennessee are all substantial manufacturing centers. The TVA also became a principal developer and producer of artificial fertilizers, another heavy power-consuming industry.

Proponents of the TVA argue that much of the significant manufacturing growth of the area over the last four decades came as a direct result of the work of the authority. This ignores the fact that other Southern areas, such as the piedmont from North Carolina to Alabama and the lower Mississippi River valley, also experienced major increases in manufacturing during the same period. Such TVA contributions as flood control, cheap water transport, and power production probably did have an impact on regional manufacturing growth. Still, it seems reasonable to assume that a substantial increase in manufacturing would have occurred here even without the TVA.

The series of dams had an obvious flood-control value. Above the dams, the TVA initiated a major program to help valley farmers control erosion at the farm. A careful inventory of the agricultural potential of the region was taken, thousands of farm ponds were constructed, and advice was offered on other types of erosion control. The goal was to hold part of the flood-waters at the farm and to slow the rate at which the lakes were filling with silt.

In addition to the water itself, the authority also owned 520,000 hectares (1.3 million acres) of land along parts of the rivers. Major public recreation areas were developed on some of this land, and recreational boating was encouraged on the water. This is now a substantial recreation facility, with one of the finest inland water-recreation potentials in the country.

The TVA must be viewed as having easily reached its basic goal to change the economy of the area. Frequently criticized by outsiders, it is generally well regarded within the Tennessee Valley itself. With over 40,000 workers, the TVA is also one of the region's more important employers. Tennessee, once one of the South's poorest and least industrial states, is now a regional leader in manufacturing.

Appalachian Regional Commission

In 1965, Congress passed the Appalachian Redevelopment Act, an extension of the Area Redevelopment Act of 1961. It created the Appalachian Regional Commission (ARC). Responsible for an area that extends from New York to Alabama, the commission has spent several billion dollars in a program to improve the region's economy. Its explicit goals are quite different from those of the TVA, its primary thrust being to improve highways in Appalachia in the hope that this will decrease isolation and encourage manufacturers to locate in the region.

The principle of the government investing in transport development to encourage regional economic growth is well established in the United States. The Erie Canal, opened in 1826, was funded by the state of New York, and canals were built by the federal government in Ohio, Illinois, and Indiana during the 1830s and 1840s. More than 40 million hectares (100 million acres) of federal land were given to western railroads to help finance their construction. Virtually the entire road and highway pattern of twentieth-century America was built by local, state, and federal agencies. The costs involved in transport development are great, and economic returns resulting from construction are usually long term. Thus, private investment in transport development is unusual. It has been more usual in other forms of communication—most notably, telephones—but only because of governmental restrictions that guarantee a monopoly over the area served.

The ARC is different from the TVA in several other important ways. First, it involves federal-state cooperation, with each level of government paying a portion of the total costs. This, incidentally, helps explain the large area contained within the project. The states were given the task of defining the target area for the commission. No one wanted to be excluded; eventually the entire physiographic region was included.

The ARC also worked to improve public and vocational education and regional economic planning and to maintain the quality of the physical environment.

A NEW DIRECTION

Once an area of substantial out-migration, much of Appalachia is now gaining more people than it is losing. Between 1970 and 1980, for example, eastern Kentucky was one of the country's faster growing rural areas. The Ozarks has experienced a population boom for a decade. Recently, no part of the country has had as great a shift from out- to in-migration as Appalachia and the Ozarks. This is surely part of the broader movement west and south that is so noticeable in the migration patterns of Americans over the past few years; nevertheless, it also represents a large shift in the region's fortunes. Certainly, Appalachia is not likely to become part of the manufacturing and urban core of the country and few in the region really want that. Poverty will remain a problem especially in mining areas. Still, there seems to be substantial reason to hope that, economically, the worst is over for Appalachia and growth in income and living standards is offered by the future.

Additional Reading

Appalachian Regional Commission. *Appalachia.* Washington, D.C.: ARC, 1967– .

Arnow, Harriette L. *Hunter's Horn.* New York: Macmillan, 1949.

Batteau, Allen, (ed.). *Appalachia and America: Autonomy and Regional Dependence.* Lexington: University Press of Kentucky, 1983.

Caudill, Harry M. *Night Comes to the Cumberlands.* Boston: Little, Brown, 1963.

Ergood, Bruce, and Bruce E. Kuhre (eds.). *Appalachia: Social Context, Past and Present.* Dubuque, Iowa: Kendall/Hunt, 1983.

Shackelforld, Laurel, and Bill Weinburg (eds.). *Our Appalachia: An Oral History.* New York: Hill & Wang, 1977.

Ulack, Richard, and Karl Raitz, "Appalachia: A Comparison of the Cognitive and Appalachian Regional Commission Regions," *Southeastern Geographer,* 21 (1981), pp. 40–53.

Watts, Ann DeWitt. "Does the Appalachian Regional Commission Represent a Region?" *Southeastern Geographer,* 18 (1978), pp. 19–36.

Whisnant, David E. *All This Is Native and Fine: The Politics of Culture in an American Region.* Chapel Hill: University of North Carolina Press, 1983.

Wigginton, Eliot, (ed.). *Foxfire,* books 1–7. Garden City, N.Y.: Doubleday (Anchor Books), 1975–82.

CHAPTER NINE

THE DEEP SOUTH: A CULTURE REGION

New England is provincial and doesn't know it, the Middle West is provincial, and knows it, and is ashamed of it, but, God help us, the South is provincial, knows it, and doesn't care.

THOMAS WOLFE

The South is a region singular in the depth and persistence of its regionality. It has maintained within itself a deeper regional "sense" than perhaps any other part of the United States. Other regions are identified as distinct cultural territories, of course. The Down East of upper New England, the Middle West, the Southwest, the West Coast, and other such regions (see Figure 1-4), possess cultural characteristics and views of themselves and each other that contribute to the sense of territorial distinctness prominent in the many published self-examinations.

If we were to map for the entire United States the many expressions of habit, attitude, and behavior relevant to a definition of culture, "the South" would often appear as a characteristic territory. As Thomas Wolfe observed, Southerners are acutely aware of their region; some resent it, but most seem proud of the identification and gain sustenance from it. The South's regionality is reinforced by the nearly uniform opinion of people living elsewhere in the country that this region is distinctive, an opinion also held by international observers of the United States.[1] The view from outside the region—until recently, largely negative and critical—is partly a function of the region's distinctness; that, in turn, is largely a function of past socioeconomic characteristics and the institutions that developed to maintain those characteristics.

In spite of the consistency with which the South is seen as a distinct, uniform culture region, this view is far from accurate. Strong differences exist within the region as well. The Gulf Coast, the Southern Highlands, the Georgia-Carolinas Piedmont, and many portions of the northern interior South possess their own versions of "Southern" culture, each at variance with the others. True Virginians know how they differ from Georgians, and both recognize their differences with Southerners from Tennessee or

---

[1] For an interesting example, see Roger L. Thiede, "The Soviet Portrayal of the United States South in Secondary School Geography Textbooks," *Southeastern Geographer*, 15 (1975), pp. 56–62.

Mississippi. But all are also clear about, and find comfort in, the special "Southernness" they share. Newcomers to the region (and outsiders) do not often recognize the basis for the cultural comfort, but those reared within the region know. Peninsular Florida and coastal Texas, locationally the most southern territories of all, are probably the least imbued with the region's cultural flavor and might well be excluded from the region. The source of this flavor is not easy to identify geographically, although its association with a regional core is reflected in the label, Deep South.

The list of cultural characteristics that give the South its flavor is long, as it would be for any region. The flavor cannot be appreciated by such a list however; only by passing through the landscape and living among its people with one's senses open can any region's flavor begin to be comprehended. Throughout much of the Deep South, one can still stroll down a dusty red-dirt road in the bright summer haze with an open field on one side and a pine woods on the other. From the road, a green wave of kudzu vine splashes itself against the pines. Thirty years ago,

the single-story, broad-porched frame house with the painted metal roof might have contained a family preparing a customary evening meal of ham hocks and snap beans, sweet potatoes, and turnip greens, followed by corn bread and molasses. The house appears pretty much the same today, but higher incomes and the spread of grocery-store chains have greatly reduced the dietary distinctiveness of this region.

References to the Deep South are references to whatever is felt to be the source area for Southern culture, with depth implying intensity. Non-Southerners often mistakenly use the term geographically, with "deep" implying a more southerly location. The states of Georgia, Alabama, Mississippi, Louisiana, and South Carolina, however, are not the only area within which Southern culture is found. In addition, many elements of this culture originated, or continue to be stronger, elsewhere in the larger region. One style of Southern country music, for example, is largely focused on Nashville, Tennessee; another style carries the name of the Bluegrass Basin in Kentucky; and a third, Old-Timey, is associated with the southern Appalachians. The

Regionally distinctive house-types may suggest some aspects of a region's cultural underpinnings. The older house on the left was enlarged by its owners as changing needs were given form in additions to the original simple boxlike structure. Note how the newer house on the right, different as it may be in detail, has many features similar to the appearance of the "developed" older house. (Stephen S. Birdsall.)

blues are a legacy of Memphis, Tennessee. Gospel music is prevalent throughout most of the South, although there are several styles. Stock-car racing, as a second example, is a highly popular, typically "good-old-boy" Southern tradition that flowered within the South and continues to be more intensely related to that region than any other in the country. Even though it has been exported widely to other regions, this part of Southern culture is almost entirely a North and South Carolina production.[2]

The region of Southern culture (the Deep South) can be viewed as a geographic composite of beliefs, attitudes, patterns, habits, and institutions associated with the entire group of southeastern states (Figure 9-1). Some of this composite is absent in parts of the territory, but much of what is seen as Southern is an outgrowth of early patterns. Many of these patterns and the changes in them that are currently underway are explicitly geographic; many others have definite geographic consequences. In comprehending the South as a culture region, therefore, it is useful to examine first the source of the early patterns and the ways in which they changed and became clearly defined and then to discuss the external pressures on the old cultural patterns and the implications of these new alterations. In other words, when considering the South as a culture region, we need to understand the origins of the culture and the modifications to it that have occurred if we are to make sense of what exists now and to guess what the region is likely to be in the near future.

THE HERITAGE

It can be argued that much of Southern culture has its roots in the region's location and the economy generated in that location. The earliest European colonization in North America was commercial and exploitive. The first century of European settlement was almost entirely meant to provide the means of, and support for, that exploitation. The coastal plain of North America south of Delaware Bay, especially that south of Chesapeake Bay, contained many areas that appeared ideal to agricultural exploitation. The long hot summers, regular rainfall, and mild winters permitted settlers a selection of crops complementary to those grown in Great Britain and northern Europe. The large number of rivers that crossed the plain, navigable to small boats at least, allowed settlement to expand freely between the James River in Virginia and the Altamaha River in Georgia. Population densities remained low throughout most of the region, with urban concentrations larger than the village size limited to port cities (Norfolk, Wilmington, Charleston, and Savannah) or the heads of navigation on the main rivers (Richmond, and later Columbia and Augusta). The strong agrarian element of Southern culture, therefore, developed from the beginning of white settlement in the region and established a pattern that did not weaken until late in the nineteenth century and remained significant until after World War II.

There is more to early Southern agrarian culture, however, than relatively low average population densities, an absence of large cities, and the presence of a supportive physical environment. European exploration during the fifteenth century and later was motivated by economic purpose more than by human curiosity and a search for knowledge. The greatest return for the effort expended in settling the Southern coastal plain was through highly structured cash-crop agriculture. The plantation organization, already well established throughout the Caribbean, came gradually to dominate the early Southern colonial economy. Production of tobacco along the James River and to the south in northeastern North Carolina and production of rice and indigo in and around the many coastal swamps in the Carolinas and Georgia were important from 1695 onward. Cotton production grew slowly in importance until about 1800; it

[2] Richard Pillsbury, "Carolina Thunder: A Geography of Stock Car Racing," *Journal of Geography,* 73 (January 1974), pp. 39–47.

FIGURE 9-1

THE DEEP SOUTH.

spread rapidly inland thereafter, from the initial concentrations on the Sea Islands between Charleston and Spanish-held Florida. The plantation form of organization was successful enough to be carried westward with cotton production, and it reached its most prevalent form in Georgia, Alabama, Mississippi, and Louisiana during the first half of the nineteenth century. Tobacco was similarly carried westward into Kentucky and Tennessee by settlers migrating from Virginia and North Carolina.

In contrast to the individually owned and family-operated farm, a plantation was a business enterprise. Its purpose was production for cash return. Because production was for export, the South's spatial organization was weakly developed with small market centers serving as collection and transshipment points; the largest cities, containing a wide variety of economic activities, were few in number. The transportation network accompanying this pattern was one that simply allowed the inland products to be moved

MAJOR METROPOLITAN AREA POPULATIONS IN THE DEEP SOUTH, 1980			
Atlanta, Ga.	2,029,710	Charlotte-Gastonia, N.C.	637,218
New Orleans, La.	1,187,073	Richmond, Va.	632,015
Memphis, Tenn.-Ark.	913,472	Greenville-Spartanburg, S.C.	569,066
Nashville-Davidson, Tenn.	850,505	Raleigh-Durham, N.C.	531,167
Birmingham, Ala.	847,487	Baton Rouge, La.	494,151
Greensboro-Winston–Salem-		Mobile, Ala.	443,536
High Point, N.C.	827,252	Charleston-North Charleston, S.C.	430,462
Norfolk-Virginia Beach-		Columbia, S.C.	410,088
Portsmouth, Va.-N.C.	806,951	Little Rock-North Little Rock, Ark.	393,774
Jacksonville, Fla.	737,541	Shreveport, La.	376,710

most directly to the coastal export centers; interconnections between the smaller market places remained few (see Figure 5-2). A major consequence was relatively high rural isolation for most of the region's population and a continuation or development of distinctively local allegiances. This isolation is common to rural populations, but it was especially pervasive throughout the South and contributed to the region's unashamed provincialism.

In additon to a strong rural provincialism, the plantation system generated two other equally significant components of Southern culture. Large-scale plantation agriculture required a large investment. Until recently, much of the annual investment was in the form of labor. The combination of low population densities, cheap land, and ample opportunities for the application of slave labor led to its early use in Southern agriculture. Once established, this practice continued to restrict population immigration because potential settlers and urban workers found freer opportunities in the North. Since early in the nineteenth century, therefore, the South has had a lower proportion of its population that was foreign born than any region of the country. Because significant immigration to the United States and Canada from countries outside the British Isles did not occur until the 1840s, the overwhelming majority of Southern whites are of English and Scots-Irish descent. The accompanying stimulation of new skills and knowledge from non-British cultural sources that al-

tered the mix of habits and attitudes elsewhere in the country throughout the nineteenth century was also lacking from Southern culture and contributed to an entrenchment of cultural expressions distinctive to this region.

There are several important ethnic variations on the dominant black and Anglo white theme found within the South's population. The two long-term resident populations that are neither Anglo nor African in ancestry are the Cajuns of southern Louisiana and several Indian (native American) groups. The name "Cajun" is probably a derivation of Acadian, for this Catholic, French-speaking population is thought to be descended from French exiles from Acadia (now Nova Scotia and New Brunswick). The rural Cajun population stayed in southern Louisiana and remained culturally distinct in spite of France's sale of the territory in 1803 and the gradual integration of the remainder of the state into the Deep South culture. Most Indian groups were removed from the South in much the same ruthless manner and at the same time as in the Midwest, but several significant exceptions remain. The Lumbee in southeast North Carolina, remnants of the once powerful Cherokee along the Blue Ridge in southwestern North Carolina, the Choctaw in central Mississippi, and the Seminoles in southern Florida managed to elude removal until federal efforts to this end decreased.

A third group, although not strictly within the Deep South, has had a major impact on local culture, especially since the end of the 1950s. Southern Florida has for many decades contained minority populations with cultural roots in Latin America. Whether from Puerto Rico, Mexico, or some other portion of Middle or South America, people from Spanish-speaking areas closer to southern Florida than are the Great Lakes moved to the subtropical portions of the peninsula. With the change of government and political system in Cuba in 1959, many thousands from the island fled to the United States. Most settled in the Miami metropolitan area. This influx gave support to the subsequent concentration of Spanish-speaking immigration to the area. By 1980, the three urban counties along the southeast coast of the state contained almost 650,000 persons "of Spanish origin," as the Bureau of the Census described the group. This is more than 20 percent of those counties' total population. The concentration is higher in Miami itself, with more than 55 percent of Spanish origin in 1980. In addition to their modifications to the visible landscapes of the major metropolitan areas in this portion of the state, southern Florida has experienced a strengthened orientation away from the Deep South and toward the Caribbean Basin and the countries that border the basin.

Although not directly tied to the plantation system, another strong element of Deep South culture took root in the agrarian, isolated communities and homesteads that gradually spread across the region. The South's population has long been characterized by its persistent adherence to various evangelical Protestant religions. Small, unpretentious church buildings still dot the countryside, consistently drawing their congregations every Sunday from the scattered rural and small-town populations. Although Methodist, Episcopal, and other Protestant congregations are numerous across the region, it is the Baptists that have numerical dominion in the South (Figure 3-7).

Free-wheeling, energetic Baptist evangelists entered the Southern piedmont and hill country hard on the heels of the mid–eighteenth-century settlers and found them eager listeners. The religion was adopted through an intensely emotional, but personal, conversion experience that was compatible with frontier hardship, isolation, and individualism. As a group, Baptists were also resistant to the rigidities of formal institutional organization; this too permitted easy expansion of the religion across the sparsely settled South. Any individual who was spiritually "born again" and felt the "call" to preach could do so without either the formal approval or denominational constraint of a church hierarchy. And because most nineteenth-century immigrants settled elsewhere in the country, the religious pattern that was established by the early 1800s was not affected by later arrivals.

The heavy use of African slaves in the Southern colonies lies at the crux of both additional components of Southern culture. One impact, heavily discounted until recent years, was the transfer of some elements of the African cultures to the region and the amalgamation of these elements with those of the white population. The first Africans arrived in Virginia in 1619, only 10 years after the initial James River settlement had been established and a year ahead of the Mayflower's arrival at Plymouth. Although slaves were not imported in large numbers until the early eighteenth century, a few blacks were present in the region and were part of its organization and social environment from the beginning. By the time of the first U.S. census in 1790, blacks comprised 39.1 percent of the population of Virginia, 43 percent in South Carolina, and 25.5 and 35.4 percent in North Carolina and Georgia, respectively. Even Maryland's population was 33 percent black in 1790. As socially and psychologically isolated and disoriented as imported Africans may have been, many of their original cultural mannerisms were retained for a time and transmitted to Southern whites. Their impact on patterns of speech, diet, and music in the South is undisputed. Undoubtedly, other more subtle contributions to the dominant white culture also occurred. And Southern culture is richer for the amalgamation.

Religion is important in the Deep South. There is more diversity in this small town in northwestern Georgia than in much of the region, for Baptists predominate, but to have this many active congregations in a town of only about 600 people suggests a high degree of support. (Bruce Roberts/Photo Researchers.)

The other indisputable cultural consequence of the black slave presence in the region is less positive. For a slaveholding situation to exist effectively and over a substantial period of time in an overtly Christian society, there must be no empathetic identification of the slave owner with the slave. For the owner to "put himself" in the slave's position would be to recognize the significance of the slave's subservience in the relationship. By extension, then, to justify the enslavement of other human beings, it is necessary to consider the entire enslaved group as decidedly and perhaps permanently inferior. The accep-

tance of this view of blacks by Southern whites was no different from the dominant European view until late in the eighteenth century. By the turn of the nineteenth century, however, opposition to slavery had gained strength in those regions where it was of less importance. Slavery was a pervasive institution throughout the Southern states by this time. Changes in white attitudes toward the supposed inferiority of blacks within the South, therefore, were small; instead, justification of slavery became more intense and self-righteous as pressure to eliminate it arose from outside.

Much of the South's antiblack attitude seems to have been tied to proximity and numbers as well as to economic aspects of the master-slave relationship. More than 90 percent of all blacks in the United States lived in the South until the beginning of the twentieth century, and they were present throughout the region in large numbers, except in the Appalachians and peninsular Florida. Events outside the South during the twentieth century have revived the argument that white attitudes toward blacks were little different across the country; the opportunities for expressing these attitudes were simply greater in the South.

As a point of contrast, however, it should be noted that the long and intimate relationship between whites and blacks in the South seemed to produce a relationship between the two groups that is less apparent outside the region. Although basic animosities remained strong, as long as both sides knew what to expect from the other, under most circumstances, the day-to-day interactions between individuals could often be conducted with little apparent tension. That tensions existed and frequently flared into violence cannot be denied and underlines the basic inequality of the relationship. Nevertheless, this greater intimacy between the races, not easily comprehended by non-Southerners, was as much a part of the distinctive Southern culture as the region's long history of violence between blacks and whites.

By the outbreak of the War Between the States, the geographic pattern of population settlement and economic organization had changed dramatically from the region's colonial beginnings. As new areas were taken from the Indians in Georgia, Alabama, and Mississippi, settlers moved westward from the Carolinas and eastern Georgia. Complete plantation operations, including entire slave populations, would often be transferred to the rich new land. Families who passed through the Appalachians from Maryland, Virginia, and North Carolina settled into the more fertile farming areas of Tennessee (such as the Nashville Basin), Kentucky (the Bluegrass Basin),

and along the major rivers in these states. Arkansas, Missouri, and southern Illinois also received settlers from these areas. Many of the attitudes and behavior patterns that identify Southern culture were transplanted with the migrations. Although slave densities were generally lower in the so-called border states, blacks were present in higher proportions in the Louisville and Bluegrass Basin sections of Kentucky, in the Nashville Basin, on the *loess* plains of western Tennessee, and throughout the tobacco and hemp plantation country along the Missouri River westward from St. Louis. In a similar manner, plantations existed in the panhandle of northern Florida and on the interior highlands almost as far south as present-day Ocala. Western Louisiana and the plains of eastern Texas became extensions of the Deep South's plantation economy during the 1850s.

As the Civil War erupted, the South was composed of a strongly rural population. Urban development was limited to the village and small-town level, except for a few major collecting points for the region's agricultural produce. These larger cities were almost all located on the coast or at major transfer points along interior waterways. Transportation and communication networks were sparse, with the best connections limited to the relatively few scattered large urban centers.

The biracial population pattern was uneven. Although blacks were likely to be found in almost every county outside the Southern Highlands, they comprised a large proportion of the total population only in the original plantation areas (south-central and tidewater Virginia, northeastern North Carolina, and tidewater Georgia and South Carolina) as well as in the new lands most suited to large-scale cotton production (piedmont South Carolina, the inner coastal plain of Georgia, Alabama's Black Belt, the Mississippi Valley, the loess plains south of Memphis, and the black-soil prairies of eastern Texas).

Supported by slave labor, the cotton plantation had become so successful that the region's

The bustle of commercial activity, fundamental to New Orleans, is caught vividly in this 1851 lithograph of the city. Riverboats and ocean schooners vie with smaller vessels and rafts on the Mississippi while many dozens of each huddle tightly along the riverbank to load or unload their goods. ("Bird's Eye View of New Orleans," 1851 by John Bachmann/I. H. Phelps Collection of American Historical Prints, Print Division, New York Public Library.)

economy was dominated by this one crop. Other crops were grown but primarily as a local food supply or a decidedly secondary cash alternative. Tobacco, rice, sugarcane, and hemp, for example, continued to be grown in areas suited to their production. In 1860, cotton dominated not only the South's economy, but also—at least in terms of export income—the entire country's; slightly over 60 percent of the total value of goods exported from the United States during that year was from cotton. Currently produced in significant quantity outside the South (see map appendix), cotton still ranked fourth in value of U.S. agricultural exports in 1981.

With the loss of the Civil War, the South's economic underpinnings were badly damaged. Railroads were torn up and equipment confiscated, shipping terminals disrupted, and most of the South's scattered industrial base de-stroyed. Confederate currency and bonds were suddenly worthless. Cotton stocks awaiting postwar sale in warehouses and ports were confiscated by Union forces. Farms and fields were in disrepair, and implements and livestock were often stolen or lost. The slave-labor supply was formally eliminated and large landholdings broken up or heavily taxed. Investment funds had either been used up during the war or were drawn off during the decade after the war. Although the large black population was no longer slave, most blacks gained little economically by their emancipation. To an overwhelming degree, they continued to live near their prewar locations.

As a region, the South had lost much politically, militarily, and economically. However, the new geographic patterns that emerged and continued to evolve without significant disruption during the next three-quarters of a century were tied in large part to underlying cultural patterns that neither the war nor the subsequent economic turmoil had weakened. Having lost almost everything except the land itself, certain aspects of Southern culture became more intense. These aspects were reflected in many of the geographic consequences of the regionwide disruption of the Civil War, consequences that reverberated throughout the nation and that laid the foundations for the current geography of the South.

THE CONSEQUENCES

The first half-century following the Civil War was a period of readjustment for the South, during which the economy had to become reconciled to the changes imposed by the war. The white population proceeded through several alternative reactions to the emancipated status of the large black population before finally settling into institutionalized segregation. Blacks, for their part, experienced changes in opportunity that were largely out of their control until the beginning of the second half-century after the Civil War. This was also a period during which Southern attitudes and feelings of isolation from the remainder of the country became even more inflexible than they had been in the antebellum period. Each of these consequences interacted with the others and were reflected in geographic patterns.

Persistent Poverty
The disintegration of the antebellum economic organization led to very difficult times for most of the South's population during the 12-year Reconstruction (1865–77). Quite aside from the destruction of transportation and manufacturing capacity, the plantation economy had become refined to the point of rigidity and overdependence on slave labor. Although the North had undergone the beginnings of industrialization and was coming to expect change as nor-

mal, this adaptability had not been valued in the South before the war. After the war, a continuation of intense exploitation was necessary to meet heavy taxation and other costs of rebuilding. The resource most available for exploitation continued to be the land; thus, cotton production remained dominant in the region's economy.

The other factors necessary for production, however, were much less available. In addition to land, both labor and capital are required. Local capital was scarce, with much of it consumed by the war effort or drawn off by the North through taxation. With credit in short supply, interest rates increased sharply and farmers, both large and small, found themselves continually in debt. This tended to perpetuate the general Southern dependence on agriculture and maintained cotton and tobacco as the most important cash crops.

The availability of labor was complicated by emancipation. No longer slave, the black population was not forced to work for white farmers for social reasons. The dream of Yankee-supplied economic independence gradually faded, however. With few jobs available in the small towns of the region, most rural blacks were forced to make whatever arrangements they could with the remaining white landowners. *Sharecropping* became the means of subsistence and a way of life for these blacks, just as it was for many poor whites who had lost their land. Once this pattern was established for the black population, it was enforced with "Black Codes" that restricted Negro movements outside the agricultural areas and with a continuation of low educational opportunities for blacks. Sharecroppers were frequently in debt to the landowner (just as he was in debt to the bankers and merchants) and were not permitted to end the sharecropping arrangement until their debts were paid.

For most rural blacks, slavery was exchanged for an agrarian existence that was not far above the subsistence level or far from indebtedness to white landowners or local merchants. Some did own the land they worked, but the proportion of blacks operating farms who were also owners was small (25 percent in 1900), and this number increased only as nonowners left agriculture altogether.[3]

Even when they owned their land, black farmers were hampered by poor access to credit, farm sizes too small to be highly productive, and the antiblack aspects of the regional culture. These conditions created and maintained low annual income levels among the overwhelming majority of rural blacks in the South. When consider-

[3] For a clear and thorough discussion of this, see James S. Fisher, "Negro Farm Ownership in the South," *Annals of the Association of American Geographers*, 63 (December 1973), pp. 478–489.

ing the distribution of poverty in the region, therefore, the distribution of the rural black population has been a strong explanatory factor. Even though the South's economy and population distribution have experienced major changes during the last three or four decades, this association between the proportion of a county's population that was black (Figure 9-2) and the median family income (see Figure 1-3) remained clearly visible in 1980.

Poverty has not been exclusively an experience of rural blacks in the South. Even during the antebellum period, a majority of the white population were small farmers; those who had the capital to create a large landholding and the expertise to operate it successfully were rela-

FIGURE 9-2
PERCENTAGE OF COUNTY POPULATION BLACK, 1980. In spite of considerable out-migration, the present distribution of racial balance in the South is very similar to that in 1860.

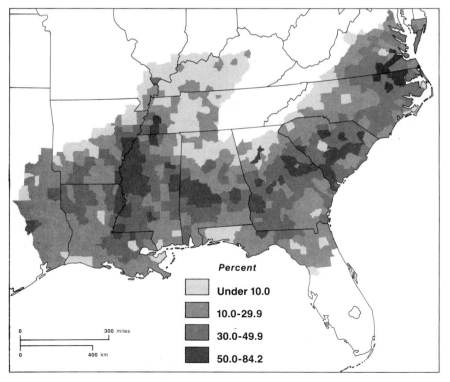

Percent

Under 10.0

10.0-29.9

30.0-49.9

50.0-84.2

0   300 miles

0   400 km

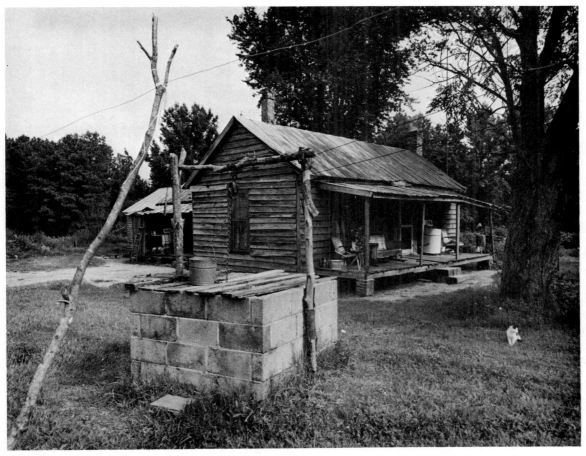

Although this house's clean exterior and yard suggest great care and effort by its occupants, complete maintenance and a full rehabilitation are just as clearly beyond the means of this low-income, rural, Southern family. Compare the form of this house with that of those shown on page 183. (U.S.D.A.)

tively few. Cotton plantations were established on the best agricultural land during the first few decades it was available for settlement. The pattern of black population reflects the location of this land. By the same token, poor whites were unable to compete for land in the main cash-crop areas or for employment with the slave population. Most poor whites, therefore, settled in the lower quality agricultural areas north and south of the Black Belt; in peripheral southern Appalachia; among the large plantations, but on locally poor land; or in nonagricultural employment. After the Civil War, hard times pushed many of the rural poor whites into sharecropping arrangements little different from those affecting blacks.

About 1880, the patterns began to change. The distribution of poverty was changed little, but the economic environment in the South entered a new phase. During this decade, manufacturing experienced rapid growth, perhaps more rapid than in any prior 10-year period. The growth of the cotton textile industry led this period of regional economic development. Largely a New England industry prior to the turn of the century, cotton textile manufacturing began to shift southward several decades earlier. It grew much more rapidly in the South after 1880 than anywhere else in the country. By 1929, 57 percent of the nation's cotton textile spindles were in the South, over 2½ times the share existing in 1890.

Within the region, textiles and the related apparel industries grew most strongly throughout the piedmont of North and South Carolina and Georgia. The shift of these industries into the region can be attributed to changes in labor availability, resource proximity, technology, and the inertia costs of location. The low-skill, labor-intensive industries originally situated along the northeastern margins of the continent's manufacturing core region began to experience a gradual loss in labor availability as workers shifted toward higher-skill, higher-paying employment. The high levels of *underemployment* in the Southeast offered as an alternative a large pool of available labor for the textile and apparel industries. Natural and synthetic fiber industries produce the raw material for cotton and synthetic textile manufacturers, just as the textile industries provide the raw material for apparel manufacturing. Taking advantage of proximity, the growth in textile and apparel manufacturing across the Carolina Piedmont and in northern Georgia was followed by an increase in the number and output of fiber industries.

The tremendous increase during manufacture in the value-to-weight ratio for wood-processing industries also makes these industries sensitive to raw material location. In addition to all this, as technological improvements occurred in these activities, New England plants became increasingly obsolescent. Internal costs of production and locational-inertia costs grew for the Northern manufacturers as taxes, land, and power costs were driven up by high-value industries. This had not yet occurred in the Southeast.

There was much more to the relocation of

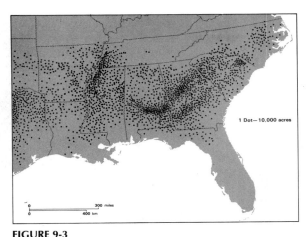

FIGURE 9-3

COTTON PRODUCTION, 1909. Cotton acreage remained high in most Southern states well into the twentieth century, a pattern reinforced by heavy regional debt, cultural inertia, and the apparent absence of alternatives.

textiles, therefore, than a desire to be close to the raw material source. Much of the new industry located outside the primary cotton-producing regions (Figure 9-3). Resistance to industry in the best agricultural areas combined with a large surplus labor force and active campaigns by local leaders to draw most of the cotton textile industry to northern Alabama, northern Georgia, and the piedmont sections of North and South Carolina. Most counties in these areas had predominantly white populations and showed their determination to keep it that way. The two chief results of this were: (1) only some parts of the South experienced the economic and cultural stimulus of early industrialization, and (2) it was the black portion of the region's population that benefited least from these changes. Although it would be a mistake to overestimate the impact of this development on general income levels (textiles are a low-wage industry, thus the appeal for cheap labor regions), the fact that there was *any* manufacturing growth in the South during this period is frequently overlooked.

Cotton textile manufacturing was not the only new source of industrial opportunities during the

1880s and 1890s. Reconstruction of the region's railroads and other public improvements stimulated the flow of money and the development of railroad towns. Cigarette manufacturing began to be focused in the tobacco regions of North Carolina and Virginia following technological advances. With the establishment of a new federal land policy and a strengthened railroad network, the South's very large timber resources began to be exploited. Much of the timber was taken out as a raw material, but furniture manufacturing in North Carolina and Virginia and (after 1936) pulp and paper manufacturing throughout the South also were an outgrowth of the exploitation. These industries continue to be important in the South today (Figure 9-4).

It was also during the last quarter of the nineteenth century that the South's most important iron and steel industry received a series of stimuli. An iron industry had existed in the region for many decades in small and dispersed smelters. By the mid-1870s, however, technological

The abundant pine forests of the South produce about two-thirds of U.S. pulpwood. Pulpmills such as this one at New Bern, North Carolina, transform wood waste to pulp through a process that requires wastewater aeration, as can be seen in the foreground. Note the heavily wooded, flat landscape. (Courtesy Weyerhauser.)

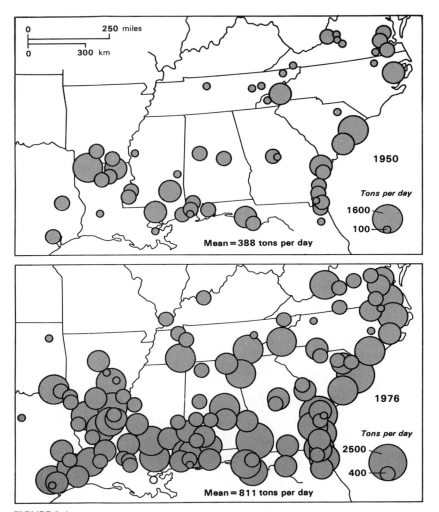

FIGURE 9-4

PULP MILL PRODUCTION, 1950 and 1976 (courtesy of J. F. Hart). The tremendous growth in the exploitation of Southern forests is indicated by changes in pulp production. Most increase in production has occurred on the Atlantic and Gulf coastal plains.

improvements in iron making had led to the rise of Chattanooga, Tennessee, as an important center of iron production. At the same time, a large deposit of high-quality coking coal was discovered near Birmingham, Alabama, and exploitation of the seam was begun before the end of the decade. Numerous iron-making compa-

nies and iron- and steel-using industries accumulated in and around Birmingham and Chattanooga. These two cities combined with the transport focus and subsidiary industries in Atlanta to form an important industrial triangle by the end of the century.

This development was significant in the economic geography of the South, although the critical importance of this triangle's appearance was weakened by the economic and political context of the period. First, the development was significant because of the way in which iron and steel production tends to draw other manufacturers dependent on steel to the region. Second, these industries are also not as low skill and low wage as textile- and tobacco-product manufacturing. Finally, this centrally located region of nonagricultural economic development could have been an industrial focus for the South as a whole, stimulating increases in labor skills, income levels, and general economic welfare through each city's connections with other major urban centers.

This did occur to some degree, but discriminatory shipping rates imposed on Birmingham-manufactured products dampened the effect considerably. Because of low labor costs and the fortunate proximity of raw materials in the Birmingham area, steel could be produced there at much lower cost than in Pittsburgh, the North's primary center of steel production. Without governmental interference, however, the U.S. Steel Company imposed and enforced what came to be called the Birmingham differential, or a pricing policy of "Pittsburgh plus," on all users of Alabama steel. Under this policy, consumers were required to pay the price of steel at Pittsburgh plus $3 per ton ($2.72 per metric ton) plus the regular freight cost from Birmingham. Even

Although most Americans mentally associate iron and steel industry with the major centers of production in the manufacturing core region, this steel plant in Birmingham, Alabama, represents the long-established heavy industry present in a few sections of the South. (Leroy Woodson/EPA-Documerica.)

though this discriminatory pricing practice was eventually ruled illegal and ended, the policy severely restricted the competitive cost advantage of Alabama steel during the rapid economic expansion decades of the early twentieth century and undoubtedly contributed to the slow growth of Southern industry. One way to view this would be to consider Birmingham and the South in general as having been much farther away from the national economic core region than in fact they were. Atlanta firms, for example, found it to their advantage to order steel from Pittsburgh rather than Birmingham, even though a Pittsburgh-Atlanta shipment had to travel more than four times the distance of a shipment from Alabama.

For more than half a century after the Civil War, changes in Southern economic development were irregular and often slow. The region was far from resource-poor, yet its population continued to have an average annual income far below the national level. Gradual industrial growth helped raise income levels somewhat, but agricultural incomes were low and most industrialization was of the low-wage variety. Geographically, the greatest growth in manufacturing employment occurred in the piedmont section of Virginia and the Carolinas and in the Atlanta-Birmingham-Chattanooga triangle. The Gulf Coast also began to experience rapid growth as local resources were developed. The number of white families who remained in poverty was large, but an even larger proportion of the black population lived at these levels. Most blacks not only resided outside the areas experiencing industrial growth and found themselves locked into restrictive agricultural arrangements, but they also experienced increasing racial repression at the same time that economic opportunities were increasing for whites.

The Institutionalization of Racial Segregation

A good deal of variation in interracial attitudes undoubtedly existed in the South prior to the Civil War. Even under the assumptions of inherent inequality required by the institution of slavery, it is clear that most whites were not comparable to Simon Legree, just as most blacks would not fit the Uncle Tom image. Following the Civil War, however, anti-black feeling gradually spread and intensified among whites until it pervaded almost every aspect of Southern life.

During the early years after the war, anti-black activities were often localized. Some former slaves were able to gain title to good agricultural land; to enter professions as educators, lawyers, and doctors; and to be elected to political office. The bulk of the population struggled under the economic constraints, but the rapid achievements of many belied the label of inferiority placed on blacks as a group.

By the late 1880s, segregation laws began to be passed requiring the racial separation of more and more aspects of Southern life. This followed by about a decade the departure of federal troops. Restrictive laws, called Jim Crow laws, were passed in each Southern state in rapid succession during the 1890s, until virtually total legal segregation was established.

Formal segregation had many geographic expressions. It was by definition the separation of the total population into two or more societies. It was sometimes necessary for members of these two groups to meet, communicate, work near one another, and in other ways have their lives overlap, but segregation's purpose was to maintain a distance between members of the different groups wherever possible. Two sets of schools were operated. Two sets of restaurants, recreation facilities, park benches, drinking fountains, rest rooms, and other points of potential contact had to be constructed and maintained. Housing was separated into white areas and black areas. Entry into certain occupations was restricted, with the second-class group finding only the less desirable employment available to them. In the South, both overt and covert restrictions were placed on black efforts to vote with almost total disenfranchisement achieved well before World War I. The geography of the Deep South, therefore, became much more complex because there were two very different human regions in the South at the same time, one black and one white. In some cases, such as residential patterns in the western South, the two areas of human activity hardly overlapped at all in the urban places. But in other cases, such as employment, the regions overlapped greatly in a geographic sense, even if they did not come into contact in social and economic ways. When individual resistance to the discrimination arose, violent repression was an effective method of removing the resister. A propensity for lynching became an ugly component of white Southern culture during the early decades of this century.

In spite of a nineteenth-century Supreme Court ruling permitting segregation as long as equality was ensured, *de jure segregation* did not lead to facilities and opportunities for whites and blacks that were "separate but equal." Basically, the desire and commitment to make the facilities equal were not present. Whites presumed themselves superior to blacks. Although this presumption differed little from that of whites outside the region, Southerners used the force of law to institutionalize the belief. The South was not a wealthy region and, in allocating scarce financial resources, black facilities received much less than half of the budget for many years. By 1930, the difference between white schools and black schools, for example, was striking (Table 9-1). Every Southern state, except Tennessee and the border states of Kentucky and Maryland, spent more than twice as much per pupil on white schools as black schools, with white schools in South Carolina and Mississippi receiving well over *five* times as much per pupil as black schools. These figures suggest some of the pressures that led to the eventual reversal of the "separate but equal" doctrine by the U.S. Supreme Court in the 1954 decision *Brown* vs. *Board of Education of Topeka.*

Out-Migration

A third geographic consequence of the war and

TABLE 9-1
Selected Characteristics of Racial Inequality in Southern States, 1930

State	Percentage Black	Annual School Expenditures per Pupil ($)		Number of One-teacher Schools per 1000 Pupils		Percentage Illiterate	
		White	Black	White	Black	White	Black
Alabama	35.7	27.43	8.31	2.03	5.04	4.8	26.2
Arkansas	25.8	22.75	10.42	4.89	4.95	3.5	16.1
Florida	29.4	34.59	10.56	n.a.	n.a.	1.9	18.8
Georgia	36.8	24.64	6.53	1.48	6.49	3.3	19.9
Kentucky	8.6	22.22[a]	21.90[a]	7.17	8.99	5.7	15.4
Louisiana	36.9	36.41	8.78	0.79	4.02	7.3	23.3
Maryland	16.9	49.41	33.89	1.74	4.52	1.3	11.4
Mississippi	50.2	30.60	4.79	1.51	6.72	2.7	23.2
N. Carolina	29.0	28.76	12.70	1.22	3.13	5.6	20.6
S. Carolina	45.6	37.53	6.65	1.07	4.27	5.1	26.9
Tennessee	18.3	28.97	15.02	3.42	5.02	5.4	14.9
Texas	14.7	38.76[b]	16.02	1.75	4.19	1.4	13.4
Virginia	26.8	29.44	13.06	2.89	4.59	4.8	19.2

SOURCE: Charles S. Johnson, *Statistical Atlas of Southern Counties.* Chapel Hill: University of North Carolina Press, 1941.

[a] Does not include Louisville.

[b] Includes salaries of superintendents.

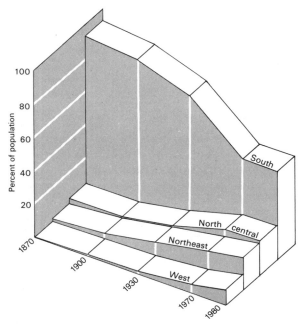

FIGURE 9-5

CHANGE IN REGIONAL DISTRIBUTION OF THE U.S. BLACK POPULATION, 1870–1980. It was not until World War I that the South's black population began to leave the region in very large numbers, a level of out-migration that continued for at least half a century.

the antebellum Southern heritage—one which had intimate connections with the patterns of persistent poverty and racial repression—was an accumulation of conditions that caused one of the country's major streams of population redistribution. For almost 50 years following the end of the Civil War, the slow trickle of black migrants from the South increased little. Lack of information about alternatives outside the region, initial improvements in local opportunities, the impoverished and low-skill characteristics of most Southern blacks of that period, and perhaps simple inertia combined to keep black out-migration below the level of natural increase. Thus, 91.5 percent (less than 5 million) of all U.S. blacks resided in the South in 1870 and 89 percent resided there in 1910, whereas

the Southern black population in that period increased by almost 4 million (Figure 9-5). During the next decade, the number of black emigrants increased sharply and subsequent decades revealed a continuation of this trend.

The numerous factors that accumulated about the turn of the century to "push" blacks out of the region have already been discussed: Jim Crow laws, violence, and near-subsistence economic conditions are the most prominent. A further reduction in economic support caused by a severe boll weevil infestation coincided with the onset of World War I. The war led to a stren-

This 1923 editorial cartoon from the New Orleans *Times-Picayune*, demonstrates the Southern viewpoint regarding Northern labor recruiters. The sharp-eyed "outsider" attempts to sway "contented" black labor with false stories. In fact, black labor was not contented, and wages *were* higher and hours shorter in the North. (Ettie Temple/Historical Pictures Service.)

THE MIRAGE
—Temple in the New Orleans *Times-Picayune*.

uous effort by the North to "pull" blacks (and poor whites) from the South. Prior to 1914, national industrial expansion had depended on millions of European immigrants to meet the large demand for labor. More than one-third of the U.S. population in 1910 was foreign born or had at least one parent born outside the country (Chapter 3). When the war ended this supply of immigrants, an alternative one was found in the large unemployed and underemployed Southern labor pool. Press campaigns and labor recruiters sent into the South extolled the virtues and opportunities of Northern manufacturing employment and Northern life in general. And once the outflow was initiated in earnest, rela-

tives, friends, and former neighbors provided the information and often the financial support required to encourage others to undertake the long move to the North.

The migration of blacks from the South was not uniform. Almost all blacks leaving the region moved to urban places. Those from the Atlantic states, however, traveled largely to Northern cities east of the Appalachians: Washington, D.C., Baltimore, Philadelphia, New York, Hartford, Providence, and Boston, for example (Figure 9-6). Blacks from the south-central states, Alabama, Mississippi, Kentucky, Tennessee, Arkansas, and Louisiana, moved north along the major transportation lines to St. Louis, Indian-

FIGURE 9-6
MAJOR BLACK MIGRATION STREAMS, 1965–1970. Migration from the South formed "streams" of population movement that flowed between states in the region and destinations in the North and West.

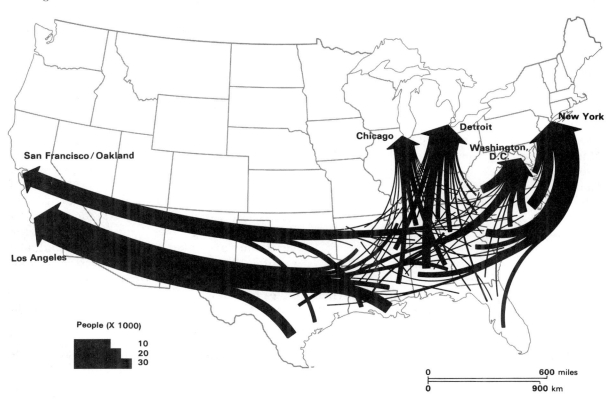

People (X 1000)
10
20
30

apolis, Gary, and Chicago or to eastern interior manufacturing cities, such as Cincinnati, Detroit, Cleveland, Pittsburgh, and Buffalo. The largest and most southerly cities tended to receive large numbers of blacks first, whereas smaller industrial centers often did not possess large black populations until after World War II. The more distant western cities, such as Denver, Portland, Seattle, and those of California also experienced a later growth in their black populations than did cities along the East Coast, with most western growth occurring after World War II; blacks migrating westward directly from the South moved primarily from Louisiana, Texas, Arkansas, and Mississippi. In each case, the factors of distance and the predominant pattern of transportation can be seen as the basic explanation for these geographically distinctive migration streams.

The Southern economy might not have suffered from this exodus if the population involved had not also been selective. After all, the average Southern black did receive very low wages if employed, and after World War II sharecropping became less necessary and much less efficient as Southern agriculture gradually mechanized. Most blacks who left, however, were between the ages of 18 and 35. Raised in the South, this group's most economically productive years were then spent outside the region. Many of those who remained behind were in their later productive years, retired, or not yet in the labor force (Figure 9-7). Racial limitations on opportunities in professional occupations also resulted in a loss of many of the most highly trained young people from the region.

Sectionalism

A fourth major consequence of the Civil War and the South's cultural heritage that preceded it was an intensification of the sectionalism already felt in the region. Partly because of the slavery issue, the South felt itself a region apart even before the war. The four years of turmoil, fought almost entirely within the South, the loss of the war, and the repressive aspects of Reconstruction all reinforced the sense of regional difference. The South is the only part of the nation to have ever suffered occupation by a conquering army, and it has taken a century and a great deal of economic growth to temper the bitterness that followed.

The war and Reconstruction were also instrumental in unifying Southern whites. The direct impact of blacks on politics during Reconstruction was not large, but politics became imbued with the race issue by the 1880s. This provided politicians and voters with an oversimplification and distraction from significant economic issues that remained until the 1960s.

The sectionalism of the South is easily visible in its political behavior. The "Solid South" was a term that indicated that the entire region voted as a bloc and often in direct contradiction to otherwise national trends. The Civil War and Reconstruction were associated with the North and the Republican Party; therefore, Southern whites became stubborn opposition Democrats. Even after blacks were disenfranchised and the Northern tenor of both parties changed, the South continued to vote Democratic, at least at the local level. When the national party became so different that Southern whites could no longer tolerate the ideological connection, the explicit sectional label of "Southern Democrats" became common. The South was clearly considered a distinctive political region, as well.

Although national political changes and Southern cultural changes have made the South far from "solid," even at the state level, the 1968 formation of the American Independent Party was a continuation of Southern sectionalism. Although receiving votes from every state, the viewpoints presented by presidential candidate George Wallace that year were basically old-fashioned Southern viewpoints, and the only states he carried in the election were Southern states. Even with a successful Southern candidate for the presidency in 1976, most political observers agreed that the white South viewed President Carter with a mixture of sectional pride for his Southern origins and Southern culture and sectional disapproval for his obvious departure from these origins and culture in appealing

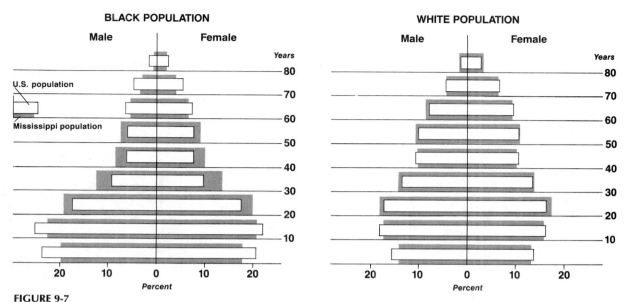

FIGURE 9-7

COMPARATIVE AGE/SEX DISTRIBUTIONS: UNITED STATES AND MIS-
SISSIPPI, 1980. It is clear that young adults of both races, but especially blacks
now in their twenties and thirties, migrated from Mississippi in large numbers,
leaving behind an inordinately large proportion of the population who, for the
movement, are too young to follow them.

to the national electorate. The non-South's near
obsession with Jimmy Carter's regional background during the first half of his presidency
did nothing to allay this sectional feeling within
the South.

The issue of sectionalism can be extremely
subtle, but it is nevertheless a manifestation of
the distinctive Southern culture. Interwoven with
the other major strands that made up the region
during the century that followed the Civil War—
widespread poverty, explicit racism, and a continued agrarian and small-town orientation—the
Southerners' sense of their region as having always been different from the rest of the country
created an isolation and regional consciousness
that reinforced whatever differences there were.
As Thomas Wolfe pointed out (see the quota-
tion at the beginning of this chapter), the Southern sense of region was strong and conscious.
Perhaps because of the strength of this regional
identification and the somewhat disdainful attention it received from non-Southerners, the

slow erosion of its potency was little noticed for
a number of years. Even after it was clear that
change was occurring, observers referred in sectional terms to the rise of the "New South." Just
what is new about the "New South's" geography?

THE ONSET OF CHANGE

The New South is a modification of the South
that preceded it. Present spatial and regional
characteristics have been built on patterns that
evolved over decades and, in some ways, over
centuries. The key to recent changes lies in the
gradual loss of regional isolation. Prior to World
War II, most of the South's population and certainly its leadership appeared to believe and react
to events as though the region had successfully
seceded from the United States, almost as though
the South was a separate country reluctantly required to continue dealing with a Northern

Located near the center of the Southeast, Atlanta's role as a major transportation focus for the region has stimulated rapid growth since World War II. (Courtesy Atlanta Chamber of Commerce.)

neighbor. The result was an accentuation of certain cultural and geographic patterns that had existed very early in the region. Since the later 1930s, however, and especially since the end of World War II, trends and pressures external to the South began to infiltrate the region and break down its isolation. Some of the changes induced in the South's geography were caused by factors that affected the entire nation's economy and have no clear relation to regional location. Some changes have followed from specific federal intervention in regional affairs. And some characteristics of the New South have come about through modifications to the distinctive Southern culture. All have eroded the isolation that set the South apart from the rest of the country.

Economic Reorganization

Changes in the South's economic structure since the mid-1930s have been dramatic and of na-

tional significance. The mid-Depression economy of the South was little different from that of 1870: dominantly agrarian, producing raw agricultural products primarily for export, capital deficient, supported by heavy use of animal power (usually mules) and hand labor, and operated through sharecropping and tenant-farming arrangements and a regionally distinctive *crop-lien system*. Such industry as existed was largely low-wage or oriented toward narrow local markets. The region's urban structure continued to reflect this orientation; small market centers, railroad towns, textile mill towns, and county seats represented the pervasive urban form in the South.

By 1985, in approximately half a century, tremendous changes have occurred. The South's economy is no longer primarily an agrarian economy. By the early 1950s, over half of the region's labor force was engaged in urban-based, nonagricultural employment, and the propor-

Throughout the tobacco regions of North Carolina and Virginia where the crop is flue-cured, the distinctive curing barns are fast becoming relic features on the landscape as they are replaced by more efficient sheds that look little different than abandoned truck trailers. Thus, landscape changes are instituted by the adoption of new technology. (Left, Courtesy: W. K. Collins, N.C.S.U. Crop Service Extension Specialist; right, Ray Wilkinson III, Raleigh, N.C.)

tion in agriculture has continued to decline. This has paralleled a sharp increase in manufacturing employment and employment in service activities. Further, the industrial mix in the South has shown a strong trend toward diversification; no longer is Southern manufacturing limited to the early stages of raw materials processing.

Within agriculture, diversification has also occurred, although that is not agriculture's most dramatic transformation. Among the traditional cash crops produced in the region, cotton remains the most important; the others include tobacco, sugarcane, peanuts, and rice. However, all of these crops, except rice, have been under strict federal acreage controls since the 1930s and 1940s, with rice under controls since about 1951. Total output has remained at high levels as growers reserved their best land for cash-crop production, but the area producing cotton, for example, is only a shadow of its former size (see map appendix; compare with Figure 9-3). This

areal shrinkage has been supported by the decay of old cotton-ginning institutions in sections of the former production area. The decline is most pronounced in the Carolinas-Georgia Piedmont and the Black Waxy Prairie of eastern Texas.[4] In addition, cotton is no longer exclusively a Southern product. Irrigated cotton is grown in the Texas Panhandle, along sections of the Rio Grande, and in Arizona and California. Within the South, soybeans have replaced cotton as the most valuable cash crop produced. By 1981, the soybean crop earned Southern farmers nearly $3.5 billion in that year, whereas cotton lagged far behind at $1.1 billion.

Although cotton dominance has declined, other crops, such as soybeans, and livestock industries have increased sharply. Beef production has im-

[4]For an excellent discussion of this, see Merle C. Prunty and Charles Aiken, "The Demise of the Piedmont Cotton Region," *Annals of the Association of American Geographers*, 62 (June 1972), pp. 283–306.

Chicken broiler and egg production require heavy investment and large-scale operations today. This laying house in Mississippi is almost identical to those found in other production regions across the South. (U.S.D.A.)

proved greatly since World War II as farmers began to improve pastures directly with better grasses and fodder crops and with higher fertilizer applications. At the same time, new cattle strains were developed to survive and thrive in the hot, humid Southern summer (see Chapter 10). Within the last 25 years, national broiler and chicken production has become industrialized and concentrated in the South. Clearly a regional industry, approximately 75 percent of commercial broilers produced in the United States were Southern in origin (see map appendix). Chicken prices have remained relatively low through the application of modern production and management techniques. Highly automated

production in large volume with opportunities to share new technology and a well-developed, spatially concentrated marketing network have combined to ensure the continued location of this agricultural industry in the South.

Even more dramatic than the diversification of agricultural production have been the transformations in the means of farm production. Agriculture is no longer an animal-power, hand-labor operation. Wherever possible, machinery has been applied to the production process, just as in manufacturing, and regional agriculture is now much more efficient than before. It traditionally has been assumed that as Southern agriculture became automated, the underemployed and marginal farmers were *pushed* from the land as unneeded excess labor. One study, however, argued the opposite.[5]

Urbanization and nonagricultural industrialization prior to 1960 were said in this study to have led to largely voluntary migrations in search of better economic opportunities; because of this labor outflow, farm mechanization was required to maintain production levels. The distinction between these alternative perspectives is important because it can affect our view of social organization and policy development. The belief that farmers were pulled from the land by greater opportunity creates quite a different viewpoint from the belief that these farmers were pushed out in spite of whatever wishes they may have had.

Associated with the mechanization of Southern farming, the traditional sharecrop system has almost disappeared since the mid-1930s. Accompanying this change has been a sharp increase in the average farm size in the South. Again, traditional wisdom has seen this increase as a reflection of sharply dropping farm *ownership*, but Professor Prunty argues convincingly for another explanation.[6] Since 1870, the U.S. census has counted tenant and sharecropped subunits as separate farms. With the end of the sharecrop system, therefore, and largely because of increased urban and industrial opportunities, there is an exaggerated appearance of reduced farm ownership resulting from farm size increases. Furthermore, much of the rural population that could leave the countryside for urban employment has already done so. Those who remain are either successful full-time farmers, part-time agriculturalists, or individuals with few formal skills who manage to eke out a living at low levels of income.

---

[5] Merle C. Prunty, "Some Contemporary Myths and Challenges in Southern Rural Land Utilization," *Southeastern Geographer*, 10 (November 1970), pp. 1–12.

[6] Ibid.

Of those who have left the rural South, virtually all have migrated to urban centers; not all of these centers, however, have been located in the North and West. Rural-to-urban migration *within* the South increased rapidly during and after World War II as the region's economy participated in the post-Depression expansion. This city-bound movement included members of both races, suggesting again that the pull of increased opportunities dominated social push factors or was, at least, important enough to partially forestall the exodus of blacks from the region. In 1940, there were only 35 cities with populations greater than 50,000 in the South. By 1950, the number had increased to 42 cities, a jump of 20 percent and by 1980 it had reached 75, an increase of 114 percent over 1940. Many other small Southern places, more accurately villages than towns at the outset of this period, have developed a certain vitality from the larger growth centers. Although still strongly rural in character, the trend in the South is toward a level of urbanization typical of the United States as a whole.

The pull to the cities was stimulated by industrial growth and a diversification that promises to match that of Southern agriculture. The proportion of the nonagricultural labor force in manufacturing jobs increased greatly and in virtually every part of the region (Figures 9-8, 9-9). The traditional industries, such as steel (Alabama), tobacco products (North Carolina and Virginia), and textiles (northern Georgia, the Carolinas, and southern Virginia), remained important but less dominant as other kinds of manufacturing activity appeared. Synthetic textiles and apparel industries, the former in the Carolinas and the latter primarily in northern Georgia, widened activities even within this broad industrial category. Chemical industries expanded rapidly along the Gulf Coast. Furniture production in the central Carolina Piedmont increased, and other wood-processing plants (paper-container plants and plywood manufacturing) became more prominent throughout the

207

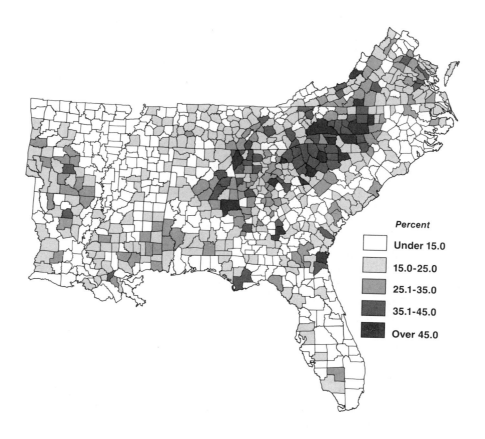

FIGURE 9-8
PERCENTAGE OF NONAGRICULTURAL LABOR FORCE EMPLOYED IN
MANUFACTURING, 1950. Except for portions of the piedmont and other scattered pockets in the South, manufacturing employment remained low across most
of the region in 1950.

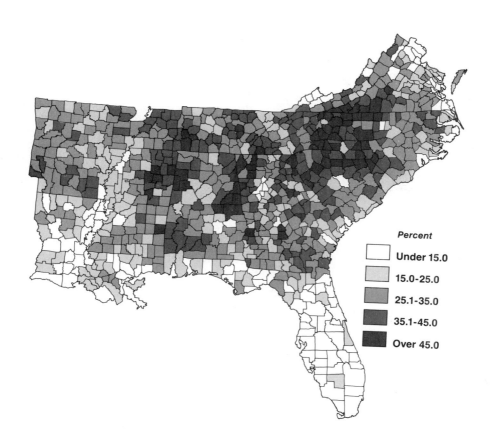

Percent

Under 15.0

15.0–25.0

25.1–35.0

35.1–45.0

Over 45.0

FIGURE 9-9

PERCENTAGE OF NONAGRICULTURAL LABOR FORCE EMPLOYED IN MANUFACTURING, 1980. Within one generation of the pattern shown in Figure 9-8, there had been a tremendous growth in manufacturing employment across the South, now a region with few areas that are without a significant manufacturing work force.

Textile and apparel industries have a high machinery-to-labor ratio, require a large amount of floor space, and demand relatively low skills from their labor force. The woman tending this modern knitting machine also represents the high female work force in such low average wage industries. (Courtesy Burlington Industries.)

eastern and Gulf coastal plains. Shipbuilding was continued at Norfolk and begun at several sites on the Gulf Coast; aircraft production at Marietta, Georgia, drew skilled labor and higher wages to the Atlanta area.

Most significant, as the average Southern consumer earned higher wages, the regional market increased enough to draw many consumer-goods manufacturers into the South. This increased the demand for nonagricultural labor,

spread the income further, strengthened the local market, and so on, into the growth cycle common to developing regions. As this trend continues, some of the low-wage industries that remain concentrated in the region (Table 9-2) are likely to be driven out by the demands of high-wage activities for the limited labor supply. Economic activity in the South, therefore, is gradually shifting into something that more closely reflects the patterns found in other regions. Even though the South continues to contain many areas with low per capita incomes, the persistent poverty of the region's population has begun to diminish.

Government Intervention

Perhaps the most significant changes in the South have been triggered by direct governmental involvement, usually from the federal level until recently. This intervention has occurred in economic as well as in social matters. Furthermore, the economic intervention has most often been in the form of a specific stimulus to development, not merely a kind of regional welfare payment to cover income deficiencies. The regional development characteristics of the Tennessee Valley Authority (TVA) are an excellent example for a portion of the South (see Chapter 8). Federal expenditures in the South have been uneven, with North Carolina, Arkansas, and Louisiana receiving less than average outlays from Washington, whereas Virginia, Texas, and Mississippi received more than the average state in per capita terms.

It has been argued that the South's rapid industrial growth since the mid-1930s is a consequence of a growing regional market, gradually demanding and able to pay for more goods and services. The question remains: Why did the market expand? Per capita incomes in the early 1930s were about one-third the national average, and the economic structure had changed little since Reconstruction, a full 60 years earlier. Professor Prunty proposed that the federal Agricultural Adjustment Acts (1935 and later)

TABLE 9-2
Index of Average Hourly Wage in Manufacturing Industries, 1952–81 (U.S. Average = 100)

State	1952	1957	1962	1967	1972	1977	1981
Alabama	79	86	86	85	85	87	88
Arkansas	69	71	69	71	73	76	78
Florida	76	79	83	84	85	82	82
Georgia	73	75	74	78	80	79	80
Kentucky	90	96	95	96	97	101	98
Louisiana	85	94	96	96	97	102	107
Mississippi	66	68	69	72	73	74	75
North Carolina	73	70	69	72	73	73	74
South Carolina	73	70	70	73	74	76	77
Tennessee	81	81	81	81	81	83	84
Virginia	81	79	80	81	81	83	85

SOURCE: Calculated from U.S. Bureau of Labor Statistics, *Employment and Earnings, 1939–1974* and *U.S. Statistical Abstracts,* 1978 and 1982.

provided the main stimulus to the market growth.[7] Although aimed at farmers throughout the country, the consequences for the largely rural South were especially strong.

Before the Agricultural Adjustment Acts took effect, the prices that farm products could demand were set to a great extent by supply and demand in the international market place. To the South, this meant that cotton prices, for example, fluctuated partly according to the production success or failure in other cotton-growing areas of the world. More important, farm labor in the cotton South (basically tenant and sharecrop farmers) was essentially in competition with hard-pressed cotton producers along the Nile or Indus rivers or elsewhere in what was still largely a colonialized world economy. When agricultural wages and prices were adjusted upward under the Agricultural Adjustment Acts to reflect more accurately national industrial wage differentials, the sharply improved market in the South for manufactured goods in-

itiated the upward development spiral currently affecting the region.

In an act of federal intervention much more widely recognized as significant to the South's social structure, in 1954 the U.S. Supreme Court struck down the racial separate-but-equal doctrine permitted almost 70 years before. Changes in the South's social geography were initiated by this decision, changes that have reverberated to every other part of the country where race affected opportunity, and the repercussions are far from settled more than 30 years after the decision itself.

The geographic and institutional consequences of racial segregation were not quickly affected by simply declaring the practice illegal. The antiblack component of white Southern culture had been consciously nurtured for too many years for change to occur rapidly, but even in this the South has shown considerable change. In some parts of the South, new schools were built for blacks during the 1950s to show that their facilities could be equal. But public facilities were gradually required to prove compliance either through direct testing by blacks or by serving the public according to a specific black-to-white ratio.

[7] Merle C. Prunty, "The Agrarian Contribution to Recent Southern Industrialization," in *The South: A Vade Mecum,* James R. Heyl (ed.). Washington, D.C.: Association of American Geographers, 1973, pp. 108–118.

Throughout this period and accompanying the South's economic revitalization, many blacks migrated from rural areas but stayed within the region. By 1980, more than two-thirds of the South's black population were living in the region's major metropolitan areas. Although the black metropolitan population has been increasing, migration by whites into the South from other regions, also mostly into the metropolitan areas, has kept the proportion of metropolitan population that was black from growing even more rapidly.

Out-migration continued at a high rate, as discussed earlier, but by the early 1970s there was a slight decline in the overall level of black outflow. This may be a result of increased economic opportunity—somewhat less limited than previously by discrimination—in the South's urban areas. The 1980 census confirmed that more blacks were entering the South than were leaving it. This suggests that regionally based racial push factors have become insignificant in national population redistribution. The South, in other words, may no longer be seen as any more threatening racially than any other region. It also would confirm the growing economic maturity of the South, for migration in the United States has been associated traditionally with the attraction of economic opportunity. Finally, it suggests that Southern culture, already containing a strong flavor introduced by its black population, may continue to benefit from greater-than-average black cultural contributions.

Cultural Integration

With the possible exception of this last speculation, a thread common to many of the South's changes since the mid-1930s is the gradual decline of its regional distinctiveness. Simply, much of what has been occurring in the South is making it more and more like the rest of the eastern half of the country. Economic diversity is replacing simple dependency on agriculture. There are indications that the region's supply of low-wage labor is almost exhausted; new industry and service activities will have to compete more actively

and may continue to force wages upward slowly. A significant infusion of Northern migrants, especially to regional metropolitan growth centers, has made some of these cities less distinctively Southern in culture and more clearly just urban. Racial discrimination has never been peculiar to the South, although its expression was often more blatant and institutionalized than elsewhere. With the end of many overt and all legalized forms of racism, the South has, at least, reached the level of interracial attitudes found through the rest of the country, although some of the old institutions occasionally resurface.

As the South has experienced these changes, the region's sense of isolation, its sectionalism, has also declined. Not only have some regional characteristics become more "national," but the North and West have also demonstrated an affinity for what Southerners recognize as their own. Improved communications have broadened the national exposure of many Southern writers, poets, artists, and others who draw deeply on the region's distinctive cultural heritage; and often the regional origin is not recognized by non-Southerners. Black cultural contributions are also more widely dispersed and less distinctly regional. Politically, Southern elections and administrations are no more tumultuous than elsewhere and no more race oriented. The full range of the political spectrum is represented among Southern elected officials, although the majority tend to continue some of the traditional orientations. Since the Civil Rights Act of 1964, blacks have had less difficulty in registering to vote and in exercising that right. All of this has reduced the defensiveness of Southerners somewhat and, therefore, reduced the sectionalism of the region's population.

Agricultural Case Study: The Yazoo-Mississippi Delta

The Yazoo-Mississippi Delta is a microcosm of several aspects of the Deep South, even though the smaller region has a history modified by its physical geography. The Delta, as it is known in Mississippi, occupies an eastern portion of the

Value of crops sold, 1974

1 dot = $1,000,000

FIGURE 9-10

YAZZO-MISSISSIPPI DELTA. With rich soil and flat land, subject to flooding and used intensively for agriculture, the delta continues to have very few urban places.

alluvial floodplain of the central Mississippi River (Figure 9-10). There are small local variations in relief, but the overwhelming visual impression is that of an almost perfectly flat landscape. Filled in since the last Ice Age by glacial outwash and flood-deposited alluvium, the Delta drops only about one-third of a meter (1 foot) of elevation for every 3.2 kilometers (2 miles) of the 320 kilometers (200 miles) between Memphis and Vicksburg.

Poor internal drainage, frequent floods by the Mississippi, and the presence of Choctaw and Chickasaw Indians kept European settlement tentative in the Delta for many years. However, after the 1830s, the lure of fertile soil, proximity to river trade, and the profits to be made from cotton gradually drew plantation agriculture into this rather hazardous region. As elsewhere in the South, with cotton production came large numbers of slaves. This was so exclusively plantation country that by 1860 blacks outnumbered whites in the entire Delta by more than 3 to 1 with some areas exceeding a 10 to 1 ratio. Development of the land was slow however; ravages of the Civil War and frequent floods kept the region largely wilderness. The Delta was inundated by floods, for example, in 1858, 1862, 1865, 1867, 1868, 1871, and 1874. The tremendous productive pressure on the South's land during the decades after the war and political pressure by railroads and land speculators during the last quarter of the nineteenth century and the first quarter of the twentieth century finally attracted federal efforts to control the flooding. Although only gradually successful, the Delta attracted population again (both white and black) as more land was cleared, swamps were drained, and new areas were planted in cotton. Cotton acreage increased almost 500 percent between 1880 and 1920. Towns within the Delta remained small, with Greenville the largest at 11,560 in 1920 (the only town larger than 8000 population). Although no longer slaves, the Delta's majority black population continued to labor in the fields as sharecroppers.

An especially severe flood in 1927 followed shortly by the national distress of the Great

213

Land in the delta is very flat and most is intensively used. In the fall, mechanical harvesters scuttle across the large fields gathering in the year's crop. This may be soybeans, as shown, or with different equipment, it may be cotton or rice. (U.S.D.A.)

Depression led to more decisive governmental intervention in the Delta economy. Large landowners recovered financially with the Agricultural Adjustment Acts and expanded acreages began to be developed, but the amount of land that could be planted in cotton was controlled from Washington. It was soon found that rice and soybeans could be most profitable additions to the traditional cotton economy, especially as both crops did well on land too wet for cotton.

Agricultural expansion during the 1930s and 1940s, however, occurred just as Delta blacks increased their migration out of the region. During the half-century after 1920, the black population in the Delta declined 33 percent. Landowners responded with increased mecha-

nization as elsewhere in the country, but as cotton had always been an especially labor-intensive crop, the transformation was more extreme. Because mechanized cultivation, harvesting, and the application of fertilizer and pesticides require considerable capital and large-scale operations, the most successful landowners often expanded at the expense of the smaller farmers.

In addition to this array of features—dependence on agriculture; a large and poor black sharecropper population; an absence of major urban centers; a decline in the relative importance of cotton; and changes stimulated by federal intervention and mechanization—the Delta also highlights Deep South patterns in its grudging industrial growth. Later to appear and less intense—even by 1980—than elsewhere in the larger region, the Delta received a few new manufacturing activities after World War II. The success of agriculture (plus the valid fear of flood) probably combined to retard industrial entry for many years. Successful farm mechanization eventually generated a surplus labor force, and, in turn, stimulated modest industrial development, most of it related to processing agricultural products; manufacturing employment grew modestly from 6491 in 1955 to 15,079 in 1970, but then grew to more than 25,000 by 1980.

Rich agriculturally, increasingly diverse economically, possessing a great deal of residual poverty (especially among its majority black population), in spite of the presence of both old and new wealth, the Yazoo-Mississippi Delta is a distinctive region, but just as clearly it is a part of the larger Deep South.

Additional Readings

Carney, George O. "Country Music and the South: A Cultural Geography Perspective," *Journal of Cultural Geography,* 1 (1980), pp. 16–33.

Comeaux, Malcolm L. *Atchafalaya Swamp Life: Settlement and Folk Occupations,* vol. 2. Baton Rouge: Louisiana State University, School of Geoscience, 1972.

Gottmann, Jean. *Virginia in Our Century.* Charlottesville: University Press of Virginia, 1969.

Hartshorn, Truman A., et al. "Metropolis in Georgia: Atlanta's Rise as a Major Transaction Center," in *Contemporary Metropolitan America,* vol. 4. Cambridge: Ballinger, 1976.

Kovacik, Charles F., and John J. Winberry. *South Carolina.* Boulder, Colo.: Westview Press, 1985.

McKee, Jesse O. *Mississippi.* Boulder, Colo.: Westview Press, in press.

Reed, John. *Southerners: An Essay in the Social Psychology of Sectionalism.* Chapel Hill: University of North Carolina Press, 1982.

Rubin, Louis D. *William Elliot Shoots a Bear: Essays on the Southern Literary Imagination.* Baton Rouge: Louisiana State University Press, 1975.

Shuptrine, Hubert, and James Dickey. *Jericho: The South Beheld.* Birmingham: Oxmoor House, 1974.

Winberry, John J., and David M. Jones. "Rise and Decline of the Miracle Vine: Kudzu in the Southern Landscape," *Southeastern Geographer,* 13 (1973), pp. 61–70.

CHAPTER TEN

THE SOUTHERN COASTLANDS: ON THE SUBTROPICAL MARGIN

The southern margins of the United States can be divided into two approximately equal sections. One half, the Southwest Border Area (Chapter 14), shares a long land boundary with Mexico. The other half traces the coastline eastward from the mouth of the Rio Grande to North Carolina and includes Florida's peninsula (Figure 10-1). Although both stretches are southerly in latitude and share a small area of overlap in southern Texas, the Southern Coastlands are as distinct from the Southwest Border Area as are any other two adjacent regions on the continent.

The Southern Coastlands comprise a distinctive region for two primary reasons (although a third factor is also an important component of the region's character). First, the region has a humid subtropical environment. The southern tip of Florida lies only 160 kilometers (100 miles) from the Tropic of Cancer, with the chain of coral islands called the Florida Keys extending another 65 kilometers (40 miles) south. The warm waters of the Gulf of Mexico also contribute a strong maritime influence to the coastlands' climate. The region has a clear appeal to visitors and potential residents, and its agriculture is distinctive specifically because of this environment.

Second, the region derives much of its character from its location at the continental margin. The environment, of course, is related to this aspect of location. But the region's role in generating American trade patterns with the rest of the world, and the distinctive industrial pattern that has arisen along this coastal region also help define the region. As in other regions, there is a good deal of potential conflict between activities that depend directly on the unblemished character of the natural environment and those that treat the environment merely as a setting within which the activity must be carried out. Thus, the two distinctive components of the Southern Coastlands' regional geography and the seeds of conflict they contain provide the theme for this region.

The third factor relevant to the character of the region is its position within the Deep South.

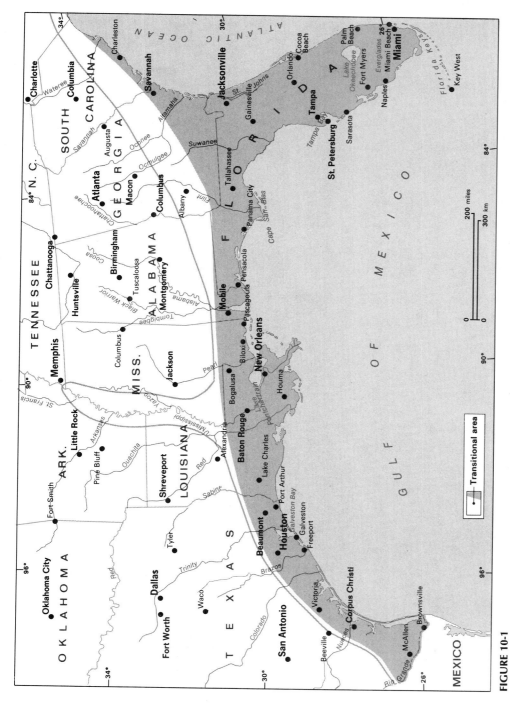

FIGURE 10-1

THE SOUTHERN COASTLANDS.

This locational component has influenced some of the cultural attributes of the Southern Coastlands and creates difficulty in attempts to define sharply the inland margins of the region. Here we restrict our view of the region to its coastal influences after having discussed the larger region's culture in Chapter 9. This southeastern margin of the continent simply does not "fit" well economically and, to some extent, environmentally with the rest of the South.

THE SUBTROPICAL ENVIRONMENT

Of the several components that make up a physical environment, the region's climate has the greatest impact on the human geography of the Southern Coastlands. A humid subtropical climate, the long growing season, mild winter temperatures, and warm, humid summers all contribute to the patterns of human activity associated with the region.

The average length of the growing season in the Southern Coastlands has an equivalent elsewhere in the United States only in southern California, Arizona, and Hawaii. Measured as the number of days between the last killing frost in the spring and the first frost in the fall, almost the entire region experiences 9 months or longer of potential growth for its agricultural crops (Figure 10-2). In much of the region, in fact, the average growing season is in excess of 10 months, and the southern half of the Florida peninsula is unlikely to experience frosts each year. In addition to the long warm season, almost the entire region receives precipitation abundant enough for most agricultural activities. An average rainfall in excess of 125 centimeters (50 inches) is received by all of the Southern Coastlands east of Houston, Texas, with most of the rain falling in the summer between April and October. Thus, abundant rainfall can also be expected in the period during which sunlight is plentiful and warm temperatures support most plant growth.

FIGURE 10-2

LENGTH OF THE GROWING SEASON. The combination of a southerly location and maritime proximity leads to a long growing season and, across most of the region, sufficient precipitation for reliable specialty crop farming.

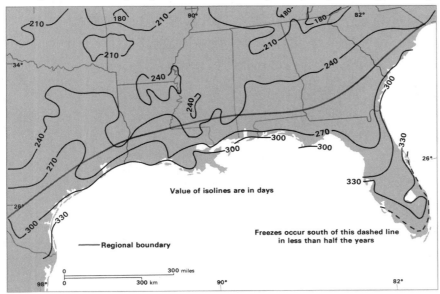

Value of isolines are in days

Freezes occur south of this dashed line in less than half the years

Regional boundary

300 miles
300 km

MAJOR METROPOLITAN AREA POPULATIONS IN THE SOUTHERN COASTLANDS, 1980	
Houston, Tex.	2,905,353
Miami, Fla.	1,625,781
Tampa-St. Petersburg, Fla.	1,569,134
New Orleans, La.	1,187,073
Ft. Lauderdale-Hollywood, Fla.	1,018,200
Jacksonville, Fla.	737,541
Orlando, Fla.	700,055
West Palm Beach-Boca Raton, Fla.	576,863
Baton Rouge, La.	494,151
Mobile, Ala.	443,536
Charleston-North Charleston, S.C.	430,462
Beaumont-Port Arthur-Orange, Tex.	375,497
Corpus Christi, Tex.	326,228
Lakeland-Winter Haven, Fla.	321,652

Agriculture

There are two primary consequences of the presence of this set of climatic conditions in North America. First, as long as other agricultural conditions are met, such as fertile soil, appropriate ground drainage, and control of insect pests, farmers can grow their crops without fear of an early frost until late in the fall. In a few locations, with careful selection of complementary crops and judicious planting, it is possible to harvest successfully two crops in one growing season, and some vegetable farmers are achieving even more. Second, however, and even more important, is the opportunity for production of specialty crops that can be grown in few other locations in the United States and Canada. Citrus fruits (produced in significant quantities outside the region only in southern California and Arizona), rice, and sugarcane are the most important specialty crops in the Southern Coastlands. All three contribute greatly to the local economies of the region. Citrus is especially important in Florida, with significant production in southern Texas as well. Only in the produc-

Most first-time visitors to central Florida are struck by the immense orange orchards encountered for mile after mile across the gently rolling landscape. (Florida News Bureau, Department of Commerce.)

TABLE 10-1

Production of Citrus Fruits, 1978, in Florida and California (Thousands of Tons Harvested)

Citrus Fruits	Florida	California
Oranges (all)	7438.9	1447.4
Tangerines and mandarins	193.3	23.2
Tangelos	177.2	24.3
Grapefruit	1953.8	208.3
Kumquats	0.4	0.2
Lemons	38.8	638.9
Limes	30.6	4.1

SOURCE: U.S. Department of Commerce, *1978 Census of Agriculture*. Washington, D.C.: U.S. Government Printing Office, 1980, vol. 1, part 51, Table 33.

tion of lemons and navel oranges does California exceed Florida's citrus output (Table 10-1).

Citrus production has been an important contributor to Florida's economy since it was first introduced to the region by the Spaniards during the sixteenth century. Although the areas of production have gradually shifted southward along the peninsula's interior, most citrus is produced south of 29° North latitude and north of Lake Okeechobee. This southward movement is the response of growers to the yearly threat of frost, but the swampy soils and standing water of the Everglades have prevented most citrus production from moving farther south into the effectively frost-free margin of Florida. About half of Florida's citrus production remains in the region between Tampa and Orlando (see map appendix).

Oranges and grapefruits are the most important of the seven major citrus fruits grown in the state. Orange production has increased slowly but steadily in recent decades; more than 7.4 million tons of oranges were harvested in 1978. Since 1945, an increasing share of the total orange crop has been processed rather than sold as fresh oranges. The proportion processed had reached about 80 percent of the total output as early as 1970. By processing the crop (mostly into frozen concentrate), a sizable industry has developed in

Florida. This spreads the benefits of this specialty crop among a larger proportion of the state's population than if only fresh fruit were shipped northward. In addition, processing allows year-round sales instead of limiting returns to the immediate harvest period. Grapefruits are produced in an area almost coincident with orange production, but total demand is lower and output is only about one-fourth that of the more important orange crop. Significant orange and grapefruit farming in the Southern Coastlands is also found under irrigation in the extreme southern portion of Texas between Brownsville and McAllen.

Because citrus are tree crops, a major share of the production costs are associated with the harvest. The fruit must be hand picked, often from the top of a long ladder. A large amount of short-term labor is required during the harvest season, a feature of citrus farming that annually draws thousands of migrant laborers to the dense, parklike groves for periods of intense physical effort. Housing facilities for temporary workers are often cheaply constructed, and although a day's labor may pay a respectable amount, the workday itself may be long. The conflicting goals of producer and worker have not been resolved, and, so far, attempts to develop mechanical pickers for citrus crops have been unsuccessful. The results are a substantial labor force whose members cannot earn enough at one location to satisfy their annual living requirements and an industry that must devote more than one-third of its total production expense to hired labor.

Although citrus crops are grown in the Southwest as well as in Florida and Texas, sugarcane production is exclusive to the Southern Coastlands in the mainland United States. Cane is the tropical source of sugar and exceeds beets as the primary source of the world's sugar supply. Within the United States and Canada, however, the temperate-climate beet crop has been a more important source of domestic sugar production. Sugarcane is, in fact, a perennial crop requiring more than 1 year for full maturity, and it is not at all tolerant of frost. In addition, water requirements for cane are high, needing a mini-

Florida's citrus harvest remains labor intensive. Whatever the fruit, it is picked by hand, one at a time, from the small trees as are the oranges shown. (Courtesy Florida Department of Citrus.)

mum rainfall of about 125 centimeters (50 inches) per year. Both temperature and water requirements would seem to preclude growth in the United States, except under irrigation, but moderately large quantities of cane are grown in Louisiana and Florida. The necessary water is available naturally in Louisiana and supplemented by irrigation in Florida; and immature cane is harvested early, especially in the Louisiana locations. In spite of the reduced sugar content in cane before it matures fully, federal import controls and price supports ensure that returns are usually high enough to justify the agricultural effort. In Florida alone, the 1981 value of sugarcane production was approximately $234 million. Even though federal controls on cane production were removed in 1974, higher levels of production are not anticipated in the Southern Coastlands because cane farming there is not competitive on the world sugar market.

Rice is less demanding than sugarcane in its climatic requirements. Given sufficient water, rice will mature within one growing season at a rate roughly in proportion to the amount of heating it receives during the summer. Within the Southern Coastlands region as we have defined it, irrigated rice is grown in Louisiana (1.1 million tons in 1978) and in Texas near Houston (1.4 million tons). The transitional nature of our regional boundary is emphasized by a rice-producing region far to the north of the Gulf Coast in the Mississippi alluvial valley sections of Arkansas and Mississippi. These portions of Arkansas and Mississippi are dissimilar to the Southern Coastlands in most respects, but large volumes of this warm climate crop are grown in those states: 2.4 million tons in Arkansas; 458,000 tons in Mississippi in 1978 (see map appendix).

In addition to the speciality crops, the Southern Coastlands are also among the continent's primary regions of vegetable production and have developed a strong beef industry since World War II. The region's subtropical envi-

221

Growing conditions for sugar cane are far below optimum along the Gulf Coast, but it can be grown. The use of machinery in large-scale operations permits profitable production where smaller scale and hand-labor would not. (U.S.D.A.)

ronment is complementary to the cold winter continental climate experienced by the large population living in North America's urban manufacturing core region. This has led to production of a wide variety of vegetable crops in the coastlands. Most fresh vegetables reaching the urban markets during the winter are grown in Florida and the southern margins of the other Gulf coastal states, although some fresh vegetables are shipped from the extreme southwestern United States. Proximity to the northeastern markets undoubtedly provides the Southern Coastland producers with a distinct locational advantage over California vegetable growers.

In a somewhat different manner, the environmental possibilities for beef production have been realized only after a good deal of effort and innovation. The long growing season and abundant rainfall do suggest the potential for high-quality pasture once the land has been cleared. The combination of high temperatures and high humidity, however, produces poor natural grasses. More important until rather recently was the impact of cattle tick fever. Even with improved pasture, the better European cattle breeds succumbed to tick fever, but the native fever-resistant cattle did not yield good beef even on the better grasses. Success of the industry was finally achieved after a long series of efforts between World War I and 1950 to eradicate the ticks that cause cattle fever (hopefully), improve pasturage, and crossbreed hardy Brahman bulls (imported to Florida from Texas) with improved domestic cattle (Table 10-2). The Florida beef industry has become an important contributor to that state's economy. Because these changes are still recent, some observers argue that the beef industry in Florida is still capable of an "enormous" increase "whenever the landowners decide to upgrade the remaining old native pastures."[1] Although some of these improvements have been adopted in other,

---

[1] W. Theodore Mealor, Jr., and Merle C. Prunty, "Open-Range Ranching in Southern Florida," *Annals of the Association of American Geographers*, 66 (September 1976), p. 376.

TABLE 10-2

Number of Cattle and Calves Sold, Florida

Year	Number Sold
1944	216,236
1954	658,617
1964	798,615
1974	966,002
1978	1,240,163

SOURCE: U.S. Department of Commerce, *Census of Agriculture*. Washington, D.C.: U.S. Government Printing Office, various.

noncoastal sections of the Southeast, it is in the Southern Coastlands, especially central and northern Florida, that the benefits are most strongly felt.

Although climatic conditions in the region are, in the balance, favorable for agricultural activity, soil conditions and quality are much more variable and less clearly beneficial. The soils range from the fertile but poorly drained muck of the Louisiana coast and Mississippi Delta to the extremely sandy soils in north and central Florida. To complicate the pattern, sections of Gulf coastal Florida and the extensive Everglades region are largely swampy muck soils or poorly drained sands, whereas the coasts of Texas and Georgia and South Carolina may have either marsh soils or sandy soils, depending on local conditions. Areas of muck in Louisiana, once they have been drained, have proven to be extremely productive, especially for sugarcane and rice. But attempts to drain Florida's swamps have met with more failure than success and have, in any case, been strongly opposed by conservationist groups. The economic (and technical) failures in Florida are primarily a result of extremely low-relief variations. Conservationist opposition is based on the widespread disruption to the complete ecological system of the vast marshlands that would result from the draining; a lowered *water table* permits extensive encroachment of salt water, exposes the grasslands to fire, and upsets the delicate life-support balance for wildlife, shellfish, and vegetation.

In sharp contrast to a need for drainage in many parts of the region, much of the remainder of the Southern Coastlands benefits from heavy irrigation. The central highlands of Florida, for example [at about 30 to 46 meters (100 to 150 feet) above sea level] are underlain by sandy soils with moderately poor to very poor water-retention capacity. It has been discovered that the productive citrus "belt" and vegetable areas have yielded annual output as much as ten times more valuable when the crops are irrigated than when precipitation is depended on as the sole source of moisture. With this degree of improvement possible and the technological capacity available for its achievement, the distinctive subtropical environment of the coastlands has been developed agriculturally beyond the levels found in much of the interior Southeast.

Amenities

In addition to the economic benefits derived through agricultural specialization, the Southern Coastlands' subtropical climate also benefits the region by its contrast with the more northern and continental climate of the North American interior. As personal disposable income and old age financial security increased in the United States and Canada during the decades following World War II, recreation and retirement became major industries in those areas that could attract large numbers of people. In an article calling attention to this trend and studying its early significance, Professor E. Ullman[2] argued that as economic reasons for migration become *relatively* less important, people will choose to live where they find it "pleasant"; he suggested that an outdoor climate most similar to a comfortable indoor climate would be most preferred. The two regions in the United States and Canada best approximating this climate are coastal southern California and the Southern Coastlands, at least in winter months. The widespread adoption of

---

[2] Edward L. Ullman, "Amenities as a Factor in Regional Growth," *Geographical Review*, 55 (1954), pp. 119–132.

air-conditioning provides summers with nearly equivalent appeal.

Even by the early 1950s, when Professor Ullman wrote, the importance of amenity factors in stimulating the regional growth of Florida and the Gulf Coast was clear; since then their effects have increased. Statistics show that Florida is viewed as one of the prime states for retirement by many approaching that stage in their lives. In 1950, 12.4 percent of Florida's population was over 60 years of age; by 1980, this proportion had increased to more than 23 percent. In comparison, Ohio and Connecticut, located in the continental manufacturing core, contained populations in 1950 of 13.7 and 13.3 percent, re-

Greater general affluence and reduced transport costs permit more of us to enjoy the environmentally pleasant places. However, this greater use has almost always meant deterioration of the environment. Thus far, the prime alternatives to this dilemma have been to place restrictions on individual "freedom of choice" or to do nothing and let the desired places become less pleasant than they had been. Many areas in the southern half (and western region) of the United States are currently facing this dilemma. (Berry's World © 1977 by NEA, Inc.)

© 1977 by NEA, Inc.

spectively, above the age of 60, but these proportions rose only slightly by 1980 to 15.3 percent for Ohio and 16.7 percent for Connecticut.

Although the amenity of a mild winter climate draws large numbers of retired persons to the Southern Coastlands, a larger direct economic advantage is generated by tourist activities in the region. New Orleans has been a strong attraction for vacationers because it mixes the opportunities of a metropolitan urban area with a culturally distinctive Afro-American flavor, all set within the mild subtropical climate of the Southern Coastlands. Between New Orleans and Mobile, Alabama, the coastal section of Mississippi, locally referred to as the Mississippi Riviera, has recently experienced a tourist boom, with construction of numerous hotels, motels, restaurants, and artificial beaches.

The most dramatic tourist magnet in the region, however, has been Florida. With long beaches facing both the Atlantic and the Gulf of Mexico, this state has drawn winter vacationers for decades from the urban centers of Megalopolis. Luxury hotels lining the oceanfront in Miami Beach and luxury estates along the coast at Palm Beach proved to be only the cores of an expanded tourism-based urban development extending over 160 kilometers (100 miles) along the southeastern coast of the peninsula. Spatially less continuous pockets of recreational development have sprouted from Palm Beach along Florida's Atlantic coast as far north as Jacksonville. And although the Gulf side is less well endowed with attractive beaches, considerable growth has occurred from Tampa Bay southward to Naples. The demand for subtropical amenities has become so strong that the development of recreation resources has spread well north along the Atlantic margin into coastal Georgia and coastal South and North Carolina. Sections of the Georgia and South Carolina coasts, such as Georgia's Sea Islands and Hilton Head, Charleston, and Myrtle Beach in South Carolina, are experiencing equivalent development booms.

Not all tourist attractions depend on re-

The coastal resources of peninsular Florida are well suited to recreation, as illustrated in this section of Fort Lauderdale where the landscape has been modified to permit maximum access to the water. (Florida Department of Commerce Division of Tourism.)

sources provided by nature. The decision by Walt Disney to construct Disneyworld, an eastern equivalent to his highly successful Disneyland in southern California, has drawn millions of out-of-state visitors to Orlando and south-central Florida. Many other new attractions were drawn to this part of the state by the promise of tourist traffic and the possibility of tourist expenditures, a promise that became self-fulfilling. The magnet provided by this central Florida recreation complex has stimulated considerable spill-over growth in the region. It is expected to be the inland link between the east-coastal and west-coastal urban clusters. By the end of the century, a small megalopolis—or coalescence of urbanized areas—is expected to extend from Miami to Tampa-St. Petersburg via Orlando and contain approximately 13 million people. Although not the only reason for this expectation, the more than $21.5 billion spent by tourists in Florida during 1982 are a clear indication of economic power.

Hazards

Although the subtropical environment of the Southern Coastlands provides a variety of advantages through its complementarity with the rest of the continent, this environment is not entirely beneficent. Agriculturally, successful vegetable production has encouraged growers to attempt to produce crops year-round. Thus, when an occasional midwinter frost does reach into southern Florida, considerable crop damage can occur. Similarly, Florida citrus is harvested between October and late spring, and a winter freeze can hurt the ripening fruit. Lake County, Florida, citrus growers, for example, usually expect the first frost about December 15 each year but may be faced with a hard freeze as much as three weeks earlier. Smudge pots, oil heaters, and other temporary (and partially effective) devices are used in attempts to protect the growers' investment. Less well publicized is the potential damage of these untimely wintery blasts for sugarcane growers in Louisiana. Ironically, the shortness and general mildness of the frost period also creates problems for agriculture because, although controlled by expensive chemical means, insect and parasitic pests thrive during the long growing season and create a constant threat to farm production. A recent example is

the late-1984 appearance of citrus canker in Florida citrus groves, an outbreak that threatened to damage production for years to come.

More erratic, more sporadic, more dramatic, and locally more destructive along the Southern Coastlands are the region's hurricanes. These large cyclonic storms are generated by intense solar heating over large bodies of warm water. When the proper set of conditions has been created, a large low-pressure atmospheric pocket results. Warm, moist air flows into the low-pressure cell and is lifted vigorously upward until the air pressure sink is very deep and creates a powerful pull on more warm, moist air. The earth's rotation and spheroidal shape mean that air flowing toward the hurricane's center will establish in the storm a broad counterclockwise rotational pattern. Depending on areal extent and the atmospheric conditions involved, the result

The Gulf Coast is especially vulnerable to the tremendous destructive power of hurricanes. Although the effects of these large storms may be felt across very large areas, as is shown by this 1972 hurricane, the most severe damage is done to the coastal region as the storm moves ashore. (NOAA.)

of all this might be a large but local thunderstorm, a gale, or a full-fledged hurricane.

Hurricanes can affect settlement far north of the Southern Coastlands, but their chief concentration is in that region. When a storm does move onshore in some section of the heavily populated region of Megalopolis, severe destruction can occur, as in 1955 when hurricane Diane struck New England. But because of their exposure toward the tropical waters of the Caribbean Sea and the southern Atlantic, Florida and the entire Gulf coastal lowlands are extremely vulnerable to hurricane damage. The same features that provide the region's amenities and make it attractive to tourists and retirees—subtropical climate and sea coast orientation—are those that create hurricanes and permit such great damage when they strike land. Because these storms are an accepted feature of the region, preparations to withstand the greatest force of the wind and rain can be made early. And because the heaviest damage is usually limited to a narrow swath onshore, many portions of the region have not been affected for years. On the other hand, because hurricanes are so variable in occurrence and strength, settlements that have spread in spite of warnings into coastal areas are very much exposed to the dangers of a large storm. And the long period between direct coastal strikes at any specific site by hurricanes tends to make newcomers complacent about the tremendous force of major storms.

The region is also gradually succumbing to what might be called a slow-acting hazard. If ignored or not dealt with, this problem will pose a considerable hazard to many in the region. In some respects, portions of the southern coastlands provide an example of the long-term environmental impact of rapid and widespread development.

As surprising as it may sound, major sections of the region will soon be faced with a situation in which water demand exceeds water supply. Even though most of the region receives abundant rainfall and no portion of the region is more than a short distance from a seacoast, there is clear evidence that water is already being withdrawn from some local resources faster than they can be recharged.

The potential problem has begun to appear most obviously where there are major urban developments at some distance from rivers that might be used as a water supply. Consider for a moment the options available to those who must provide water to growing municipalities. If rivers are too distant and the land is too flat to make constuction of large reservoirs feasible, the remaining possibility is to withdraw water from the water table by drilling wells and pumping it as needed.

In central Florida, this is the primary method used. The region's rains are usually sufficient to recharge the water table. However, even the normally abundant rainfall of the region is not always enough. The tremendous growth in population and business between Cape Canaveral and Tampa—with the greatest concentration in Orange County (containing Orlando)—has meant that roads, parking lots, homes, motels, amusement parks, and all the other facets of tourist development have spread across the land at an amazing pace. Water that falls in these built-up areas is carried away through storm drains and does not percolate down to the water table. When combined with the increased withdrawal, the land surface loses some of its support from below (as the water table drops), and it can settle, sometimes quite suddenly. The abrupt appearance of a *sinkhole* that swallowed an automobile dealership in Winter Park, Florida, brought national attention to this potential hazard, if not its underlying cause. Other, smaller cavities have appeared in urban areas of central Florida since the Winter Park experience. Although we do not suggest that the state will disolve in some catastrophic manner, this does demonstrate that the repercussions of development for an environment are not always immediate or obvious.

ON THE MARGIN OF THE CONTINENT

In a way, the subtropical character of the Southern Coastlands is a function of its location along the southern margins of North America. But there is more to the significance of location than the physical environment found at that place. Just as urban places can be studied in terms of their absolute and relative locational characteristics or in terms of site and situation (see Chapter 3), analysis of other kinds of places can be enhanced by examining both types of locational characteristics. The Southern Coastlands' absolute location is probably best reflected in the subtropical nature of the physical environment. The region's relative location, however, its position with respect to, and spatial relations with, other places may be more effectively understood in terms of its character (1) as an exchange area between the continental interior and overseas territories, and (2) as a geological region of considerable economic importance. Both characteristics are direct consequences of the continental marginality of the region. The first is a function of trade, whereas the second is more related to factors of resource location and exploitation. Both have generated consequences that help define the variability that exists within the southern coastlands.

Trade

The coastline of the Gulf of Mexico has only a few good-quality harbors suitable for large-volume trade activities. A shallow, emergent coastline containing many *high-action beaches,* much of the coast itself is either backed by extensive swamps or partially shielded behind offshore bars. If passage can be made through gaps in the sand bars, they can provide protection from rough seas. Coastwise shipping uses this protection in the *intracoastal waterway system.* However, most bays behind the bars are too shallow for good anchorage for the large ships engaged in transoceanic trade. Therefore, most of the larger ports have developed on the edge of large river estuaries along the coastline or a short distance inland from the mouths of rivers emptying into the Gulf or the Atlantic (see Chapter 3). Pensacola (Florida), Mobile (Alabama), and Galveston

and Corpus Christi (Texas) have all developed in the shelter of large bays, each of which also bears the name of the port city. Jacksonville (Florida) on the St. Johns River, Brownsville (Texas) on the Rio Grande, and New Orleans (Louisiana) on the Mississippi River share riverine port characteristics. The largest single city in the Southern Coastlands, Houston, was not originally a port city but has become one through construction (beginning in 1873) and repeated improvement of the Houston Ship Channel across shallow Galveston Bay.

Perhaps more important than physical facilities, a coastal location must be accessible to inland sources of exchange items before trade can be generated. Each of the bays providing good harbor facilities is the outlet for a river draining a part of the interior, but these rivers vary greatly in navigability. All assisted early settlement expansion and some are still navigated by small barges. All have had their access to the continental interior strengthened by rail connections with major inland markets. Jacksonville, for example, was an early terminus for railroads entering Florida from Georgia. It also was the main focus for a hinterland that extended westward into the state's panhandle and south into the agriculturally rich central highlands. Other ports were too distant or too low quality to be attractive alternatives. A result was that Jacksonville was well established even before highway connections reinforced its locational bases for growth.

When discussing the accessibility of Southern Coastland cities to their hinterlands, New Orleans is in a class by itself. The city was founded early in the European colonial period by the Spanish, soon afterward succeeded by the French, as a southern counterpart to Quebec and Montreal on the St. Lawrence. New Orleans was both a control point and a shipping focus for the entire Mississippi River system. The Mississippi was navigable (with proper caution) by shallow draft paddle-wheel steamers far to the north, well into the agricultural heartland of the continent. The river's main tributaries were also navigable along much of their length and extended the single waterway system both westward into the Great

The high-altitude vertical photo of New Orleans dramatically demonstrates the city's intense urban development within a large meander and the trade and transfer point advantages along the lower Mississippi River. (NASA.)

Plains (see Chapter 12) and eastward into the continent's manufacturing core region (Chapter 5). New Orleans' site, within a large river meander and on the low-lying river delta only a few feet above mean sea level, meant that flooding was an annual threat often realized. But the city's situation was of such tremendous advantage to all associated with trade that its population grew early in the nineteenth century and has remained large. New Orleans, in fact, was easily the largest urban place south of the Ohio River and east of the Rocky Mountains until after World War II.

New Orleans lies midway between the recreation region of the eastern Gulf and the industrial centers of the western Gulf. The French colonial heritage is consciously maintained in the city's French Quarter, the *Vieux Carré*. A strong non-Anglo cultural imprint is apparent outside the Quarter as well, but the distinctive mix of local and European cuisine, a wealth of jazz and dixieland performances, legitimate eighteenth-century architecture, and an anything-goes reputation has drawn millions of tourists to the city. A casual visitor to New Orleans, however, may be surprised by the heavy barge and ship traffic

The fog-enshrouded spires of St. Louis Cathedral overlook Jackson Square in the very heart of New Orleans' French Quarter. The French cultural inheritance has been preserved here and is one of the attractions that draws many thousands of visitors to the city each year. (Courtesy of the Greater New Orleans Tourist and Convention Commission.)

mineral resources and the Gulf Coast access have been the primary geographic factors in Houston's rise to urban prominence.

Resources

The true margin of North America as a continent does not coincide with the seacoast. A shelf of land below sea level extends beyond the coastline. In some cases the shelf extends only a few miles, but along much of the Atlantic coast and in the Gulf of Mexico, the edge of this shelf may lie more than 80 kilometers (50 miles) from shore. Mineral exploration along the coast from the Rio Grande to the Mississippi River mouth led to the discovery of an extensive series of petroleum and natural gas deposits, both onshore and offshore (Figure 10-3).

When the Gulf Coast oil field was brought into production during the early 1900s, Houston was still a moderate-size city of less than 75,000. By 1930, the city's population had multiplied four times to almost 300,000. Through the Depression years and World War II, Houston continued to grow, doubling its population again by 1950. And by the time the 1980 census was taken, the city itself had grown to 1,595,138 to rank fourth in the United States behind New York, Chicago, and Los Angeles. Located about midway along the long coastal arc beweeen the Mississippi and the Mexican border, Houston also lies at the coastal apex of the Texas Triangle joining Dallas-Fort Worth and San Antonio. With Dallas ranked seventh in population in 1980 and San Antonio ranked 10th, connections with these major inland growth centers and the large cotton exports that originate in eastern Texas have also contributed to the locational strength of Houston.

The search for additional petroleum deposits along the Gulf Coast was extended seaward before midcentury. The petroleum companies' success created new problems, of a political nature, even while economic problems were temporarily alleviated by the discoveries. In overcoming the technological difficulties of drilling for, and extracting petroleum from, platforms

on the river (it is the busiest port in North America) and by the heavy industry supported by this traffic.

The other major city in the western portion of the Southern Coastlands, the largest in the entire region by 1970, provides a contrast with New Orleans. Houston is a "new" city in many ways, whereas New Orleans has a long history of economic importance. Not actually located on a river or even on Galveston Bay, Houston's port connections became more important to growth after World War II with the rise of a local petrochemical industry. Houston possesses a high degree of focality for most of lowland Texas (see Figure 3-4), but these overland connections also followed after the city's economic growth, stimulating it further; they did not precede it. Regional

FIGURE 10-3

PETROLEUM AND NATURAL GAS DEPOSITS: WESTERN GULF COAST-
LANDS. The long arc of petroleum and natural gas deposits paralleling the
Louisiana-Texas Gulf Coast has, with the coastal location of these resources, gen-
erated a large refining capacity at Texas and Mississippi River ports.

far beyond sight of land, conflicting claims between state and federal governments over jurisdiction of continental shelf resources flared for
the first time. Texas and Louisiana had immediate stakes in the argument, and it was clear
that other coastal states, especially those bordering the Gulf, would also be interested before too
long. The federal form of political organization
in the United States allowed a series of litiga-
tions between the individual states and the national government. One result of this complex
series of court cases has been variable jurisdiction over continental shelf resources in the Gulf
of Mexico by the Southern Coastland states;
Florida and Texas are sanctioned to claim up to
9 miles seaward, whereas Louisiana, Alabama,

231

Highway, rail, and water transportation come together to support the industry around the turning basin at the end of the Houston Ship Channel. Currently the fourth largest city in the United States and still growing, Houston's CBD can continue to appear small and lost on the vast coastal plain of Texas. (Sam C. Pierson/Photo Researchers.)

and Mississippi are limited to 3 miles. The claims by Texas and Louisiana were initially the most significant because of the already known resources off their coasts.

The continuous rapid increase in domestic consumption of petroleum, natural gas, and petroleum products led the national government, during the early 1970s, to open commercial bidding on offshore tracts between Biloxi, Mississippi, and Tampa Bay, Florida, and during the early 1980s off the Atlantic coast. Opposition to exploratory drilling in the eastern section of this region has been strong from environmentalist groups and from economic interests already

present but based on the coast's amenity characteristics. Florida, naturally, has led the opposition, a reflection of its long Gulf coastline and its extremely strong recreational resources orientation. Although oil spills or leakage could have disastrous short-term effects on certain tourist activities, many conservationists argue that even more serious consequences are possible. Extensive mangrove swamps, coastal marshlands, and commercial fishing beds are in a delicate ecological balance, and they could be badly damaged for long periods if oil-drilling mishaps or shipping accidents polluted them. The classic disagreement between those who would develop resources as they are needed and those who wish to preserve the balance of the natural environment is not likely to be resolved easily along the Gulf Coast. The dispute also illustrates the regionalization within the Southern Coastlands, for the western and eastern portions of the region each represent opposing interests.

So extensive are the petroleum deposits along the coast between northern Mexico and the Mississippi River that the resource importance of the Southern Coastlands would be a national priority even if these deposits were the only mineral resource in the region. Texas and Louisiana are currently the two leading petroleum-producing states, far exceeding the two next most important, California and Oklahoma. Although Texas and Louisiana possess large producing fields well inland from the Gulf, the coastal fields are major contributors to both states' totals.

Natural gas, often found in conjuction with petroleum deposits, is another resource in the Gulf. For many years, gas was burned off as an unwanted by-product of petroleum production. But the energy potential of this resource was eventually realized and generated a search for, and utilization of, natural gas independent of petroleum exploration. The numerous scattered Gulf Coast deposits of natural gas are spatially intermingled with the region's long arc of petroleum deposits. Pipelines to carry the gas radiate from the main coastal production centers to the primary consumption points across the continent and within the manufacturing core.

Although a less common mineral association, the geologic formations of the Texas and Louisiana coastlands that contain petroleum and natural gas contain two additional minerals of economic value—sulfur and rock salt. The subsurface rock contortions found where petroleum and natural gas have been trapped in economically recoverable deposits were formed in this region by the gradual upthrust of large salt *domes*. Far less valuable than either of the mineral fuels, rock salt is nonetheless mined in large quantities in southwestern Louisiana. More valuable than salt, although not so much as petro-

leum or natural gas, is the sulfur that is also found in the *caprock* covering many of the salt domes. The large deposits at Beaumont, Texas, and across the state boundary near Lake Charles, Louisiana, supply all U.S. needs. Additional deposits both inland and beneath the continental shelf indicate an abundance of this mineral for many years. Also of national significance is the major phosphate production from substantial deposits in Florida.

Industrial Development

Petroleum and natural gas extraction do not usually generate strong local urban and industrial growth. The process of exploration and drilling requires specialized and expensive equipment but not the large variety of support materials or manpower of many mining operations. Once the crude oil or gas is pumping, only routine attention is required. However—as was dramatically illustrated again in the 1970s by the Persian Gulf states for any Americans who had forgotten their own earlier experience—large-volume petroleum production does generate tremendous quantities of capital in a short time. Although both mineral fuels of the region can be transported cheaply and reliably through pipelines and by ocean-going tankers, locally accumulated wealth has attracted a wide variety of industries able to use the minerals near their production sites. Petroleum refineries were built outside all major ports from Corpus Christi, Texas, to Pascagoula, Mississippi, with the most intensive concentration around Houston, Beaumont, and Port Arthur.

Of broader development impact than the refining process itself are the indsutries that depend on the refinery output for their own existence. The petrochemical industry is one of the most important and rapidly growing enterprises in highly industrialized countries, such as the United States and Canada. Natural gas and petroleum products are used as chemical components for a great many products in everyday use. Items ranging from plastics, paints, and antifreeze to fertilizer, insecticide, and prescription

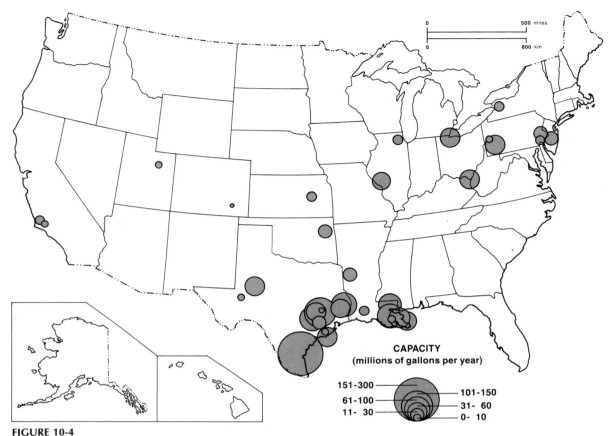

FIGURE 10-4
BENEZENE PRODUCTION CAPACITY, 1981. Dependent on large stocks of
coal or, preferably, petroleum, the production of this basic petrochemical is
heavily concentrated along the western half of the Southern Coastlands.

drugs have their origin in chemical plants located along the western Gulf Coast (Figure 10-
4). In addition, other chemical industries that are
not tied to petroleum and natural gas production, such as those producing sulfuric acid, su-
perphosphate fertilizer, and synthetic rubber, are
major consumers of sulfur and salt (sodium
chloride). The coincidence of the basic minerals
within a region possessing large capital investment capabilities has produced rapid economic
and population growth, especially in urban centers such as Houston.

There is more to the location of industry,
however, than the availability of capital and
proximity to raw materials, even petrochemical
industries. The two large oil fields extending from
southeastern New Mexico to central and eastern
Kansas, for example, have produced vast quantities of petroleum, natural gas, and capital but
with much less industrial development than along
the Gulf Coast. The difference is largely a matter of accessibility. The Southern Coastlands,
again, are on the continental margin and as such
constitute the line of exchange between water

transport and land transport. Further, because water transportation is cheaper per ton-mile than land transport, even though slower, shipment of the finished products from the Gulf Coast can be accomplished efficiently by ocean carrier to the ports of Megalopolis and by barge via the intracoastal waterway and the Mississippi River system to the manufacturing core.

Just as coastal location can facilitate shipment out, products and raw materials can be moved more efficiently to Gulf Coast industries. The most significant raw material import along this coast is bauxite. This ore, from which aluminium is eventually refined, is originally of low aluminum content. Ideally, the ore is refined near the site of its extraction, but the refining process requires large amounts of electricity. Because most bauxite imported by the United States originates on the energy-poor Caribbean island of Jamaica or the South American counties of Surinam and Guyana, the ore can be shipped efficiently by freighter to Gulf Coast ports. Once there, abundant local fuel permits ore refinement to levels of aluminum content that can be transported economically to various markets across the continent. Only at Arvida in Quebec and Kitimat in British Columbia is imported bauxite ore refined in significant quantities in North America outside of the Gulf Coast.

The distinctive character of the Southern Coastlands, then, is strongly divided. Although the entire region possesses a subtropical environment, the eastern half of the coastlands has been developed largely with emphasis on the advantages of such an environment. The rapid decline in precipitation west of the Sabine River has limited the agricultural possibilities of the Texas Gulf region. Recreation and retirement are possible along the coast west of the Mississippi, but these possibilities have not been exploited as strongly as along the eastern portion of the coast. Mineral exploitation and the consequent industrialization have provided more obvious economic returns in the western portion of the coastlands.

The three major urban centers also illustrate the region's internal contrasts. Houston and the Miami area have both grown rapidly in recent decades, but one is based on manufacturing and trade, the other on recreation, travel and a Latin American commercial connection. Locationally between the two, New Orleans can claim both industrial and recreation importance, although, by not possessing either entirely and because of its constrained site, it has not grown as rapidly. By 1975, the industrial and mineral exploitive character of the western Gulf began to encroach on the eastern Gulf. As petroleum costs remain high, pressure on the recreational character of coastal Florida will also mount. Changes in the eastern half of the Southern Coastlands cannot yet be anticipated. The rate, degree, and kind of change, however, will be only partially in the control of those presently living and vacationing in the region.

Additional Readings

Balseiro, Jose Agustin (ed.). *The Hispanic Presence in Florida.* Miami: E. A. Seeman, 1976.

Blake, Nelson Manfred. *Land Into Water—Water Into Land: A History of Water Management in Florida.* Tallahassee: University Presses of Florida, 1980.

Cortes, Carlos E. (ed.). *The Cuban Experience in the United States.* New York: Arno Press, 1980.

Fernald, Edward A. (ed.). *Atlas of Florida.* Tallahassee: Florida State University Foundation, 1981.

Hilliard, Sam B. (ed.). *Man and Environment in the Lower Mississippi Valley,* vol. 19. Baton Rouge: Louisiana State University, School of Geoscience, 1978.

Lewis, Peirce F. *New Orleans: The Making of an Urban Landscape.* Cambridge: Ballinger, 1976.

Longbrake, David B., and Woodrow W. Nichols, Jr. *Sunshine and Shadows in Metropolitan Miami.* Cambridge: Ballinger, 1976.

Palmer, Martha E., and Marjorie N. Rush. "Houston," in *Contemporary Metropolitan America,* vol. 4. Cambridge: Ballinger, 1976.

THE DEEP NORTH: THE AGRICULTURAL CORE

The interior of North America is vast and nearly flat. Between the Appalachians in the east and the Rocky Mountains to the west, there are relatively few pockets of landscape that vary from the gently rolling swells of this dominant countryside. So large is the area encompassed, however, that other characteristics become the means of subdividing the territory into numerous regions. Low average temperatures, a short growing season, and related environmental factors separate the Northlands (Chapter 17) from the remainder of the interior plains. The distinctive cultures of the Deep South (Chapter 9) and the Southwest Border Area (Chapter 14) clearly set these regions apart from each other and from adjacent portions of the interior plains.

The balance of the interior is characterized by gradual *longitudinal* differences in the amount of annual precipitation and the ways in which the land is used in light of this rainfall transition. The western portion of this territory is a drier region dominated by grassland vegetation. It has a grand openness to it that has led these plains to be called the Great Plains and Prairies (see Chapter 12). The eastern half of the interior is no less a plains, but the more humid growing season, different vegetation, higher *carrying capacity* of the environment, and more numerous natural transportation routes have supported the development here of the continent's agricultural core region (Figure 11-1).

In most directions, the thematic distinctiveness of the agricultural core is not troublesome. The agricultural core does not overlap strongly with Appalachia, with the Northlands, or with the cultural influences of the Deep South. The core's western boundary is more clearly a transition zone, however. Many would argue that the far-flung wheat belts are too significant to both the United States and Canada to be excluded from a region designated the agricultural core of North America. We agree that wheat is tremendously important, but we also consider the two portions of the large agricultural interior too different to treat them usefully as a single region.

FIGURE 11-1

THE AGRICULTURAL CORE REGION.

There remains a possibility of confusion over differences *within* much of the territory occupied by the agricultural core. As we discussed earlier, the agricultural core overlaps territorially with several other thematic regions (Figure 1-5). Most prominent is the eastern portion of the agricultural core, which appears to occupy the same area as the western half of the continent's manufacturing core region. It bears repeating that this is no geographic accident; much of the urban development and manufacturing capacity of the region was stimulated initially by the demands of agricultural producers or the tremendous volume of agricultural output. From

the eastern cities of the region, such as Buffalo, Detroit, and Cincinnati, to the western regional centers of Minneapolis-St. Paul, Omaha, and Kansas City, these marketing and shipment centers gradually adapted their activities to include processing and supplying the needed technological hardware demanded by the region's farmers.

The in-between location of this territory is reflected in the variety of names that have been applied to the region. Most common, perhaps, is the "Middle West." Widely used and probably a permanent fixture among America's geographic place names, this term is basically a carry-

237

over from the days of eastern seaboard dominance. It implies that "the West" begins at the Appalachians with the closer Middle West gradually merging into the Far West somewhere in the general area of the Rocky Mountains. The Great Plains are clearly as much in the continent's east-west "middle," and many students, when asked to identify the Midwest, point to Kansas, Nebraska, and the Dakotas. Only the sequence of European settlement and habit prevent Oregon residents, for example, from referring to the region as the Middle East. In spite of this, the Middle West (or Midwest) will probably continue to be the most widely used and understood name for the U.S. portion of the continent's agricultural core region.

Even though both themes are not expressed across identical territories, the Midwest is clearly a region of farms *and* factories, of a white, Protestant, dispersed, agricultural population *and* a black and immigrant, clustered, urban population. These two aspects of the Midwest are essential to an understanding of the entire region. The urban-manufacturing contribution has been discussed in Chapter 5; the agricultural component deserves equivalent consideration here.

It may be helpful to think of the regionalization of this section of the United States and Canada by picturing each thematic region as a separate plane, or flat surface, lying over the appropriate portion of the earth's surface. The manufacturing plane extends eastward toward Megalopolis and contains islands of industrial clusters west of the main region where regional centers support more local, but still significant, manufacturing activities. The agricultural plane can float at a different level over the same portion of the earth's surface and extend eastward across Ohio and around Lake Erie into southern Ontario. This agricultural plane may be seen to have holes in it where urban and manufacturing functions crowd out almost all direct agricultural production. These holes are largely a matter of scale and definition, for even the most urban counties contain some agricultural production. Cook County, Illinois, for example,

is almost entirely filled by Chicago and its suburbs and was the second most populous county in the United States in 1980, yet it contained over 20 thousand hectares (49,482 acres) of farmland yielding almost $28 million in farm products in 1982. By placing our emphasis on the agricultural functions, however, the region is essentially rural and small town. As such it possesses a distinctive economic and cultural orientation, an orientation that lends strong support to a distinctive regional label—the Deep North.

The Deep North is a culture region based, as strongly as the Deep South, on an accumulated mix of habits, attitudes, and reactions to the traditional opportunities for livelihood and contact with other groups within the region. The term "Deep North" refers to an area of cultural intensity and pervasiveness, but it does not refer to an area that is farther north than other places. Basically, the Deep North is small-town and rural America that is specially flavored with the agricultural patterns of this region. The Deep North should not be confused with the urban America of the Midwest, for in spite of their proximity, they are quite different regions. The population of the Deep North is politically and socially conservative, with an underlying what's-in-it-for-me suspicion, cautiously secure in what is and not strongly exposed to the pressures for change found in major urban centers or in the transition zones between regions. "Middle America" is a popular term applied to the region's population. "Middle" means in between, and, as such, this population shows neither extreme of America's social, economic, or political spectra. Geographically, "middle" also may imply a degree of isolation, or at least buffering, from the forces of change. By being rural and small town on the one hand, and located in the continent's interior on the other, the Deep North possesses many of these features and helps define the character of Middle America.

Consider the contrast between the Deep South (Chapter 9) and the Deep North in terms of their population composition. The South's population had few immigrants from Europe or else-

MAJOR METROPOLITAN AREA POPULATIONS IN THE DEEP NORTH, 1980[a]	
St. Louis, Mo.-Ill.	2,356,460
Minneapolis-St. Paul, Minn.-Wis.	2,113,533
Kansas City, Mo.-Kans.	1,327,106
Louisville, Ky.-Ind.	906,152
Omaha, Neb.-Iowa	569,614
Davenport-Rock Island-Moline, Iowa-Ill.	383,958
Fort Wayne, Ind.	382,961
Peoria, Ill.	365,864
Des Moines, Iowa	338,048
Madison, Wis.	323,545
Lexington-Fayette, Ky.	317,629
Evansville, Ind.-Ky.	309,408
Appleton-Oshkosh, Wis.	291,369

[a] For other large metropolitan areas, see table for North American Manufacturing Core, p. 98.

where after the nation's independence; the white population was almost entirely of English or Scots-Irish descent (with a sprinkle of Germans and the notable exception of the French influence in southern Louisiana). The large black population possessed considerable potency in the formation of Deep South culture. The Deep North population is predominantly white. It had received contributions of foreign-born migrants until late in the nineteenth century. Most of these immigrants originated in northwestern Europe—Germany, the Netherlands, the several Scandinavian countries, and the British Isles. Later immigrants, from eastern Europe and some of the Mediterranean countries, found the better agricultural land already occupied, so they settled in the nearby metropolitan areas of the manufacturing core. The agricultural core region encloses many large cities; but because their urban function is by definition nonrural, the contributions to Deep North culture by the cities' populations have been indirect or through political confrontation with agrarian interests and attitudes at the state level.

The proportion of local population that is black is small in the Deep North. Perhaps more im-

portant, local blacks are nearly "invisible" to many residents of the region. This is not primarily a function of race tolerance by Deep North whites, although that may exist. Instead, it is a result of the combined effects of low black numbers in small town and rural areas, large and increasing clusters of blacks in the major urban centers, and a job-opportunity structure that relegated the region's few blacks to occupations that do not expose them to the rest of the population. During the half-century period of northward migration by southern blacks, labor demand was higher in metropolitan areas and new jobs were limited in the Deep North; few blacks were attracted to settle here. In fact, much of the agricultural core has declined in population since the 1930s. Most new jobs that did become available were filled by local people.

The pattern of Deep North culture, then, is partly reflected in its stability, a resistance to some kinds of change, and a degree of isolation from many of the forces that have created change. In this, there are strong similarities to the Deep South. Also, as in the South, however, recent changes have disrupted this stability, and modifications to the Deep North way of life have become as insistent as elsewhere in North America. The underlying culture in this region is strongly and explicitly tied to the environmental setting. The geographic structure of the Deep North and current alterations in that structure clearly demonstrate this interdependence.

THE ENVIRONMENTAL BASE

In dealing with this region as the agricultural core of the United States and Canada, the physical environment is most important. To some extent, technology has been used as a modifier of the natural environment in many human activities. Urban and industrial activities use technology intensely and are, therefore, less subject to influences of the local environment. Unless their purpose is specifically to tap some aspect of the environment (e.g., recreation centers or port cities or hydroelectric power sites), many of the activities associated with urban location and manufacturing are carried on wherever the nonenvironmental conditions for these activities are met.

In spite of heavy application of technology, agriculture remains intimately associated with the physical environment. Industrial fertilizer, irrigation, and development of new plant varieties have enlarged the areas in which crops can be grown and increased yields attained in old areas. New tools and farm machinery have changed the productivity of individual farmers. Controlled environments, such as greenhouses, and research in *hydroponics* indicate that if we are willing to pay the price, agriculture can be almost as independent of nature as urban activities. The costs, however, are immense, and in North America, it is not yet necessary to bear them outside the agricultural experiment stations.

The natural mix of environmental characteristics in the Deep North is excellent for certain kinds of agricultural pursuits. After a brief period of wheat dominance, settlers of this region selected a variety of corn-based farming systems that changed little for more than a century. The success of these farming patterns contributed greatly to the stability of this region's economic and cultural forms.

The suitability of the environment for corn production does not mean that it is the most desirable climate on the continent. As one long-term observer of this region has stated:

Geographers have long extolled the virtues of the Middle West for growing corn, as well they should have, but for anyone whose aesthetic requirements transcend those of a cornstalk the climate is pretty darned miserable, winter or summer. . . . It's a great climate for making a living, but it's no place to have to live, unless you consider making a living the principal purpose of living.[1]

---

[1] John Fraser Hart, "The Middle West," *Annals of the Association of American Geographers,* 62 (June 1972), p. 267, footnote 14.

Although Hart's comment refers to the region's climate, much the same could be said about the relief and, less strongly, about the soils of the Deep North. Even in terms of its natural transportation resources, the agricultural core is outstandingly blessed. The need to share these waterways with urban and manufacturing activities, however, has greatly altered the aesthetic benefits originally present.

Climate

The early American and European settlers who moved across the Appalachians and onto the eastern interior plains were more concerned with survival and pursuit of a livelihood than a pleasant environment. Except for the resistance of the local population (the Indians) and the usual vagaries of Mother Nature, the environment was supportive. Most of Ohio, Indiana, and lower Michigan were covered by a mixed hardwood forest. The trees encountered were familiar and indicated to the experienced easterner where the best soils were located, and they also provided considerable local fuel and building material. Near the western margin of Indiana and farther into Illinois and southern Wisconsin, the small openings and glades in the forest were larger and more frequent. Except along the rivers and in hillier country, Illinois, Iowa, and parts of southern Minnesota and northern Missouri appeared to be as much open grassland as forest. By the time settlers reached north-central and western Iowa, the dense woodlands had been left miles behind. Only along river courses and in valleys were substantial clusters of trees to be found. The truly barren prairie, essentially treeless, stretched westward from the margin of the agricultural core region to the Rockies.

This gradual east-west transition from heavily forested land to grassland encountered by waves of American settlers was an indication of the average precipitation they could expect. The relationship is not perfect, for soil and groundwater conditions also govern tree growth, even in areas of comparable temperature patterns. Small areas of prairie, for example, were encountered in

Ohio and Kentucky, and the Prairie Wedge, a zone of grassland where trees might be expected on climatic grounds, reached well into central Illinois. But in general, the presence of trees indicates adequate moisture for crop growth. Except for the northwest corner of the region and a few sections of Michigan, Ontario, and eastern Wisconsin, the entire agricultural core receives an average of more than 75 centimeters (30 inches) of precipitation each year. The southern margin of the core can expect in excess of 100 centimeters (40 inches) each year. More important, most of this annual precipitation occurs between the end of April and the beginning of November. Thus, the moisture is received primarily as rainfall during the growing season. Also important to plant growth, the variability of this rainfall over a 10-year period is low. Farmers in the more humid sections of the region are less likely to suffer from devastating droughts than are agriculturalists working in drier sections of the continent. Summer rains do often come in the form of intense thundershowers, occasionally accompanied by damaging hail and high winds, but even in this, the region's farmers are less likely to be economically crippled than those located on the open plains. Rainfall amounts, frequency, and timing, therefore, are all beneficial to the Deep North's economic structure.

Crops harvested each year also need sufficient time to grow to maturity before the first killing frost. As the in-between location of the agricultural core would indicate, the region has neither the long growing season of the South nor the short period of crop growth of the Northlands, the latter too brief for reliable production by nineteenth-century farm technology. The average last killing frost in the spring ranges from mid-April across the southern portions of the agricultural core to mid-May along its northern margins. The first fall frost normally is not expected until late September anywhere in the region and may not arrive for several more weeks in southern Illinois, Indiana, Ohio, and northern Kentucky. The growing season regularly extends over a full four to five months throughout the region, a length sufficient to permit flexibility if late spring rains postpone the customary period for cultivation and planting by a week or two.

In concert with an adequately long frost-free period, growing season temperatures must also be high enough to stimulate rapid crop growth. It is largely a reflection of the agricultural core's interior location that it possesses these temperatures. Like other interior regions, the Deep North has a climate referred to as continental. A continental climate is characterized by a wide range in its temperatures through the year. The coldest winter temperatures at a given latitude are often as low as those occurring much farther north at sites closer to a moderating maritime influence. Similarly, summer temperatures can be expected to climb as high as those found at more southerly latitudes. At Peoria, Illinois, for example, located near the center of the Deep North, the average temperature in January is −4° C (25° F), whereas the average in July is 24° C (76° F). The average annual temperature range, calculated throughout the region as the difference between the mean cold month and the mean warm month temperatures, clearly demonstrates the continental character of this region's climate (Figure 11-2). For the agriculturalist, the high summer temperatures encourage rapid crop growth. As Hart pointed out, the average non-farm resident of the region is less likely to appreciate the miserable combination of hot days, warm nights, and high humidity. Although southerners experience long summers of much the same character, southern winters are milder and refreshingly short when compared to the Deep North's long, often gray, and uncomfortably cold winters.

Relief
Just as the agricultural core's climactic mix is highly appropriate for farming—neither too hot nor too cold, neither too wet nor too dry, and with an adequate frost-free period—the region's topographic relief is properly moderate. The

FIGURE 11-2

AVERAGE ANNUAL TEMPERATURE RANGE. The effects of interior continental location and the modest moderating influence of the Great Lakes are both
evident on this map.

landscape is gently rolling with few areas of either
very flat or very hilly terrain. The low relief has
meant that a large proportion of the total area
could be used for cultivation and that fields could
be as large as practical for good management
without a high risk of erosion. As farm machinery was developed, it could be used throughout
the region. The scattered hillocks and stream
courses that broke up the unending land swells
were an obvious location to maintain a woodlot
or pasture. The persistently rolling landscape also
permitted good soil drainage and, in most cases,
restricted swamps to small areas. Where land was
especially flat, as in the Great Black Swamp region of northwestern Ohio, or especially hilly, as
in the plateau footlands of the Appalachians and
Ozarks, settlement was postponed until the
abundant "easier" land was taken. Like the climate of the agricultural core, then, the landscape relief was intermediate and highly suitable
for good farming.

The highly practical, though unimpressive,
landscape that dominates the Deep North is
largely a consequence of the same glaciation that
created the harbors of Megalopolis (Chapter 4),
and the mechanisms of landscape formation were
the same. As the heavy ice mass spread outward
from its Canadian Shield center, soft sedimentary hilltops were ground down by its weight and
movement. The debris, or glacial till, removed
in this way became incorporated in the ice sheet,
gradually settling out and partially filling the
valleys between the decapitated hills. As the glacier fronts later retreated, long low hills of this
debris, the glaciers' terminal moraines, remained to offer a few lines of slightly greater
relief for the human populations that followed.
The tremendous quantity of meltwater released
by the glacial retreat eroded several major river
outlets, such as the Illinois River west and south
of Lake Michigan and the Mohawk-Hudson river
valleys east and south of Lake Ontario. The

Almost all of the land in the Deep North is used—crops include corn, soybeans, wheat, and hay as well as pasture. Even the small woodlots are part of the transformed landscape. Note as well the scattered farmsteads and the gently rolling terrain. (Erwin W. Cole/U.S. Soil Conservation Service.)

higher surface altitude of the Great Lakes during this period submerged large areas of what is now dry land south of Chicago, south of Saginaw Bay in Michigan, and the Great Black Swamp lake plain extending from Toledo, Ohio to Fort Wayne, Indiana. Much of the land in these areas, especially the last, is a *lacustrine plain* (former lake bed). It is extremely flat and not far above the local water table. Although the soils there are deep and rich, settlement was forestalled until effective ground drainage could be developed.

Immediately south of the Ohio River in north-central Kentucky is a large basin many would argue does not belong within the continent's agricultural core. Beyond the limit of glaciation, the Bluegrass Basin, or Bluegrass Plains, nevertheless does extend the region of low relief and highly productive agriculture into the margins of the Appalachian Plateau. The low, rolling relief of this basin is primarily a residual *karst* terrain, which is one that has developed over thick limestone bedrock. The limestone is gradually

Although some of the horse farms in the Bluegrass Basin of Kentucky are owned by "new money" rather than "old," the life-style emulated by many in this region presents the strong appearance of one image of the old Southern aristocracy in modern trappings. (Josephus Daniels/Rapho—Photo Researchers.)

soluble in moving water and permits many major surface features to be worn away. The limestone is also dissolved underground and forms stalactite- and stalagmite-columned caves that can extend for miles. The Mammoth Cave complex southwest of the basin is probably the best known in this region. Small solution holes (sink holes) are often visible at the surface (see Chapter 10), but the countryside has the same prevalent rolling appearance found throughout the glaciated areas.

Soils

The soils of the agricultural core can be described in much the same qualitative terms as the region's climate and relief: they are good, often much better than average, but usually not excellent. As discussed in Chapter 2, soil research has distinguished hundreds of soil types in North America, each varying in terms of their sand-clay-silt content, their water-holding capacity, their acidity, and numerous other variables. By taking a broad view, soils of the agricultural core have been grouped into two basic

types. With the major exception of central Illinois and south-central Wisconsin, soils east of central Iowa are alfisols—almost all of which are in the udalf subcategory (see Figure 2-5). Western soils within the region and those through much of Illinois are mollisols.

Alfisols are formed under conditions of moderate moisture and usually in association with coniferous or mixed forests. Although the thin surface soil horizon is deficient in humus, the B-horizon has not been heavily leached of agriculturally important minerals. In general, the udalfs found throughout the eastern Deep North only require careful plowing (to avoid erosion), some form of crop rotation, and application of agricultural lime (to neutralize the mild soil acidity). With these relatively modest practices, the land remains productive.

Mollisols are formed under different conditions and are the most fertile major soil group in the region. These soils were formed under grasses rather than under forest cover. They range in color from dark brown to almost black, indicating a high organic content. They also tend to be rather deep, with the surface horizon between 50 and 150 centimeters (about 1.5 to 5 feet) thick. The high organic content is a result of the grasslands' dense, tangled root network. Contribution of dead roots and grass leaves to the matted soil surface keeps organic content high. Mollisols, among the most fertile of all soils, are naturally suited to grain production.

The major soil exceptions to these two broad categories are the alluvial soils found within the main river valleys and former lake beds, and the swamp soils. Both types of soil are capable of high fertility but often require special treatment. As mentioned earlier, settlement of the Great Black Swamp in northwestern Ohio occurred about two decades after land around its periphery. Today, the area is intensively cultivated and highly productive, but deep drainage ditches around the fields' rims continue to indicate the original soil moisture conditions. Glaciation strongly altered prior natural drainage patterns and led to a jumbled landscape of low relief and poor drainage. Many small swamps, lakes, ponds, and bogs are scattered throughout the region and are especially characteristic of Wisconsin and Minnesota. In most cases, they have not been drained because of the expense involved and the more than adequate quality of the remainder of the region's soils.

Natural Accessibility Network

The waterway pattern in the agricultural core region is excellent. In this factor, the natural environmental conditions provided settlers with highly beneficial transportation opportunities. The natural network of accessibility was supplemented by land connections, some of them linking the region directly with the eastern seaboard population centers. Even prior to the railroads and further development of the road network, however, the river and lake connections within the region permitted easy and inexpensive shipment of goods to these same markets as well as to the main international trade ports. Much of this has been discussed earlier in terms of the continent's manufacturing core region (Chapter 5), but it bears further emphasis because river and lake shipping were quite important to the region's agricultural development.

Movement of settlers into the region was earliest along the larger waterways (see Figures 3-1 and 3-2). The southern Great Lakes; the Ohio River; the Illinois, Wabash, and Wisconsin rivers to the east of the Mississippi; and the Missouri River westward to Kansas City all provided major routes of entry for settlers and major routes for marketing their produce. The eastern Great Lakes offered more direct shipment through the Mohawk-Hudson routeway to New York. The entire river network funneled into the Mississippi system and, because of the low relief and regular rainfall, was navigable by small boats and barges with few interruptions throughout.

Detroit grew as a military control point and focus for farm products. This city, whose name literally means "the narrows" in French, is located at the best crossing point between Ontario and Michigan and is also near the entry of the northern lakes into Lake Erie. The southern Michigan hinterland was not as rich as that of northern Ohio, however, and Cleveland remained more populous until after 1910 when mass-produced automobile industries transformed the Motor City's economic structure. Located at the Great Bend of the Ohio River, Cincinnati became the main collecting and shipping center for agricultural products from the southeastern portions of the agricultural core as early as 1820 and never relinquished its dominance over that section of the river's trade. Kansas City, whose urban area is divided politically between two states at the junction of the Kansas and Missouri rivers, also experienced early urban growth by handling large quantities of agricultural products in river transit. As discussed earlier, Chicago's location was also beneficial to rapid growth. Near the southernmost end of Lake Michigan, it also lies only a short distance from the upper Illinois River. The transshipment opportunities provided by this site were supplemented by extensive canalization projects and by land connections built west and south across the rich agricultural core and, later, eastward directly to the growing cities of Megalopolis. The opportunities for industrial growth in these centers are clear. Their initial growth, however, and the growth of many other urban places in the region was a direct consequence of the high-quality agricultural landscape and the natural network of accessibility provided by the region's waterway system.

THE AGRICULTURAL RESPONSE

As the settlement frontier gradually moved westward across the region, it was accompanied by a wave of wheat production for eastern markets. Wheat was a high-value crop, one that was in demand in the population centers and one that was familiar to both producer and consumer. The bulk of raw wheat was not a great problem for shippers as long as water transport was continuous, but flour milling soon became established at the points of embarkation (such as Cincinnati on the Ohio River) or sites at which the grain was to change from one transport mode to another (such as Buffalo at the lake terminus of the Erie Canal). Continuous wheat farming was hard on the region's soils, however, and primary zones of production moved westward with the expanding line of settlement.

For farmers who remained behind, as wheat production moved on, the next best agricultural product was meat from domestic livestock. Both cattle and hogs were raised, with the importance of one over the other varying from year to year. Hogs grow to market size rapidly if supplied with sufficient feed; even grain-fed cattle reach a "finished" stage more readily than animals raised on pasturage alone. The intensive hog-raising activities of southern Ohio farmers earned Cincinnati the nickname of Porkopolis by the 1830s. So economically reliable was feed grain and livestock farming that it quickly supplanted wheat production as the dominant farming system throughout the agricultural core. The system was also well suited to long-term operation within the limits of the natural environment, and it could take advantage of the natural waterway network. Alternative farming systems were occasionally tried, but none was as persistently successful as feed grain and livestock farming.

The grain that best met the combination of environmental requirements and high economic return was corn. Well adapted to a humid summer climate, corn thrived during the region's long hot days and warm nights. Because even corn could not be grown year after year on the same soil, a three-year crop rotation system became common as early as 1820 in southwestern Ohio; a crop of small grain (usually wheat or oats) and a hay crop (such as clover or alfalfa) were alternated with corn in each field over the rotation period. If a fourth year was needed in the system, the land could be allowed to rest in *fallow* under one of the recuperative grasses and perhaps pastured lightly. Under this system, the land is always contributing in some direct way to livestock production or cash-crop production.

Given that corn is ecologically suited to the region's climate and soils, it has additional economic benefits over other feed grains. Grain yields are high because plants can be grown close together, and each plant produces two or more ears of grain. Furthermore, the large quantity of vegetative matter produced by each plant can also be used as feed with appropriate supplements and cutting. Stalks and leaves were chopped finely and stored in large silos to permit fermentation and provide winter feed for the farm's animals. No other plant yielded such quantities of usable feed, and its dominance spread to the limits of environmental support and beyond. It may sound strange for a crop to be grown beyond its environmental limits, but along the northern margins of the Corn Belt, as the region came to be called, where summer temperatures do not permit full ripening of the grain, the plant is cut green and chopped for *silage*.

Throughout the continent's agricultural core, then, a mixed farming operation of crop/livestock production has provided farmers with economic security beyond that found in any other major

Farming in the Deep North, as here in Illinois, requires careful application of new methods with traditional crops. The alternative strips of corn and alfalfa reduce soil erosion while maintaining high productivity. Note also the new silos and machinery sheds. (Tim McCabe/U.S.D.A.)

agricultural region. Localized specialty-crop production elsewhere in the United States and Canada can yield higher economic returns, but only in the agricultural core are relatively high returns consistently achieved by so many farmers. Farming by this system is demanding, for three to four crops plus livestock must be handled correctly throughout the year. It has been suggested that the Deep North respect for, and faith in, hard work can be traced to the rewards of Corn Belt farming that were pretty well assured as long as the system's rigorous organizational and labor requirements were followed.

Although corn has until recently been the most important crop grown in the agricultural core, whether measured by value or by volume, its distribution of production is not even (see map appendix). Corn is grown in substantial quantities from central Ohio to central Nebraska. Its southern margin of production follows the irregular relief of the Appalachian foothills in Kentucky and southern Indiana and the Ozark highlands in southern Missouri. Although it could be grown farther south on the rich alluvial soils of the Mississippi River valley, it is replaced there by the higher returns of cotton, rice, and (more recently) soybeans (Chapter 9). The northern margins of the Corn Belt extend from the southern Ontario Peninsula across south-central Michigan and Wisconsin to the subhumid plains of southwestern Minnesota. The areas of most intensive corn production remain the fertile mollisol regions of Illinois and Iowa. When comparing the distributions of cash-grain and livestock farming in the region, those areas most suited to crop production appear to have fewer farms strictly engaged in raising livestock. Even in this, however, both specializations occur near one another; the livestock farms are the prime consumers for the specialized corn farmers.

A distinctive characteristic of the central Deep North landscape is a semiregular rectangular field pattern. Although not as rigid as that found across the Great Plains states, the rural geometry of the region is stubborn in its consistency.

The original 13 states had developed their internal administrative and ownership boundaries in an unsystematic manner, indicating that settlement preceded governmental organization. The *metes and bounds* system of lot designation relied on visible landscape features, compass directions, and linear measurement to define land parcels. The irregularly shaped results were often subject to confused interpretation and litigation. Through the Ordinance of 1785, the land north of the Ohio River and west of Pennsylvania,

The square township and range land survey system is portrayed dramatically in this vertical aerial photo. As roads were constructed along the edges of many of the one square mile sections, the regular survey pattern was reinforced and remains little affected by transfers of land ownership. (Aerial Photography Field Office/U.S.D.A.)

known as the Northwest Territory, was delimited according to the regular rectangular *township and range* survey before it was opened to settlement by the newly independent American farmer. A series of east-west baselines and north-south meridian lines were drawn, and township squares 6 miles (9.6 kilometers) on each side were laid out with respect to these lines. Each township was further divided into 36 sections of 1 square mile (2.59 square kilometers) each. By 1804, the 640 acres (259 hectares) in each square mile was subdivided into quarters of 160 acres (64.75 hectares). These quarter-sections were designated the minimum area that could be purchased for settlement (at $1.24 per acre/$0.50 cents per hectare), but the minimum was later reduced to 80 acres (32 hectares) and, still later, to 40 acres (16 hectares). The irresistible logic of this system remains visible in the predominantly rectangular road network of most of the United States between the Appalachians and the Rockies and in Hollywood's dramatic references to "trouble on the south forty," although the latter is usually on the Great Plains rather than in the Deep North.

Although the land survey system and the ecological and economic realities of the Deep North produced a kind of inevitable homogeneity to the landscape, there are portions of the agricultural core that lie beyond the Corn Belt. In Wisconsin and central Minnesota, north of the centers of grain production where the climate prevented feed grain maturation, farmers chose dairy farming as an economic substitute. Corn in silage as well as other grains, such as oats and barley, and abundant hay crops provided excellent support for large dairy herds. When the supply of fresh milk exceeded even the large demand of the nearby cities, it was converted into butter and cheese for more leisurely shipment to more distant markets. This pattern of farm production was not at all unfamiliar to the German and Scandinavian immigrants who settled the area. A similar belt of dairy farming is located on the Ontario Peninsula north of the corn- and mixed-farming region. Within the United States, Wisconsin continues to produce a large proportion of the nation's surplus milk and approximately half of the country's entire cheese output (see map appendix).

Another distinctive extension of the agricultural core's boundaries beyond the Corn Belt occurs around the western Great Lakes. As in the fruit-growing region of the Niagara Peninsula (Chapter 6), fruit production is possible far to the north of its inland latitude limits in a narrow band along the Lake Michigan shores of Wisconsin and especially Michigan. The moderating influence of the lake retards fruit-tree blossoming in the spring, usually until after the last frost, and it also retards the arrival of the first killing frost in the fall. Cherry, apple, and (to a lesser extent) grape production are all important. A similar effect is found along the southern shore of Lake Erie, especially the few coastal counties in Pennsylvania and New York, where grape production has been significant for more than a century.

CHANGES IN THE PATTERNS

The entire region was pretty well settled by 1890, and the mixed-farming system based on corn and livestock that had been found to work well in southern Ohio was carried west to the edge of the Great Plains with only local adjustments. Early technological improvements, such as the reaper (1831), the steel plow (1837), and other devices suited to the region's chief economic activity, tended to ensure the system's success. More recent changes, however, have stimulated modifications to the traditional geographic patterns.

Soybean Substitution

One of the more subtle changes in Corn Belt patterns lies in the rise of one crop's importance since World War II. During this period, soybeans have been grown in increasing acreage throughout America. As late as 1925, less than 200,000 hectares (500,000 acres) of soybeans were harvested in the United States. By 1949, soybean acreage had increased to about 4.5 million hectares (11 million acres), with almost 3.7 million hectares (9 million acres) of those in the agricultural core and most of the balance immediately south in the Mississippi River valley (Figure 11-3). During the next 20 years, soybean acreage exploded to 16.1 million hectares (41 million acres); plantings in the agricultural core exceeded 10 million hectares (25 million acres) and the Mississippi River valley increased to nearly 4 million hectares (10 million acres). Nationwide acreage had nearly doubled again by 1979. Market demand for soybeans remained greater than supply throughout the period of acreage growth, and soybeans began to be described, perhaps too enthusiastically, as the mircale food of the future.

The reasons for the tremendous increase in soybean production are several. First, as a *legume,* soybeans act as a soil reconditioner by increasing the nitrogen content of the soil in which they grow. Second, the environmental requirements for soybean production are rather broad. Although vulnerable to some local conditions or climatic variations, soybeans generally may be grown throughout most of the eastern United States, parts of Ontario, and even in areas receiving less than 50 centimeters (20 inches) of rainfall if irrigation is feasible. Third, soybeans can be used in a variety of ways. In this, they are similar to corn. The bean itself can be eaten directly or milled to produce an edible vegetable oil and meal low in fat but high in protein. The meal has been used primarily as a livestock feed supplement, but an increasing amount has entered human consumption patterns in place of some grains, as milk substitutes and as meat helpers. And fourth, the world food and feed situation has maintained export demand for soybeans at high levels as well. This has kept prices relatively stable, an encouraging condition to farmers. Only increased production in other parts of the world, such as recent expansion in Brazil, is likely to temper soybean's tremendous growth in the United States.

This combination of advantages has concentrated a great amount of soybean production in

FIGURE 11-3

SOYBEAN ACREAGE, 1949 AND 1978. The early decades of the expansion of soybean production are clearly evident within what, traditionally, had been a farming region dominated by corn.

the agricultural core. The traditional three- and four-year rotations are gradually giving way to a two-year corn-soybean rotation. In some cases within southern portions of the core, early maturing varieties of soybeans can be planted in the late spring after a harvest of winter wheat, giving the farmer three crops (corn, wheat, soybeans) every two years without significant loss of productivity in any year. The large, almost level fields of the agricultural core that permitted early mechanization of farming operations are well suited to soybean cultivation. Much of the same equipment can be used and the high returns per work-hour contributed to the early acceptance of the crop even before the market developed as intensely as it has. By 1971, harvested acreage of soybeans exceeded that of corn in southern Illinois and northern Missouri and was greater than two-thirds of the corn acreage in western Ohio and the Grand Prairie region of east-central Illinois. In 1979, for the first time ever in the United States, more acreage was planted in soybeans than in corn within the traditional corn-producing sections of the agricultural core. Once this is recognized more broadly outside the agricultural community, the region may be referred to as the Corn-Soybean Belt.

Mechanization and Farm Size

A more complex set of changes in agricultural core geography is built on new levels of mechanization and alterations in average farm size. As discussed earlier, the land survey in the region set the minimum farm size that could be purchased at 160 acres (64.75 hectares), and then at half or one-quarter that amount at various times.

After the initial purchase, of course, parcels of land could be broken up and sold in even smaller lots or added to previously established farms. Hart determined that the amount of land in various farm-size categories remained stable during the first third of this century, and he argued that this stability was a continuation of farm-size patterns set during the original purchase period.[2]

Whatever the case, by 1900, farm size in Deep North states showed marked variation: about one-third of the farms were of 73 to 202 hectares (180 to 499 acres), another third were of 40.1 to 72 hectares (100 to 179 acres), and most of the remainder were smaller than 40 hectares (99 acres) (Table 11-1). The amount of land in farms smaller than 73 hectares (180 acres) began to decline after 1935; the greatest increases occurred in farms larger than 105 hectares (260 acres). By 1964, more than 50 percent of the farmland in these states was in farms larger than 105 hectares (260 acres); fully 1 hectare in 5 was located on a farm larger than 202 hectares (500 acres) in area, a trend that has continued. At the same time that large farms were increasingly important in the region, small farms were disappearing rapidly. Less than 10 percent of the states'

---

[2] Hart, op. cit., p. 273.

TABLE 11-1
Proportion of Farmland in Farms by Size Category in Eight Agricultural Core States

Size Category (Acres)	1900 (Percent)	1935 (Percent)	1964 (Percent)	1974 (Percent)	1978 (Percent)
0–49	6	4	2	2	2
50–99	18	15	7	5	5
100–179	34	35	19	14	12
180–259	36	21	19	14	12
260–499		19	33	31	30
500 or more	6	6	20	34	40

SOURCE: Calculated from John Fraser Hart, "The Middle West," *Annals of the Association of American Geographers*, 62 (June 1972), Figure 16, p. 273; U.S. Department of Commerce, *1974 Census of Agriculture*. Washington, D.C.: U.S. Government Printing Office, 1978; and U.S. Department of Commerce, *1978 Census of Agriculture*. Washington, D.C.: U.S. Government Printing Office, 1981.

farmland remained in farms smaller than 40 hectares (100 acres) by 1964, and only 7 percent was in farms smaller than this by 1974. The largest farms were located in a wide band across central Illinois and northern Missouri and generally along the western margins of the region. Smaller farms were more eastern in location—primarily in Ohio, southern Michigan, and eastern Indiana and Wisconsin.

The primary reasons for these changes in farm size are economic and related to mechanization of operations. Deep North farmers have traditionally taken advantage of mechanical innovations to increase their output per work-hour. The large fields and gentle terrain in this region permitted early and continuing use of farm machinery that would have been impossible on smaller farms and erosion-prone hill farms. Individual farmers were able to operate over areas much larger than would have been possible if more hand labor had been required.

After about 1935, however, several changes occurred in the degree of mechanization. These changes had significant consequences for the form of farm operation in the region and eventually for general farming throughout the United States and Canada. The worst of the Depression years were followed by a variety of government-backed rural development schemes: rural credit was loosened, price support and cropping controls were legislated, and public services were extended into the countryside. Tractor-pow-

Throughout much of the region, soybean production is alternated with corn and, in some cases, has replaced the grain entirely. The fields remain large, however, as does the machinery used for cultivation and harvesting. (Courtesy International Harvester.)

ered machinery became more common on all but the smallest farms, many of which were already marginal.

This application of technology to the production process is no different than that which occurs in manufacturing and the results have been the same. Economies of scale favored medium and large farms during the early years of farm-size growth. A labor shortage during the early 1940s accelerated the mechanization process, and innovations became oriented increasingly toward large-scale operations. Two- and four-row equipment gave way to six- and eight-row equipment. Storage and shipment operations also became mechanized and more and more attuned to the requirements of large-volume producers. At each stage of improvement, the main benefits could be achieved by the largest producers with fewer benefits associated with medium-size operations. The squeeze on small-scale farmers was completed from the other end as well because larger operators could afford the equipment that would give them the greatest cost efficiency. One consequence was the gradual accumulation of farmland into acreages that could effectively use the new large machinery. Even here, another size advantage appears. As farm operations have become larger and more expensive, the demand for land to permit these operations has also increased. Land prices have increased, and the larger farmers are those best able to obtain credit or afford additions whenever land becomes available for purchase.

Accompanying these changes in farm size, there has also been a gradual decline in the amount of land farmed in the region. Not all land formerly owned by small-scale farmers is aggregated into larger farm operations. The proportion of county land in farms (Figure 11-4) was above 80 percent across much of the region in 1978, with most of Iowa and Illinois still showing rates above 90 percent. Even so, the overwhelming number of counties in the Deep North had experienced a reduction in land in farms across the previous two decades (Figure 11-5). Only in central Illinois and portions of Iowa

within the region and in the Red River valley in northern Minnesota were there significant clusters of counties showing growth in farm area since 1959. Granted, it is unlikely that counties already having more than 90 percent of their land in farms in 1959 could increase that proportion, but the general pattern was one of moderate decline in farmland. The greatest reduction in proportion of land in farms occurred around the margins of the region we defined as the Deep North. This supports the argument that a major factor in the overall decline is a consolidation of successful efforts onto the better land.

The Family Farm and Ownership Changes

Deep in American folklore—and to some extent in Canadian folklore—is the almost sacred notion of the family farm. Individually owned and family operated, the farm was passed down from generation to generation with only a succession of hired hands included in the farm operations. A gradual increase in nonfarm population did nothing to disturb this image, for was it not the family farm that embodied much of the independence and proximity to nature that was missed by the large numbers of urbanites? Although they knew this image to be exaggerated in some ways and downright untrue in others, farmers of the Deep North fit the pattern until World War II as well as did those in any major region. During the four decades since the war, it has become clear that the fully owned, individually operated family farm is in rapid decline. Although still dominant in the United States and Canada, it is disappearing most rapidly from the agricultural core states. The relatively self-sufficient dairy farms in Wisconsin—frequently with an orchard or two, a vegetable patch, a few hogs and chickens to complement the cows, and even a rollicking mixture of children and collie dogs—probably approach the nostalgic image better than farms in most areas. If the current trend continues, however, only a small proportion of the significantly productive farms in the Deep North will fit the cultural image by the end of the 1980s.

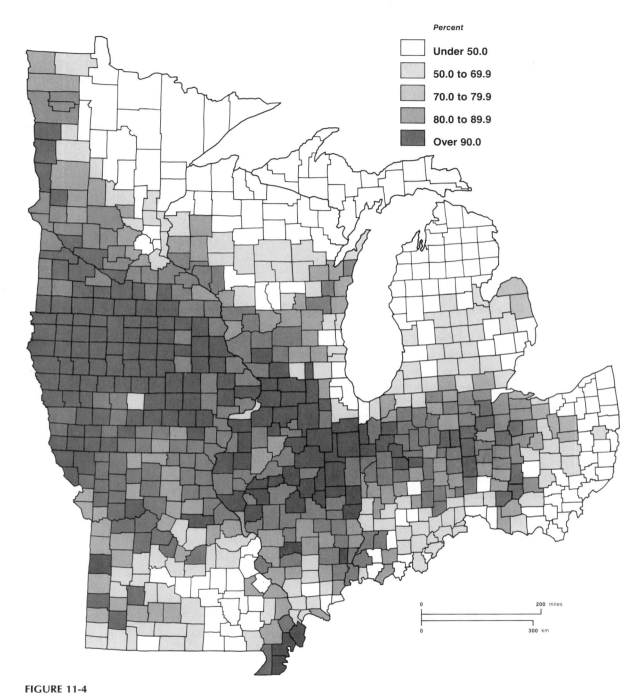

FIGURE 11-4
PERCENTAGE OF LAND IN FARMS, 1978. Throughout the core of the Deep North, more than 80 percent of each county is in farmland. Only in urban counties and toward the periphery of the region does this proportion fall below 50 percent.

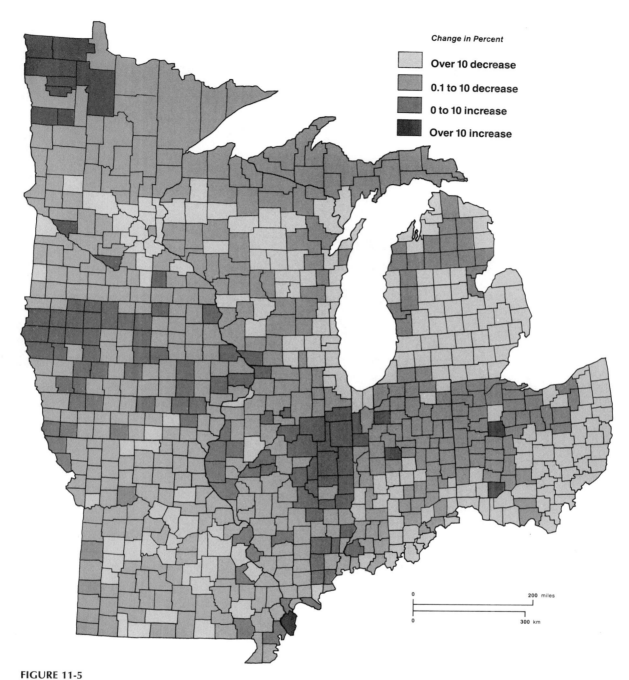

Change in Percent

◻ Over 10 decrease

◻ 0.1 to 10 decrease

◼ 0 to 10 increase

◼ Over 10 increase

FIGURE 11-5

CHANGE IN PERCENTAGE OF LAND IN FARMS, 1959–1978. The greatest growth in farmland occurred in central Illinois and central Iowa, but the largest zones of decline in farmland can be identified around the edges of the most intensively farmed portion of the region.

255

Dairy farms in the Deep North, such as this one in Ohio, provide a bucolic landscape that contains all of the elements of the region's culture. (Erwin W. Cole/U.S. Soil Conservation Service.)

The decline of the family farm is directly associated with the demands of increased farm efficiency. When it was defined by hard work, alertness to detail, and rigid adherence to traditional methods that had proven themselves reliable, efficiency was controlled directly by the person involved. As the cost advantages of large-scale farming became much more important, the ability of the farmer to use large equipment and to use it on larger farms shifted the bases on which farm efficiency was established. Personal effort and individual integrity still contribute to a farm's success, but the factor of scale is increasingly critical. It is difficult to anticipate the long-term impact of this change on Deep North culture.

There are several ways in which Deep North farmers and those in other sections of the United States and Canada have already adapted to these changes. As the need for more land per operation increased, some farmers have found it feasible to rent or lease additional land rather than purchase it outright. The rented land can be part of a nearby farm or several sections of farmland in scattered locations and still be accessible by the well-developed network of paved rural roads (Figure 11-6). The rented land may continue to appear in the census as a separate small farm operated by what is called a part owner, thus masking the intensity of small operation decline. Other farm operators may be full tenants, choosing to work for the landowner through one of several arrangements. As a U.S. Department of Agriculture study pointed out

FIGURE 11-6

ASSOCIATED PARCELS ON FRAGMENTED FARMS (reproduced by permission from the *Annals of the Association of American Geographers,* Volume 65, 1975, E. G. Smith, Jr.). This example demonstrates clearly some of the scattered, non-contiguous land parcels owned and worked by individual farmers. A variety of trends are combining to suggest that this pattern may increase.

. . . farmland investment by nonoperator landlords is similar to stockholder investment in corporations. However, the nonoperator landlord is not synonymous with the "Wall Street" type of investor. Evidence suggests the majority of these landlords are either retired farmers, members of farm families, or closely associated with agriculture through the small rural communities in which they live.[3]

In addition, about one-third of those renting farmland were leasing it from a relative, often as a means of transferring the land from one generation to another.

Within the agricultural core, tenure characteristics reflect the new economic reality. Of the eight Deep North states, the primarily dairy state of Wisconsin had the greatest proportion of its farmland in the hands of full owners in 1978 (Table 11-2). More than two acres in every three in Illinois, Indiana, and Iowa were operated under part-time or tenant-farming arrangements, a proportion greater than that found in any other

---

[3] Bruce B. Johnson, *Farmland Tenure Patterns in the United States.* Washington, D.C.: U.S. Department of Agriculture, Economic Research Service, 1974, p. 4.

TABLE 11-2
Tenure Characteristics of Farms in Agricultural Core States, 1978

	Land in Farms			Average Farm Size			Market Value per Farm		
State	Full Owner	Part Owner (percent)	Tenant	Full Owner	Part Owner (acres)	Tenant	Full Owner	Part Owner ($1000)	Tenant
Ohio	37.2	50.0	12.7	103	308	171	16.7	59.1	30.2
Indiana	31.0	56.3	12.7	103	362	203	20.0	72.9	39.3
Illinois	21.5	56.2	22.3	133	437	280	26.6	86.3	57.4
Iowa	29.0	50.1	20.9	165	425	250	41.8	105.2	56.5
Missouri	46.7	43.6	9.7	176	475	251	17.4	55.0	31.9
Michigan	39.4	55.1	5.5	104	305	142	15.4	56.4	24.5
Wisconsin	47.5	45.8	6.8	154	304	179	26.0	63.9	39.7
Minnesota	37.9	51.6	10.5	188	450	241	30.0	70.8	39.0

SOURCE: U.S. Department of Commerce, *1978 Census of Agriculture.* Washington, D.C.: U.S. Government Printing Office, 1981.

state. Average farm-size differences also illustrate the tenure patterns; farm-size averages are much lower for full-owner operatives in every state. This is repeated again in the distribution of farm-product market values per farm. The large accumulation of invested capital through part-ownership agreements has raised the average market value of owner-renter farm combinations to more than twice that of single-owner farms, even in Wisconsin. The ability of part owners to use fully the largest equipment over large total farm acreages justifies the additional tens of thousands of dollars required by such large-scale farming.

SETTLEMENT PATTERNS

The region's visible landscape contains more than large rectangular farm fields, livestock pens, and a network of straight roads. Houses dot the countryside in a distinctive pattern. Small market centers and service communities are located farther apart but still at regular distances from each other. The internal form of many of these small towns reflects the immensely practical survey system used across the rural landscape. The farm population living in the region has been declining, as elsewhere in the United States and Canada (2.6 percent of the total U.S. population in 1981 was classified as in farming, down from 8.7 percent in 1960), but a growing rural nonfarm commuter population has come out from the cities to replace those who have left the countryside. The settlement pattern of the Deep North is undergoing a process of change that promises to affect its traditional geography as strongly as increased mechanization has been altering the organization of farming in the region.

Spatial Reorganization and Transport Technology

The gradually declining farm population and the increasingly sophisticated technology at the disposal of the average rural resident have had a consistent and significant impact on village and small-town patterns throughout the United States and Canada. The consequences have been especially persistent in the agricultural core region. There has been an increase recently in the population of rural and small-town America, but the pattern for the Deep North is much less clear. Both trends, the national rural population decline (into the 1950s) and the reversal to rural growth everywhere except in the agricultural core itself since that time, are related to changes in transport technology.

During the 1930s and 1940s, rural roads were paved and individual farmers soon could afford a reliable truck to carry livestock or farm produce to market. Transport costs declined as transport efficiency increased. Similarly, the improved road network permitted more rapid movement; the farmer could carry his own produce farther than he had been able to several decades earlier, and with no increase in time or money. As a consequence, farm produce began to be trucked past the local village to one of the larger market centers or livestock yards. Deprived of at least a portion of its consumer population, the smallest communities did not grow in population and, in many cases, began to lose what small population they had.

An examination of population change in 15 rural counties in northern Ohio and Indiana between 1910 and 1960 found that there was a direct association between village size and population increase; the larger the town, the greater had been the absolute population increase. For the very smallest communities, those with 1910 populations below about 1000, actual declines in population had often occurred.

In terms of the settlement pattern in the Deep North, therefore, (1) the rural landscape was becoming less densely populated and (2) the population was increasingly clustered. Restating this in terms of time and distance, individuals of the 1960s were traveling farther but at no increase in travel time. The utility of the small rural hamlet had declined because it no longer served much direct economic purpose.

The relationship between travel time, distance, and the process of spatial reorganization was studied by D. G. Janelle in a way that illustrates these changes.[4] He observed that in the past, a demand for accessibility often led to an improvement in transportation, usually through some technological development. Further, the transport innovations often improved access between points by reducing the direct cost of travel or the time required to complete the trip. The innovations may be in either the vehicle of movement (railroad, automobile, truck) or in the routeway. By determining the average travel time between a number of towns and cities in southern Michigan, Janelle found that a trip that required almost 24 hours to complete by land transport in 1840 had been reduced to about 2½ hours by 1900 and to about 80 minutes by 1965. In effect, the *time-distance* between these places had shrunk greatly; they had become much closer together if we measure distance as travel time.

It was also determined, however, that the time-distance between places did not converge at the same rate. The larger the place, the greater the demand for improved access to other locations; small towns and villages do not generate as much interaction with other places as major metropolitan concentrations do. Also, the populations of large cities interact most strongly with other large cities. Therefore, improvements in transportation are greatest between the more populated centers. In terms of time-distance convergence, large cities have grown closer together more rapidly than small towns and villages. As discussed earlier (Chapter 5), there are economic benefits to being located close together. Thus, the benefits of time-space shortening are enjoyed primarily by those places that are already the most dynamic economically; the larger places grew, whereas the smallest communities lost their economic base and declined slowly in population.

A mild contradictory trend has appeared in much of the eastern United States and Canada but may not include the Deep North. There have been indications that since 1970, many rural areas and nonmetropolitan places are no longer losing population and some are actually growing. A mid-decade study of nonmetropolitan cities and towns concluded that there was a good deal of evidence for population growth in all but the smallest villages, even though growth was more

---

[4] D. G. Janelle, "Spatial Reorganization: A Model and Concept," *Annals of the Association of American Geographers,* 59 (June 1969), pp. 348–364.

rapid in metropolitan areas. Furthermore, sharp differences appeared between villages in North Central states and Southern states. On the one hand, "most resurgence of growth in very small places appears to have taken place in the South," whereas on the other hand, "it is essentially in the North Central states, where there are hundreds of very small incorporated places, that the notion of 'dying' small towns comes closest to reality."[5] Thus, although the Deep South and the Deep North are both strongly agrarian, they are quite different regions in this regard as well. Hardships experienced by southern farmers for a century or more may have been a prelude to regional population growth now that many of the former difficulties are slowly dissolving (Chapter 9). In contrast, the efficiency that has made Deep North farming so successful may also be the basis for continued population decline in the region.

An Urban Case Study: Clarksville, Iowa

There are a large number of small towns and villages in the Deep North that could be used to illustrate the characteristics of the region. The village of Clarksville, Iowa, is representative of many agricultural core communities (Figure 11-7).

Located in the north-central part of the state in Butler County, Clarksville has grown steadily but slowly since the beginning of this century. Its population of 895 in 1910 reached 1328 in 1960 and 1424 in 1980. The village is incorporated and has its own post office, bank, and a new high school, but it has remained basically a service and market center for nearby farmers. Even so, changes in the marketing of farm products led to the recent closure of one of Clarksville's railside grain elevators. Other ad-

vances in technology have meant that the weekly newspaper is no longer printed in town; instead, it makes better economic sense to use the more efficient press in Waverly, a somewhat larger community about 10 miles away.

Clarksville is not served by any of the region's main highways. This has kept the village's growth slow, even for the Deep North, and it contributes to the citizens' open curiosity about the presence of any strangers. Individuals arriving in town by road are recognized immediately and expected to be there for a reason; few would drive through town when traveling between major destinations. A recent visit to Clarksville elicited several friendly inquiries, "Who in town are you here to visit?" and clear surprise at the response "No one; just passing by."

Although Clarksville has continued to grow, however slowly, there are indications of a potential for decline in the ratio of female-to-male population and the age structure of that population. There were 757 females in Clarksville in 1980 compared to only 667 males. The median age of Clarksville's population (35.5 years in 1980) was more than 5 years older than the median age for the entire state (30.0 years). And almost one-quarter of the town's population was 65 years old or older in 1980 (the state proportion was only 13.3 percent). It appears that a significant number of the population migrate out following high school—probably in search of greater employment opportunities—and leave behind a somewhat elderly population, a significant proportion of which are single older women.

Clarksville's urban form is also typical of Deep North farm communities. The square and rectangular residential blocks have been laid out around the two main streets. Topography is ignored in the regularity of this pattern. The town's small business district is clustered around the intersection of the two through roads, but neither road is busy enough to require a traffic light at the intersection. Residences retreat from this center at low housing densities. With a small six-block exception to the north, the village boundaries form a neat $2 \times 2\frac{1}{2}$ mile rectangle, al-

[5] Glenn V. Fuguitt and Calvin L. Beale, *Population Change in Nonmetropolitan Cities and Towns* (Agricultural Economic Report No. 323). Washington, D.C.: U.S. Department of Agriculture, Economic Research Service, 1976, pp. 4, 6.

FIGURE 11-7

CLARKSVILLE, IOWA. In addition to the representative features of the Deep North landscape shown on this map, you are encouraged to use the aerial photograph of this same community on page 262 to enhance the information that might be identified by either source alone.

though the political area is far from filled by the village's homes. The square land survey system can also be observed well beyond the village's limits in the rural road pattern. Only the railroads and short sections of the two main through roads deviate from the straight north-south, east-west section boundaries.

Clarksville, Iowa, is representative of thousands of small communities in the Deep North. Stable or slowly growing in population, it has be-

gun to be affected in small ways by changes in technology and the region's economy in spite of its insulation from these changes. A peaceful, attractive, friendly place in which to live, Clarksville offers few of the locational advantages of communities with greater access to local centers of economic activity. It is likely to hold its own to the end of the century, but there is little, aside from its people, that suggests significant changes ahead.

Small towns of the Deep North, such as Clarksville, Iowa, obtain much of their character because they have grown slowly, if at all. The "downtown" remains a few blocks long, the small square residential blocks are occupied by only six or eight homes, and one could easily walk into the surrounding countryside from anywhere in town in less than 10 minutes. (U.S.G.S. High Altitude Aerial Mapping Photography, N.C.I.C. Headquarters.)

Additional Readings

Andrews, Clarence A. (ed.). *Growing Up in Iowa: Reminiscences of 14 Iowa Authors.* Ames: Iowa State University Press, 1978.

Carter, William. *Middle West Country.* Boston: Houghton Mifflin, 1975.

Dillon, Lowell I., and Edward E. Lyon (eds.). *Indiana: Crossroads of America.* Dubuque, Iowa: Kendall/Hunt, 1978.

McAvoy, Thomas T. (ed.). *The Midwest: Myth or Reality?* Notre Dame, Ind.: University of Notre Dame Press, 1961.

Nelson, Ronald E. (ed.). *Illinois: Land and Life in the Prairie State.* Dubuque, Iowa: Kendall/Hunt, 1978.

Platt, Rutherford H. "Farmland Conversion: National Lessons for Iowa." *Professional Geographer,* 33 (1981), pp. 13–121.

Raban, Jonathan. *Old Glory: An American Voyage.* New York: Simon & Schuster, 1981.

Rafferty, Milton D. *Missouri: A Geography.* Boulder: Westview Press, 1983.

CHAPTER TWELVE

THE GREAT PLAINS AND PRAIRIES

In a desert you know what to expect of the climate and plan accordingly. The same is true of the humid regions. Men have been fooled by the semiarid regions because they are sometimes humid, sometimes desert, and sometimes a cross between the two. Yet is is possible to make allowances for this, too, once it is understood.

C. WARREN THORNTHWAITE

Kansas was chosen because it was dull and brown, a striking contrast to the rainbow and the land beyond.

(From a discussion about *The Wizard of Oz:*
H. FRANK BAUM)

The historian Walter Prescott Webb suggested in his book *The Great Plains*[1] that the northwest Europeans who settled much of the United States faced three great "environmental encounters"—areas where climatic conditions were so unlike those of their home region that the agricultural crops and settlement patterns developed in Europe were inappropriate. Before those areas could be settled, new crops, new land-use patterns, and sometimes new technological developments were needed. The first of these "encounters" was with the high summer temperatures and humidity levels of the Southeast. The second problem environment was the arid Southwest and interior West. The third was the great continuous grasslands located astride the center of the country (Figure 12-1).

European settlers encountered several environmental problems on the grasslands of the Great Plains and the prairies of Canada. Average annual precipitation was much less than in the East. Within the grasslands, the western margins were much drier than the eastern area. There was also considerable variation in the annual precipitation, a variation that seemed to come in broad cyclic patterns. Violent storms accompanied by high winds, hail, and tornadoes were common. Blizzards with wintry blasts intensifying the cold drove the snow into immense drifts. The hot, dry winds of summer parched the soil and sometimes carried it away in great billowing clouds of dust.

---

[1] Walter Prescott Webb, *The Great Plains*. Boston: Ginn & Company, 1931.

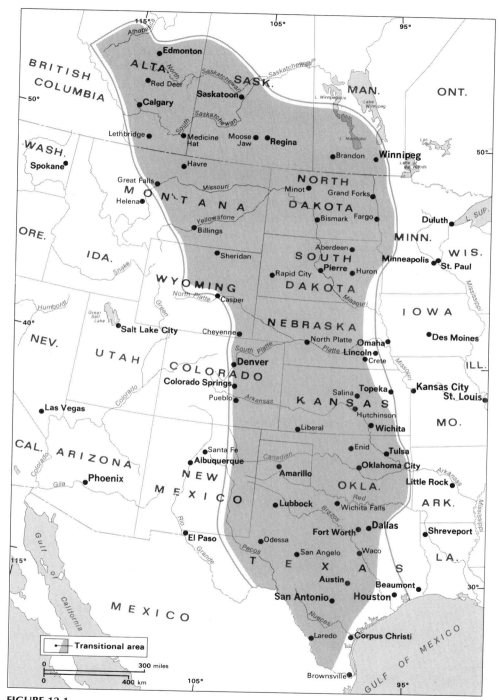

FIGURE 12-1

THE GREAT PLAINS AND PRAIRIES.

The region's sparse natural water supply would not support tree vegetation except along the stream courses. Many of these streams were small and flowed only intermittently. Eastern farmers, accustomed to a plentiful supply of water for crops and animals as well as ample wood for building, fencing, and heating, had to adapt to quite different conditions in their attempts to settle the Great Plains.

PERCEPTIONS OF THE PLAINS

When Coronado, the Spanish explorer, wandered into south-central Kansas in search of the mythical Cibola, the cities of gold, he was sadly disappointed at the rude Indian villages he found instead. In spite of this setabck, he was struck by the richness of the land. "The country itself is the best that I have ever seen for producing all of the crops of Spain," he said, adding that the land was "very flat and black" and "very well watered by the rivulets and springs and rivers."[2] The lack of gold or silver plus the great distances separating the area from the core of Spanish colonial development in central Mexico, however, overcame the attractiveness of the environment, and the Spaniards never occupied any substantial part of the Great Plains.

After the travels of Lewis and Clark from St. Louis to the mouth of the Columbia River and back in 1804 and 1805, additional American exploration was followed by gradual settlement of the Plains. The Americans found a land of seemingly endless, almost flat grasslands. (What schoolchild has not read *Around the World in Eighty Days*, which includes the tale of a covered wagon railcar propelled eastward across the Plains by a sail?) This land, so unlike the wooded lands of the East, was often branded a desert. "Almost wholly unfit for cultivation, and of course, uninhabitable for a people depending upon agriculture for their subsistence" was the reaction of Major Stephen Long in his 1821 expedition across the region.[3] The Oregon Trail, stretching from Kansas City to the Willamette Valley in Oregon, was for many a way of bypassing the unfriendly lands of the Plains for the more familiar demands of the wooded territory in the Pacific Northwest. Some American geographies published during the middle decades of the nineteenth century chose to brand the area "The Great American Desert." For many, that was an acceptable definition.

Coronado and Long were both describing roughly the same part of the continent, the area that we know now as the southern Great Plains. These men differed, however, in their environmental backgrounds, which strongly influenced their perceptions of the area. Coronado was from the dry, basically treeless plateau region of central Spain; he had spent additional time in the equally treeless, dry lands of central and northern Mexico. To him, the Great Plains vegetation, although containing few trees, seemed lush. Compared to much of his experience, the streams were abundant and flowing. Long, a product of the humid, forested eastern United States, had quite a different perspective. To him, the lack of trees was shocking; the streams seemed dry and inadequate. Only gradually did widespread attitudes similar to Long's disappear in the United States; even today, many Americans would still agree with his evaluation.

Such great variation in perception of the quality of the Plains environment is not surprising. Numerous studies of human perception have emphasized the importance of individual experience on perception of an environment. We each evaluate and filter selectively all new information, and previous experience is a major part of that filter. New situations are evaluated within the framework of those past experiences. If the new situation is basically different than those encountered previously, we may not know how to react and that environment will be avoided. Correspondingly, familiar situations require little adjustment and may be readily accepted.

The importance of past experiences on perception can perhaps be best illustrated with a few examples. Students at many universities across the United States have been asked to rank the

---

[2] In Webb, op. cit., p. 107.

[3] In Webb, op. cit., pp. 156–57.

states of the country in terms of their desirability as residential locations. Some states, such as Colorado, Vermont, Washington, and California are usually ranked high. Southern and Great Plains states are normally found at the bottom of the list. Beyond these usual extremes, marked variations occur. Students at the University of Minnesota see little difference between Mississippi and Alabama, ranking them both quite low, whereas students at the University of Alabama perceive a great difference between the two, ranking Alabama as the most preferred state but placing Mississippi well below the average. Conversely, Alabama students identify little difference between North Dakota and Minnesota, whereas those at Minnesota see a great gap separating these two states. When asked to rank the states with respect to the quality of their climates, students at the University of North Carolina ranked states across the South high, with rankings basically declining toward the North. Their peers at the University of Vermont reversed that evaluation, placing their own cooler North at the top of the list.

Along a somewhat different line, a series of studies on the perception of natural hazards have found a direct relationship between the frequency of occurrence of a particular hazard and how accurately the frequency is perceived. In a study of floodplain development, for example, it was found that inhabitants of areas that seldom experienced flooding paid little heed to its likelihood. Those in areas of frequent flooding gave that possibility considerable emphasis in development planning. In a study of drought perception in a series of counties extending from central Kansas into eastern Colorado, researchers found that farmers to the east, where drought was relatively less frequent, consistently underestimated its frequency. To the west that discrepancy declined, until in eastern Colorado perception and reality coincided.

Previous experience, then, is a major contributor to perception. It is not surprising that Spaniards found the Plains amenable; it was similar to the physical environments with which these explorers were familiar. The differ-

ences—perhaps a little more water, a little more grass—were small and positive. For the typical American, the difference was more radical, and for many, such as Major Long, the reaction was strongly negative.

THE PLAINS ENVIRONMENT

The topography and vegetation of the grasslands is among the least varied—some would say most boring—to be found anywhere in Canada or the United States. Early settlers following the Oregon Trail could reach the Pacific Coast in one season of travel, in part, because the grasslands were so easy to cross. The region lies entirely within the interior lowlands physiographic province. The underlying sedimentary beds dip gently. Elevation increases gradually, almost imperceptibly, from east to west. Along the eastern margin the elevation is only 500 meters (1600 feet), whereas in the west, Denver, at an altitude of more than 1500 meters, is the Mile-High City. Physiographically, the largest portion of the Great Plains is the High Plains stretching from the Edwards Plateau of south Texas northward to southern Nebraska. Covered by a thick mantle of sediments that are often quite sandy and extremely porous, this section is generally flat. Some areas, such as the Llano Estacado (Staked Plains) in west Texas and eastern New Mexico, are ex-

MAJOR METROPOLITAN AREA POPULATIONS ON THE GREAT PLAINS AND PRAIRIES, 1980	
Dallas-Ft. Worth, Tex.	2,974,805
Denver-Boulder, Colo.	1,620,902
San Antonio, Tex.	1,071,954
Oklahoma City, Okla.	834,088
Tulsa, Okla.	689,434
Edmonton, Alta.	657,057[a]
Calgary, Alta.	592,743[a]
Winnipeg, Man.	584,842[a]
Austin, Tex.	536,688
Wichita, Kans.	411,313
Colorado Springs, Colo.	317,458

[a] Indicates 1981 population.

The Plains and Prairies landscape is more diverse than commonly imagined.
Badlands, such as these in South Dakota, result from the erosion of sedimentary
rock by water and wind. (South Dakota Division of Tourism.)

tremely flat. Only along streams, such as at Scottsbluff on the Platte River in western Nebraska or at Palo Duro canyon on the Red River in panhandle Texas, has erosion resulted in substantial local relief. The Lake Agassiz Basin, formerly occupied by the largest of the Pleistocene lakes (see Chapter 2), is another exceptionally flat area and includes much of southern Manitoba, eastern Saskatchewan, and the valley of the Red River of the North in North Dakota and Minnesota.

Not all portions of the region are so unvarying topographically. The most obvious excep-

tion is the Black Hills section of South Dakota and Wyoming. A large, dome-shaped area of eroded igneous rock, the Black Hills are associated both geologically and topographically with the Rocky Mountains to the west. The Big Horn Mountains, in north-central Wyoming, are a major outlier of the Rockies. Other, smaller outliers, most notably the Cypress Hills in southern Alberta and Saskatchewan, dot the western margins of the northern Plains. In central and northwestern Nebraska, the Sand Hills offer a dense, intricate pattern of grass-covered sand dunes, many of which are well over 30 meters

(100 feet) high. The dunes were created by sand blowing along the southern margins of continental glaciers during the Pleistocene. *Badlands* topography—extremely irregular features resulting from wind and water erosion of sedimentary rock—is widespread on the unglaciated Missouri Plateau between northern Nebraska and the Missouri River. The best examples are found in the Badlands National Park of western South Dakota. North of the Missouri River and west of the Lake Agassiz Basin, the glaciated Missouri Plateau, although sometimes flat, is covered with ponds, moraines, and other glacial features. Many visitors to north-central North Dakota are surprised at the many lakes found there, so important to migratory wildfowl on the midwestern flyway.

To state that the characteristic vegetation of the grasslands is grass would be a deceptive oversimplification, ignoring both the variety and distribution of grasses. Although agriculture has destroyed much of the original grasslands vegetation, the moister eastern portions [areas with more than 60 centimeters (24 inches) of annual precipitation in the north or more than 90 centimeters (36 inches) in the south] were originally a continuous tall-grass prairie, where grasses grew between 30 centimeters (1 foot) and 1 meter (3 feet) in height. Big bluestem was perhaps the most characteristic vegetation type. Grasses were shorter in the drier climates toward the west. Along the western margins of the Plains, continuous prairie grasses gave way to bunch grasses, with grama grass a good example of the dry steppe grasslands. These shorter, more separated grasses could succeed in the semiarid conditions of the western Plains.

The prairie grasses have developed deep, intricate root systems that commonly extend much deeper into the soil than the grass blades reach above, allowing them to utilize available water. The tangled root system made the prairies exceptionally difficult to plow. The first settlers often had to employ "Bonanza Teams," heavy plows pulled by as many as 20 animals, to break the sod.

The Problems of Climate

Precipitation. Most precipitation on the Great Plains results from the interaction of air masses generated over the northern interior of the continent and over the Gulf of Mexico. Air masses from the Southwest also often interact with the continental air masses, but they tend to contain less moisture and create much less precipitation. The warm moist tropical maritime air flowing in from the Gulf of Mexico, the prime contributor of moisture to the Plains, commonly curves up the Mississippi Valley and then moves northeast, missing much of the Great Plains entirely. One result of this pattern is the marked westward decline in average precipitation amounts across the U.S. portion of the region. This is represented on maps by lines of equal annual precipitation, called *isohyets*, which bend more toward the east on the northern Plains. In Kansas, for example, average annual precipitation varies from a moist 105 centimeters (42 inches) in the southeast to a semi-arid 40 centimeters (16 inches) in the southwest.

The patterning of precipitation on the Canadian grasslands is markedly different. The east-to-west zonation is replaced by one that runs northeast-southwest. The driest portion of the prairies, called Palliser's Triangle after a nineteenth-century surveyor who pronounced the area unfit for cultivation, is found along the Alberta and Saskatchewan border with Montana. This dry zone stretches perhaps 300 kilometers (185 miles) north of the national border in the west. To the north and east the precipitation is adequate to support a tall-grass prairie. This Park Zone is the major area of cultivation on the prairies.

Periods of higher than normal precipitation on the Great Plains result when tropical air masses move northwestward from the Gulf of Mexico, which brings these air masses over portions of the Plains. This provident current is far from dependable, however. Annual precipitation amounts can vary widely, although they usually fall between 80 and 120 percent of the average (Figure 12-2). During severe drought years, such

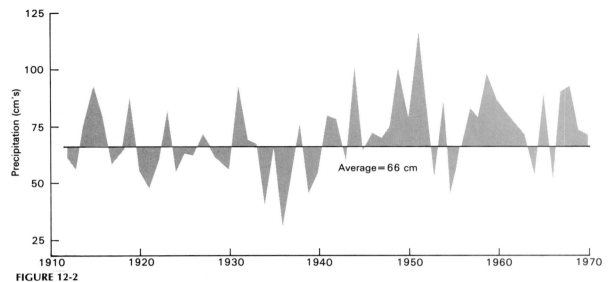

FIGURE 12-2

RAINFALL VARIABILITY IN CRETE, NEBRASKA. Crete is located on the
eastern margin of the Great Plains and thus is less affected by annual precipitation variability than more westerly sections of the region. Nevertheless, the 1930s
drought followed by several decades of greater than normal precipitation is
clearly evident.

as those during the mid-1930s, precipitation
amounts may fall to less than 75 percent of the
long-term average. Fortunately for the Plains
farmer, the bulk of the precipitation falls during
the period of more rapid crop growth, from April
to August. About three-quarters of the region's
total annual precipitation can be expected during this five-month period.

Major droughts have tended to occur in 20-
year cycles, with the last five centering on the
1890s, 1910s, 1930s, 1950s, and mid-1970s. For
farmers operating close to the Plains low-mois-
ture margins, severe long-term periodic variations in precipitation have resulted in times of
widespread optimism and expansion followed by
years of climatic and economic setbacks. During
the period around World War I, when precipitation abundance coincided with high prices for
wheat, the premier crop of the Plains, acreage
expansion was particularly noticeable. Farmers

moved into the drier southwestern margins of
the wheat-growing area in Kansas and Oklahoma. The collapse in wheat prices in the early
1920s brought economic recession to much of
the wheat belt by 1925.

The Great Depression of the 1930s coupled
with several years of severe drought on the Plains
meant economic ruin to areas of recent expansion as crop production, marginal even in good
years, shrunk to near zero. One result was the
famous Dust Bowl. With the soil-binding quality
of the land destroyed by years of repeated cropping, dunes of 3 meters (10 feet) or more in
height were built up by the dry winds, and the
sky was often laden with dust. A second consequence was the migration of large numbers of
destitute farmers from the farms of the southern Plains to the new promised land of California. Great as the soil damage was, the Soil Conservation Service found that the drought years

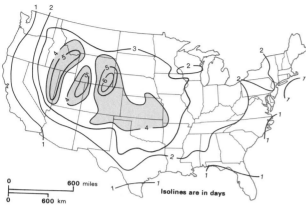

FIGURE 12-3

AVERAGE ANNUAL NUMBER OF DAYS WITH HAIL. Hailstorms each year destroy crops valued at many hundreds of millions of dollars. Damages are especially heavy on the High plains.

of the 1950s—following as they did years of ample precipitation and high demand for wheat throughout much of the 1940s—in reality resulted in the damage of far more acres by wind erosion than the 1930s drought.

Some of the region's spring and summer precipitation comes in the form of violent thunderstorms. Hail is occasionally a product of these storms. These frozen pellets, sometimes measuring more than 7 centimeters (2 inches) in diameter, which can look like golfballs or even baseballs, have the power to devastate a crop of mature, top-heavy wheat. Much of the southern and west-central Plains experiences frequent hail storms, with parts of western Nebraska and southeastern Wyoming leading the continent in average annual hail frequency (Figure 12-3).

Wind. Tornadoes, often with funnel wind speeds in excess of 350 kilometers per hour (200 miles per hour), are another violent result of the storm systems of the Great Plains. Although the area affected by one funnel is limited, the frequent occurrence of tornadoes on the central Plains makes them a significant regional hazard (Figure 12-4).

Wind has been a mixed blessing on the Great Plains. Late spring and summer wind velocities on the central and southern Plains are among the highest in interior North America. In the past, this served to maximize the efficiency of windmills in the region. However, the high winds can also mean that the amount of moisture evaporated and transpired by plants (when sufficient moisture is available) is high across much of the Great Plains. Maximum rates vary from about 165 centimeters (60 inches) annually in Texas to around 75 centimeters (30 inches) at the northern edges of the Plains in Canada; levels depend on average temperature and, therefore, on latitudinal location. Because these rates, called potential *evapotranspiration,* are high, they limit the effectiveness of precipitation.

The *chinook,* a winter wind, occurs when dry, relatively warm air from the Pacific Coast pushes over the Rocky Mountains. As it descends onto the Great Plains it warms still further and is much warmer than the cold, continental air mass commonly found over the region in winter. The Pacific air temporarily pushes the cold air from the western Plains, and a rapid, dramatic temperature rise results. Partly because of this interesting phenomenon, winter temperatures along the higher western area are slightly warmer than those along the eastern edge of the Plains, and cities such as Denver and Calgary are able to bask occasionally in balmy midwinter days not found farther east on the grasslands.

Temperature. As is common with moist continental locations, seasonal temperature variations on the Plains are considerable, and the annual temperature range increases as one moves northward. Amarillo, Texas, has an average January temperature of 2°C (35°F) and an average July temperature of 28°C (82°F); the corresponding temperature range for North Platte, Nebraska, is from −5°C (23°F) to 26°C (78°F); and for Winnipeg from −17°C (1°F) to 20°C (68°F).

In addition, the length of the frost-free season, which averages 240 days in central Texas

271

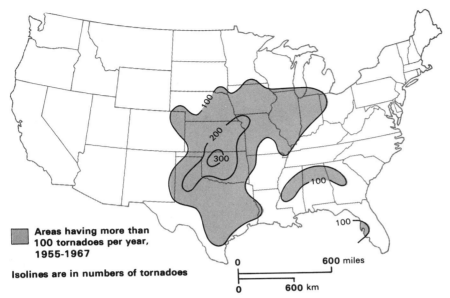

FIGURE 12-4

AVERAGE ANNUAL TORNADO FREQUENCY. Tornados are far more common on the Great Plains than in any other area of the world. Tornado warning sirens are a normal part of spring and early summer in communities in "Tornado Alley" from northwest Texas to northeast Kansas.

and less than 90 days on parts of Canada's prairies, varies widely around the average from year to year. As with annual temperature range, the variation increases as one moves northward. Along the U.S.-Canadian boundary, frost-free periods in some years are as much as 30 percent higher or lower than the area's 100- to 120-day average.

Blizzards. Snow, wind, and cold are all part of one of the most devastating weather elements on the Plains: the blizzard. A blizzard occurs in winter when a very cold polar air mass pushes southward along the Rocky Mountains and onto the Plains, breaking the usual west to east storm pattern. High winds, intense cold, and considerable amounts of snow are assoicated with these storms. Ordinarily, snowfall amounts are not high in the region, averaging perhaps 60 centimeters (24 inches) on the northern Plains. A blizzard can last for several days, and because Plains ranchers usually leave their stock outdoors during the winter, a severe blizzard may block the animals' access to food and result in high animal mortality. It is estimated that as much as 80 percent of the cattle of the northern Plains died during a series of unusually difficult blizzards from 1887 to 1888. The impact of these storms was even more severe because of a widespread decline in the quality of the steppe grasslands, the result of a decade of overgrazing.

EARLY SETTLEMENT OF THE PLAINS

Plains Indians
The pre-European occupation of the Plains by the American Indians was limited. Hunting, particularly for buffalo, was the primary economic activity. Most tribes lived along the streams

in semipermanent settlements. With no means of rapid long-distance overland movement (the dog was the only domesticated animal in pre-European North America), the Indians could not leave the dependable water supplies of the streams for any long period. This was a substantial problem, for the migration of the great buffalo herds often took this food source far away from the settlements for many weeks.

When the Spanish departed from the southern Plains following their initial explorations, they left some of their horses behind, a "gift" that dramatically altered the life-style of the Plains Indians during the following two centuries. By the time Americans reached the Plains in the early nineteenth century, they found what many have called the finest light cavalry in world history. The horse had diffused throughout the grasslands, and the Plains Indians, no longer restricted to the waterways, freely followed the buffalo migrations. This new resource enabled the Indians to thrive as never before. A few tribes, like the Dakota (Sioux) in the north and the Apache and Commanche, had used their well-developed warrior abilities to dominate large areas.

The American Indians were gradually pushed from most parts of the Plains during the nineteenth century (see Chapter 14). To some extent, this was caused by European hunters who hunted the Plains buffalo to the brink of extinction. Mostly, however, the Indians were simply swept aside by the flood tide of American settlement.

American Settlement

The early American perception of the region as an unpromising and difficult place to settle was not totally wrong. The widespread lack of trees meant that farmers had none of the traditional material used for the construction of houses and barns, for fencing, or for fuel. Water sources were scarce; often rivers and streams had only a seasonal flow. Those who arrived early settled along these waterways. Later arrivals had no ready access to flowing water because they were bound by the existing system of water law that gave all *riparian rights* to the land owner. This application of English common law in North America had been satisfactory in the more humid areas of the East, but it was totally inadequate on the Plains. The crops that settlers brought with them to the Plains often failed, and crop success varied greatly from year to year as precipitation amounts fluctuated widely—this was an unfamiliar climatic pattern from that which had been found in the East. Agricultural production rates were also generally lower, and the 160-acre (64.75-hectare) farm size that seemed so adequate farther east proved to be too small on the Great Plains.

The settlement frontier hesitated along the eastern boundary of the Plains partly as a result of these problems. Settlers tended to bypass the Plains for the Pacific Coast, until technological and land ownership changes made Plains settlement more inviting.

Grasslands ranching. During this hesitation in settlement expansion, an alternative economic system swept across the region with the speed of a prairie fire. An extensive ranching economy had been introduced into south Texas by Spaniards and into east Texas by frontier settlers from the U.S. South. This spread from Texas northward as far as the grasslands of Canada's Prairie Provinces during the period from 1867 to 1885.

A series of events and changing conditions permitted this rapid expansion of ranching. The American Civil War years resulted in a large unbranded cattle herd in south Texas as most ranchers went off to war. With the end of the war, the South's economy was in shambles, and the ranchers were broke—but millions of cattle were wandering around unclaimed. In the North, the war had spurred urbanization and increased the commercial demand for food. Railroads were given the right to construct routes from the settled Mississippi Valley to the West Coast, but they found little potential for profitable freight orig-

This feed yard near Amarillo, Texas, has a capacity to feed 70,000 head of cattle at one time. Cattle raised on the grasslands of the Plains are now commonly taken to a feedlot such as this one for fattening on a special mix of feed before they are marketed. These lots dot the region. They are an important factor in the widespread relocation of the beef cattle slaughter industry from major cities in the Midwest to smaller facilities in the towns of the Plains. (U.S.D.A.)

inating in the territory in between. The U.S. federal government aided the railroad companies with land grants of tens of millions of acres along the railroad right-of-ways, and the railroads actively encouraged agricultural settlement along their routes to generate transportation demand. Even so, until farmers arrived and until rail lines were completed to the Pacific Coast, business was limited.

Great herds of cattle were driven northward from south Texas to railheads in Kansas both for shipment east and to stock the huge, relatively unsettled Plains region. The possibilities of open-range cattle ranching were soon obvious. Within a decade the system spread northward across the Plains and into Canada. By 1880,

perhaps 5 million head of cattle had been moved from south Texas to the Kansas railheads or to the ranges of the central and northern Plains.

The open-ranching economy collapsed rapidly in the late 1880s. Widespread overgrazing, competition from the superior beef of expanding cattle-raising operations in the Midwest, a slipping national economy, the disastrous winter of 1887–88, and a rapid influx of farmers onto the Plains combined to end this short period of American history. (At least one television western series, "Gunsmoke", had a longer tenure than the period it attempted to portray.) The open-range, unimproved ranches were pushed to the drier western side of the Plains or were forced into a more restrained, fenced operation.

The agricultural frontier. A series of innovations during the two decades following the Civil War also gradually eliminated, or lessened in severity, the problems encountered by new settlers. Barbed wire, developed commercially in the 1870s, provided an effective alternative fencing material to take the place of the missing wood supply. For a time, dwellings constructed of sod provided adequate housing. The prairie sod was difficult to plow because of its intricate system of grass roots, but when lifted off the soil like peat and used as a building material, it would resist erosion for years. Nevertheless, these "soddies" were relatively short-lived, and most settlers replaced them as soon as possible with frame homes. Lumber was brought in by the railroads, which were under construction all across the Plains by the 1870s. The development of a simple windmill and mechanical well-drilling devices meant that sufficient water could be obtained locally for humans and animals, if not for irrigation. The technology for deep-well drilling had long been known, but neither in northwestern Europe nor in the eastern parts of North America had windmills ever been used much. It was the widespread adoption of this technology on the grasslands that led to its subsequent acceptance across most of rural America. Grain farming also became increasingly mechanized during this period, and an individual could cultivate far more land than had been possible previously. This enabled farmers to operate larger farms and, thus, compensate for lower yields.

A series of major modifications in U.S. land laws increased the acreage that an individual could obtain from the federal government. The Homestead Act of 1863 granted 160 acres (64.75 hectares) to each settler. The Timber Culture Act several years later allowed settlers to own an additional 160 acres (64.75 hectares) if a portion of that land was planted in trees. It was hoped that this would give farmers the additional farmland they needed and would also introduce wood production to the nearly treeless environment.

Finally, crops that were better adapted to the growing conditions of the region were introduced into the agricultural system, and farmers began to improve their understanding of how to use the Plains environment. Hard winter wheat is perhaps the best example. First brought to the United States by Mennonite immigrants from Russia, it was far better adapted to the dry growing conditions of the Great Plains than were the wheat strains grown there earlier.

Settling the Canadian Prairies

The Canadian prairies were owned by the Hudson's Bay Company until 1870 and company policy discouraged settlement. The agriculturally uninviting Canadian Shield between southern Ontario and central Manitoba also was a barrier to westward movement. The Canadian government began to foster settlement of the area in the 1870s after it acquired the prairies. The completion of the Canadian Pacific Railway in 1885 made the move much easier. Still, much of the early agricultural settling of the prairies came from the American Middle West. In 1890 good agricultural land was no longer available for frontier settlement in the United States, but there was land in Canada. By World War I over a half-million Americans had moved to the prairies.

One important result of this settlement hesitation was that the Canadian government and settlers on the prairies learned from the mistakes of American Plains settlement. Farmers were encouraged to increase their land holdings by purchasing railway land grants or other privately owned land. Ranchland was only leased and could be reacquired by the government if needed for agriculture. The government encouraged (and still encourages) mixed farming in the Park Belt in hopes of reducing the climatic hazard to the agricultural economy.

PLAINS AGRICULTURE

The agriculture of the Great Plains is large scale and machinery intensive. In the United States especially, it is dominated by a few crops particularly suited to the Plains climate. As men-

Intricate wind breaks of trees and bushes such as these in North Dakota are
common on the northern plains and protection against wind and blowing snow.
(U.S.D.A.)

tioned, the most important is wheat (see map appendix).

Winter wheat is planted in the fall. Before the winter dormant season sets in, the wheat stands several inches tall. Its major growth comes in the spring and early summer, when precipitation is at a maximum and before the onset of the dessicating winds of summer. Today winter wheat is grown across much of the United States, but its zone of concentration is the southern Plains from northern Texas to southern Nebraska. Production of winter wheat spread steadily

northward in recent years with the development of new strains that can better tolerate cold weather. In fact, it is now more important than spring wheat in Montana.

Spring wheat, grown primarily from central South Dakota to the south-central Prairie Provinces of Canada, is planted in early spring and harvested in late summer or fall. It is suited to areas of severe winters and somewhat cooler summers.

Today the Great Plains remains the continent's premier wheat-producing region. The three Prairie Provinces grow most of Canada's wheat, with Saskatchewan easily outstripping its two neighbors combined. Kansas usually leads in the United States, followed closely by North Dakota, and then, in distant order, by Oklahoma and Montana. Of the top seven wheat states in the United States, only Washington (ranked fifth) is not a plains state. Thus, it is largely on the abundance of Plains agriculture that the United States and Canada rank as the world's two leading wheat exporters (see Chapter 1).

A special kind of migrant labor has long been important in the harvesting of wheat. Beginning in the south around June 1 with the winter wheat harvest in Texas, custom combining crews gradually follow the harvest northward, finishing some four months later in the Prairie Provinces. Unlike migrant farm laborers harvesting other crops in the United States, these people, often in large crews that use many combines and trucks, have traditionally been well-paid agricultural workers. The increasing size of wheat farms—average farm size in most of the wheat belt now exceeds 400 hectares (1000 acres)—has meant that more wheat farmers can now afford their own combines, and custom combining is less prevalent today than it was before World War II (Figure 12-5).

The rapid, continuous increase in farm size in the region has resulted largely from the purchase by other farmers of the farms of those who have chosen to leave the area or the task, leading to an increasing number of noncontiguous farms. This, in turn, has helped foster a growing number of "sidewalk farmers," farmers who live in town and travel to their various acreages. Although demanding when performed, fieldwork on a wheat farm must be done only two or three times a year, so there is little need for the farmer to live on his or her land. In fact, some so-called "suitcase farmers" live far from their fields and visit them only occasionally. The corresponding processes of farm fragmentation and part ownership also have become increasingly common as a result of land sales (see Chapter 11).

Besides the freedom to live in town, many see a second advantage in this type of disrupted land ownership. A single hailstorm usually affects only a small area. By dividing the total wheat crop over several widely separate areas, the grower is, in effect, ensuring that in any one year, it is unlikely that an entire crop will be lost to hail or to a particularly severe thunderstorm.

A major problem with profitable wheat production continues to be the necessity, and difficulty, of moving the harvest rapidly to storage in the large grain elevators that dot the Plains. In Canada especially, many a small prairie town is dominated by a series of small, often wooden, grain elevators to which grain is brought by truck and then sent on to larger facilities by rail. The small elevators can hold only a part of the local crop; if all of the wheat is to be successfully removed from the fields before it is subjected to rain and destroyed, many grain hoppers must be made available at the right times by the railroads. Occasionally, the harvest outstrips the railcar supply and wheat is piled in large, temporary mounds near the elevators. Most Canadian wheat goes first to Winnipeg and then to Thunder Bay for transshipment on the Great Lakes. An increasing amount is now moving through Pacific coast ports, reducing somewhat the role of Thunder Bay. All western wheat presently goes to Vancouver. Canadian authorities are considering the establishment of new facilities at Prince Rupert, British Columbia. Most U.S. export wheat moves through the Great Lakes or in barges down the inland waterway system and the Mississippi River.

The soft international market for wheat during much of the 1960s led to a definite interest in crop diversification in the Canadian Prairie Provinces, and barley acreage increased rapidly. Today, Alberta has more acreage in barley than in wheat. Total wheat acreages in the Prairie Provinces, which totaled 12 million hectares (30 million acres) in 1967, declined to 7.6 million hectares (18.7 million acres) in 1971, with the government's Low Inventories for Tomorrow program playing an important role in the re-

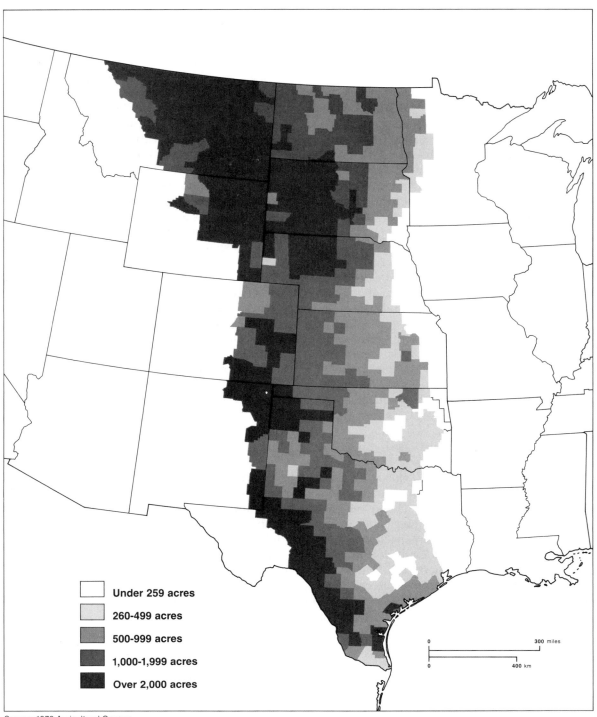

Under 259 acres

260-499 acres

500-999 acres

1,000-1,999 acres

Over 2,000 acres

0   300 miles

0   400 km

Source: 1978 Agricultural Census

FIGURE 12-5

AVERAGE FARM SIZE, 1978. Farmers on the western margins of the Great Plains need larger farms to compensate for lower per acre economic returns that result from sparse precipitation.

duction. Harvested area increased later in the decade, primarily in response to a growth in international demand. Only in Saskatchewan does wheat continue to dominate total farm production.

Sorghum has emerged as a major crop on the southern Plains in recent decades. Able to withstand dry growing conditions, this African grain now equals winter wheat in importance on the hot, dry southwest margins of the Plains. Most of the grain sorghum crop is used as stock feed. On the northern Plains, barley and oats are major second crops, with most of the continent's barley crop coming from the Lake Agassiz Basin. Nearly all flaxseed produced in North America also is grown in the northern Plains. Sunflowers are rapidly increasing in importance in the Red River valley of Minnesota, North Dakota, and Manitoba.

WATER CONTROL AND IRRIGATION

Irrigation in the United States and Canada is usually associated with the dry region between the Rocky Mountains and the Sierra Nevada and with the southwestern United States. This is understandable, for crop agriculture is made possible in much of this area only by irrigation. Yet, the benefit derived from irrigation may be higher in many semihumid or even humid areas—in terms of the level of increased production per dollar invested—because irrigation water may be used either as a supplement in dry times to maximize yields for crops already grown in the area or to grow crops for which the available moisture is not quite sufficient. Evaporation rates from irrigation systems are often lower in more humid areas than in drier regions, and it is not as necessary to flood large quantities of water onto the land to provide an off-flow sufficient to

remove the salts that might otherwise build up in the soil. Generally, increased financial security for the Plains farmer—through an assured crop in drought years—has been the most important reason for irrigation, not the production of crops that would fail to survive on rainfall alone (Figure 12-6).

There are a number of Great Plains areas where large-scale irrigation developments are important. Perhaps the most noteworthy of these is on the High Plains from Colorado and Nebraska to Texas. The High Plains are underlain by the Oglala aquifer, a vast underground geologic reservoir beneath 250,000 square kilometers (100,000 square miles) of the area containing an estimated 2 billion acre-feet of water. This is "fossil" water, much of it deposited more than

FIGURE 12-6

EXISTING AND PROJECTED WATER DEFICITS. This map, prepared by a presidential commission on population and resources, suggests that much of the Southwest and interior West is already fully using available water supplies. The same problem is expected to become widespread on the Great Plains in the next few decades. Deficit use usually results in a declining water table.

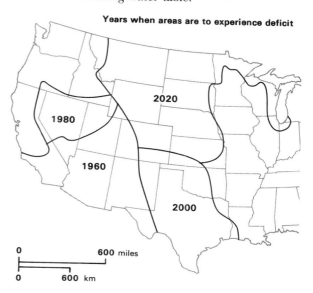

Years when areas are to experience deficit

279

Large, expensive center-pivot sprinkler irrigation systems have become common throughout the Great Plains and West where subsurface water is used for irrigation. Each circle is ½ mile (840 meters) in diameter. A problem for the farmer is how to use the remaining nonirrigated areas. (Earl Otis/U.S.D.A.)

1 million years ago. About a quarter of the aquifer's area is irrigated, almost entirely with Oglala water. The High Plains is a major agricultural region, providing, for example two-fifths of the country's sorghum, one-sixth of its wheat, and one-quarter of its cotton. Irrigated lands here produce 45 percent more wheat, 70 percent more sorghum, and 135 percent more cotton than neighboring nonirrigated areas. Groundwater withdrawals have more than tripled since 1950 to more than 20 million acre-feet annually. Recharge replenishes only a small percentage of this amount. Groundwater withdrawal rates are also quite uneven.

Early in this century, the area centered on Lubbock, Texas, became a significant region of cotton production. Irrigated farming, using water from wells drilled into the water-bearing sands that underlie much of the southern High Plains, gradually replaced the early *dry-farming* approach. Today the region is the most important area of cotton production both in Texas and in the United States. More than 50,000 wells currently supply irrigation water in the area. One negative result of this massive mining of the groundwater has been a rapid depletion of subsurface water. The average well depth is currently more than 50 meters (155 feet) and increasing. This represents a serious problem for the area because deeper wells are much more expensive. A recent estimate suggests that by the year 2000 wells will be so deep that irrigation will be too expensive on half the Texas High Plains. And the resource can be exhausted, meaning there may soon be insufficient water available to meet the demand. Even if use is reduced, replenishment of groundwater reserves is a slow process in this sub-humid environment.

New Mexican laws concerning the conservation of underground water are stricter than those in Texas. Thus, the areas in New Mexico adjacent to Texas have little irrigated land, and most of this area in the more western state is still composed of ranches. The effect of this legal difference is clearly visible from the air; the subdivided farms of Texas stand in sharp contrast to the extensive, unbroken landscape of New Mexico's ranches. It is also reflected in the higher *water table* on the New Mexico side of the boundary.

The second major irrigated area on the Plains is in northeastern Colorado, with sugar beets the primary specialty crop. The area has long been irrigated from wells and from the waters of the Platte River in both private and federal irrigation districts. The National Reclamation Act of 1902, passed largely at the instigation of Plains and Western farmers and politicians, provided for the development of federal irrigation districts in these areas. The federal government covers the cost of construction, and those who use irrigation pay for water at what many nonfarmers argue are uneconomically low rates. This act has provided one basis for a complex pattern of federal dams and intricate water-movement systems scattered widely over the West and constructed at a cost that has reached billions of dollars. Even so, these waters are no longer adequate to meet demands. In recognition of this, the government has funded the recently completed Big Thompson River project, which is designed to carry water from the west slope of the Front Range of the Rocky Mountains to the east slope and the irrigated lands beyond. The most striking technological feature of this project is a 30-kilometer (13-mile) tunnel, lying 1200 meters (3950 feet) below the Continental Divide in Rocky Mountain National Park.

The largest of the water impoundment projects on the Plains, however, is the Missouri Valley project, originally called the Pick-Sloan Plan (Figure 12-7). The plan was an outgrowth of two different sets of needs of the residents of the Missouri River valley. People living at the lower end of the valley, including those in Kansas City and St. Louis, desired above all to have an effective system of flood control. An annual average of about 100 centimeters (40 inches) of precipi-

tation falls on this area, and its residents are usually more concerned with an overabundance of water than with drought. In contrast, people in the upper Missouri Valley, especially the Dakotas and Montana, hoped for a system that would provide ample water for irrigation. The resultant multibillion-dollar system, composed of a series of large earth-fill dams on the upper Missouri as well as numerous dams on many of the tributaries of the river, is meant to satisfy both hopes. Irrigation has expanded considerably in many upstream areas, whereas residents in the lower valley hope (with some justification) that major flooding is less likely now. An added result has been a vast boom in water recreation, particularly power boating, throughout an area that previously had almost no standing water. Many argue that the costs, both monetary and in terms of damage to the natural environment, have been too great, but benefits of the plan have been considerable.

FIGURE 12-7

THE MISSOURI VALLEY PROJECT. The project, to meet the need for irrigation water in the western portion of the Missouri River basin and flood reduction downstream, is an example of large-scale federal intervention to help address regional problems.

ENERGY RESOURCES OF THE PLAINS

The sediments of the Plains contain major reserves of energy resources—petroleum, natural gas, and coal (see Figures 2-6 and 2-7). To the south, major petroleum and natural gas fields are traditionally among the country's leading suppliers of these products. The Panhandle field, encompassing western portions of Texas, Oklahoma, and Kansas, is the world's leading supplier of natural gas. The same three states are also major petroleum producers, and recent developments have also added Wyoming to the list of leading U.S. petroleum suppliers. Alberta has long been the predominant producer of both petroleum and natural gas in Canada. Coal has been mined for many years at scattered locations in the region, from southern Alberta and Saskatchewan to Texas.

Despite this past energy predominance, it is the future of energy production that excites a mixture of anticipation and concern in many Plains residents. Increases in energy cost and demand have focused attention on the region's rich coal and petroleum resources. North Dakota can boast of sizable energy resources, mostly in the form of soft coal. Wyoming anticipates becoming the leading coal-producing state in the United States before 1990. Denver has become a focus of considerable petroleum-based wealth, and Colorado's governor has expressed apprehension about his state's readiness for the development that is being thrust on it. Alliance, Nebraska, has nearly doubled in size since 1975 because of its location on the Burlington Northern rail line, which carries coal eastward from the Wyoming fields. Emerging from decades of economic stagnation, scores of other communities are experiencing similar growth spurts.

This energy boom did not happen earlier because of a combination of factors: distance from major markets, technological problems, and abundant alternative energy sources. With substantial energy price increases in recent years, the economic barriers to exploitation have diminished. Although Plains coal is often found in thick, easily mined seams, it is expensive to ship to Eastern markets. Some had considered shipping the coal eastward by pipeline, first transforming the coal into "slurry," a fine coal powder mixed with water, then pumping it inexpensively like liquid petroleum. However, the great demand this would place on the region's limited water supply doomed the plan.

Technological problems have also limited petroleum production from the Athabasca tar sands in Alberta. Located about 400 kilometers (250 miles) north of Edmonton, it is estimated that these sands contain at least 300 billion barrels of petroleum, an amount that approximates the entire world's current known reserves of conventional oil—and some estimates amount to more than twice that. (Although the actual tar sand area is beyond the margins of the Great Plains and Prairies region, its development focuses on Edmonton; hence, it is included here.) Unfortunately, because this petroleum saturates the sands in a solid form, it must be mined somewhat like coal, and technology for its refining is still in the development stage. As energy prices rise and the applicable technology improves, this reserve will be of tremendous importance to Canada.

It is difficult to anticipate the scale of these developments and their possible consequences. At least 100 billion tons of low-sulfur subbituminous coal that meets strict antipollution laws can be found near the surface in the northern Plains of the United States, an amount equivalent to that needed for 125 years at current national levels of consumption. Within 2000 meters (6000 feet) of the surface, the total is perhaps 1.5 trillion tons. Already the structure of the regional economy is shifting, with agriculture and ranching declining in importance. Despite strict regulations, environmentalists fear that future exploitation will result in soil erosion, water and air pollution, and rapid population growth. Whatever form the future development takes, the northern Plains are beginning a period of rapid change.

POPULATION PATTERNS

Population decline, or at best stagnation, has become the accepted standard across much of the Great Plains during the past 50 years. The Canadian provinces, especially Alberta, have fared somewhat better than the U.S. Plains states during this period. Where substantial population increases are found, the areas are almost invariably associated with significant recreational or natural resource developments, or with urbanization. These three factors reflect a pattern found generally in both countries. The difference is that this region has a decided lack of urban centers,

major recreational potential is minimal, and there were few important natural resource developments until recently.

Much of the American portion of the region is served by major urban centers that are found somewhat beyond the peripheries of the Plains. Chief among these are Kansas City and Minneapolis-St. Paul. The largest American cities on the Plains, Denver, Dallas-Fort Worth, and San Antonio, are all peripheral. Denver is a major regional center for much of the central Plains, but it also plays the same role for much of the interior West. It is a regional office center as well as the focus of financial activity for energy-re-

Sited along the Red River near the eastern edge of the Canadian prairies, Winnipeg, Manitoba, possesses strong focality in the northern grasslands. (Courtesy Manitoba Government Travel, Department of Tourism and Recreation.)

source development on the northern Plains and in the interior West. The Dallas-Fort Worth urbanized area, with a population approaching 3 million—easily the region's largest—seems to straddle the margin between the Plains and the South. Overall, oil, the general effects of the Texas boom economy, and this urban area's emergence as a dominant regional office center for the Southwest mark its growth. Still, Dallas seems more of a city of the humid east, whereas the smaller Fort Worth—50 kilometers (30 miles) to the west—is a ranching and stockyard center that is clearly part of the Plains. San Antonio, third in size on the Plains, is the largest commercial center in south Texas plus the home of several major air bases. The city's population is over 50 percent Hispanic, placing it clearly in the Southwest Border Area as well (see Chapter 14).

Many of the somewhat smaller, but still quite important, centers serving the area are also peripheral—cities such as Tulsa and Omaha. The service areas of the cities grouped around the edges of the Plains tend to be elongated east-west zones that cover the region. Within the United States, the eastern peripheral centers are located along the far more densely settled lands found from about the 98th meridian eastward.

In Canada, by comparison, all of the major cities of the Plains are separated from other heavily populated areas by broad expanses of only sparsely settled territory. Winnipeg, for example is nearly 1600 kilometers (1000 miles) from southern Ontario. One result is that the Canadian prairie cities, although peripheral to the region in location, all clearly focus on it. Winnipeg, especially, is also a complex manufacturing center that produces goods for the region and processes agricultural products before they are shipped eastward. Calgary has prospered from Alberta's energy boom. Some call it the most American of Canadian cities, at least in part a result of the impact of American energy firms. Edmonton also shares in the energy development but maintains a more complex commercial and governmental function.

Most towns on the Plains began as transpor-

tation centers, commonly strung out along the railroads. Those that have prospered maintain some transport service function, but they have also become established regional market centers. Some are also supported by special local conditions—Oklahoma City and Tulsa, for example, are important petroleum centers. Wichita is a manufacturing center for small aircraft. And, as mentioned, many cities are now growing because of recent energy developments.

The beef-processing industry has expanded into many smaller Plains communities in the United States during the last three decades. Formerly, the industry had been concentrated in the large cities of the Midwest, east of the Plains, where facilities were large and complex. Changing technology in the slaughter industry, the growth of feedlots on the Plains, and more diversified marketing patterns gradually made smaller plants located near the new feedlots more economical. The vast stockyards of midwestern cities, such as Chicago, are now gone, replaced by production from scores of smaller Plains towns.

Transportation routeways on the Plains were originally built to cross the area rather than to serve it (Figure 12-8). Thus, most major highways and railroads in both countries pass east-west across the Plains, with few lines running north-south. Of the seven interstate highways passing through substantial parts of the region in the United States, six are east-west routes. In Canada, the only substantial north-south highway follows the western edge of the prairies in Alberta. The same overwhelming predominance of east-west movement is also true for railroads in both countries.

Whatever the cause and justification of this pattern, one important result for American residents of the region has been a continuation of the economic and perceptual orientation toward areas off the Plains instead of toward areas within it. North and South Dakotans, for example, focus strongly on the Minneapolis-St. Paul area. The financial and trade institutions of the Twin Cities play a major role in these two states; the

FIGURE 12-8

MAJOR RAIL AND AUTO ROUTES. The north-south and especially east-west linearity of major overland transportation routes on the Great Plains can be frustrating for anyone hoping to travel diagonally across the region.

vently; and the joys of that urban area become, vicariously, at least, those of the Dakotans as well. Typical residents of the Dakotas are likely to be more aware of the Twin Cities than of the bordering Plains areas to the north and south, and the same pattern is repeated elsewhere across the Plains. Again, this is far less the case in Canada.

IDEAS OF THE PLAINS

Interestingly, the concept of the Great Plains as a region is a relatively recent addition to the popular terminology. Even today, most residents of the Plains, if asked to locate themselves in terms of the names of the physical region in which they live are unlikely to answer with the term Great Plains. Instead, a more localized term will usually be given. Thus, a Texan may live on the Staked Plains, a Kansan in the Gypsum Hills, and a Nebraskan in the Platte Valley or Sand Hills. These and several score other distinctive areas are the basic reference regions of the Plains population. The Great Plains as a broader region is, to a substantial extent, an academic invention of the twentieth century. The widespread concern over the problems of the region, both environmental and economic, especially as they developed during the depression years of the 1930s, may have been the major contribution to this conceptual development.

One major and, in a sense, negative result of the development of the Great Plains idea has been the continuation of a belief that the Plains environment is a monotonous one of flat uniformity. This sterotype is often encouraged when travelers cross portions of the Plains by car. Two of the most traveled U.S. routes, I-40 and I-80, pass through the Staked Plains and the Platte River

principal Twin Cities' newspaper is read widely across the region; the activities of the various professional sports teams there are followed fervalley, respectively. These areas are flat and, by some measures, the least attractive portions of the region. In fact, the physical environment of the Plains possesses substantial variation. The Sand Hills of Nebraska, the Badlands of South Dakota, the Alberta uplands, the Manitoba lowland—these areas are certainly different in both geologic structure and physical appearance.

One other misconception held by many is that the population of the Great Plains, like the physical environment, is ethnically homogeneous. This is not the case. In fact, many parts of the region have striking diversity as the norm, especially in Canada where recent immigration has played an important part in twentieth-century population growth. The Scandinavian settlement of the northern United States is well known. East European settlement, less widely recognized, was important in many areas, as was illustrated by Willa Cather's fine novels of pioneer settlement in Nebraska. Mexican-Americans are a major component of the population of the Colorado Piedmont. Winnipeg is today one of Canada's more ethnically diverse cities; restaurants there offer food from at least a dozen countries. Large numbers of American Indians are also located on the Plains, especially in South Dakota and Oklahoma. In spite of its unifying features, the Great Plains is a region of great diversity in physical and cultural landscapes.

GRASSLANDS LITERATURE

Yes, we buried him there on the lone prairie,
Where the owl all night hoots mournfully,
And the blizzard beats and the winds blow free
O'er his lonely grave on the lone prairie.

From the folk ballad,
"The Ballad of the Dying Cowboy"

It is in the songs, folktales, and literature of a

The dark billows of a dust storm are approaching a small Plains town in this photograph from the mid-1930s. Improved farming practices have lessened the likelihood of such storms, although smaller dust storms still develop, especially during the region's cyclic drought periods. (Library of Congress.)

region that we often find the truest measure of the appearance and nature of a place. The regional novelist and balladeer often recognizes the environment as an ever-changing stage for his or her characters. The landscape is, therefore, important as backdrop—as scenery. However, its role is usually not passive. It can influence and mold the thoughts and actions of the characters, in turn, it can be influenced by those actions. The landscape, thus, becomes a vibrant, important, and even a dominant element in good regional fiction and semifiction.

Like the Deep South, the Plains has produced an uncommonly large volume of such works. Through most of them run a few common threads that seem almost to bind them to the grasslands environment. Hamlin Garland, raised in South Dakota in the late nineteenth century, introduced one of these themes in his novel *Main-Traveled Roads* (1891):

How poor and dull and sleepy and squalid it seemed! The one main street ended at the hillside at his left and stretched away to the north, between two rows of the usual village stores, unrelieved by a tree or a touch of beauty.[4]

This barrenness, much the result of treelessness, is a theme repeated in Plains novels. Beret, the wife in O. E. Rolvaag's magnificent tale of Norwegian frontier settlement on the Minnesota-Dakota border, *Giants in the Earth,* would look at the treeless horizon of her new home and realize "the full extent of her loneliness, the dreadful nature of the fate that had overtaken her." She wondered: "How will human beings be able to endure this place? Why, there isn't even a thing one can hide behind." In the story her husband eventually dies in a blizzard, and the isolation and barrenness of her life drives Beret insane.[5]

In *The Grapes of Wrath* (1939), John Steinbeck viewed an Oklahoma wracked by a decade of economic depression and years of severe drought:

The right of way was fenced, two strands of barbed wire on willow poles. The poles were crooked and badly trimmed. Beyond the fence, the corn lay beaten down by wind and heat and drought, and the cups where leaf joined stalk were filled with dust.[6]

To emptiness, then, is added the theme of a mercurial, often devastating climate.

Mari Sandoz developed this idea of climatic hazard in *Old Jules,* a biography of her father in the Sand Hills of Nebraska published in 1935. Her father was proud of an orchard he had planted; then a hail storm struck.

. . . suddenly the hail was upon them, a deafening pounding against the shingles and the side of the house. . . . One window after another crashed inward. . . . Jules came in from the garden and sat hunched over a box before the house. All the trees were stripped, bark on the west and south, gone. Where her garden had been Mary planted radishes and peas and turnips as though it were spring. The corn and wheat were pounded into the ground, the orchard gone, but still they must eat.[7]

Despite this frequent malevolence, the Plains also offers beauty in its starkness and opportunity in its emptiness. Per Hansa, Beret's husband, rushed to his new field to begin breaking sod with a "pleasant buoyancy" and felt "the thrill of joy that surged over him as he sank the plow into his own land for the first time."[8] Willa Cather also focused on this love for the farmland in her *O Pioneers* and *My Antonia.*

A long-term resident, born and raised on the Great Plains, when asked to comment on the beauty of the mountains of central Pennsylvania responded, "Someone needs to cut down the trees

[4]Hamlin Garland, *Main-Travelled Roads.* New York: Macmillan, 1899, p. 70.

[5]O. E. Rolvaag, *Giants in the Earth.* New York: Harper Bros., 1927, p. 39.

[6]John Steinbeck, *The Grapes of Wrath.* New York: Viking Press, 1939, p. 24.

[7]Mari Sandoz, *Old Jules.* Boston: Little, Brown, 1935, p. 250.

[8]Rolvaag, op. cit., p. 47.

first so I can see them [the mountains]." His response was a bit tongue-in-cheek, but its point was important. The openness of the Great Plains, the feeling of distance that distressed many people of the woodlands when they moved onto the grasslands, is now perceived as a positive aspect of the environment by many of its current residents. This, they proudly state, is "Big Sky" country.

Additional Reading

Alwin, John A. "Jordan Country: A Golden Anniversary Look," *Annals of the Association of American Geographers,* 71 (1981), pp. 479–498.

Blouet, Brain, and Frederick Luebke (eds.). *The Great Plains: Environment and Culture.* Lincoln: University of Nebraska Press, 1979.

Jordan, Terry G. *Trails to Texas: Southern Roots of Western Cattle Ranching.* Lincoln: University of Nebraska Press, 1981.

Kraenzel, Carl F., (ed.). *The Great Plains in Transition.* Norman: University of Oklahoma Press, 1955.

Lawson, Merlin L., and Maurice E. Baker (eds.). *The Great Plains: Perspectives and Prospects.* Lincoln: University of Nebraska Press, 1981.

Luebke, Frederick C., (ed.). *Ethnicity on the Great Plains.* Lincoln: University of Nebraska Press, 1980.

Morgan, Dan. *Merchants of Grain.* New York: Viking Press, 1979.

Pearson, Jeff, and Jesica Pearson. *No Time but Place: A Prairie Pastoral.* New York: McGraw-Hill, 1980.

Rogge, John (ed.). *The Prairies and Plains: Prospects for the 1980s.* Winnipeg: Department of Geography, University of Manitoba, 1981.

Smith, P. J., (ed.). *The Prairie Provinces.* Toronto: University of Toronto Press, 1972.

Smith, P. J., and Davis B. Johnson. *The Edmonton-Calgary Corridor.* Edmonton: Department of Geography, University of Alberta, 1979.

Weir, Thomas R. *Manitoba Atlas.* Winnipeg: Department of Natural Resources, Province of Manitoba, 1984.

CHAPTER THIRTEEN

THE EMPTY INTERIOR

Stretching from the eastern slopes of the Rocky Mountains westward to the Sierra Nevada of California, the Cascade Range of the Pacific Northwest, and the Coast Mountains of British Columbia is found the largest area of sparse population in subarctic North America (Figure 13-1). This low average population density is the key identifying feature of this region. Indeed, there is much variation in other elements of the territory's geography. Portions, notably the Rocky Mountains in the east and the Sierra Nevada-Cascades-Coast Range-Coast Mountains systems along the western margins, have some of the most rugged terrain in the two countries. Between these mountain chains lies a series of plateaus, many of which contain extensive flat areas. Annual precipitation ranges from more than 250 centimeters (100 inches) in the Cascades of Washington to less than 25 centimeters (10 inches) in the plateau country. The population of the region is mostly of northern European origin, although Hispano-Americans and American Indians are in significant propor-

tions in the south (see Chapter 14). Irrigated agriculture is important in many areas, as is ranching, whereas, in other areas, lumbering, tourism, and mining are dominant.

This massive expanse of land extends nearly 1600 kilometers (1000 miles) from east to west in its wider southern sections and some 600 kilometers (375 miles) in parts of British Columbia. The region contains some of the most strikingly scenic portions of the continent. This vast scenic resource and its variable character have created the dominant perception of the Empty Interior held by most Americans. The impact of humans on the region, although locally important, has been overshadowed to a great degree by the varied splendors of the natural environment.

A DIFFICULT ENVIRONMENT

The Mountains

As discussed in the chapter on the Great Plains, the impressions that any individual or group has

289

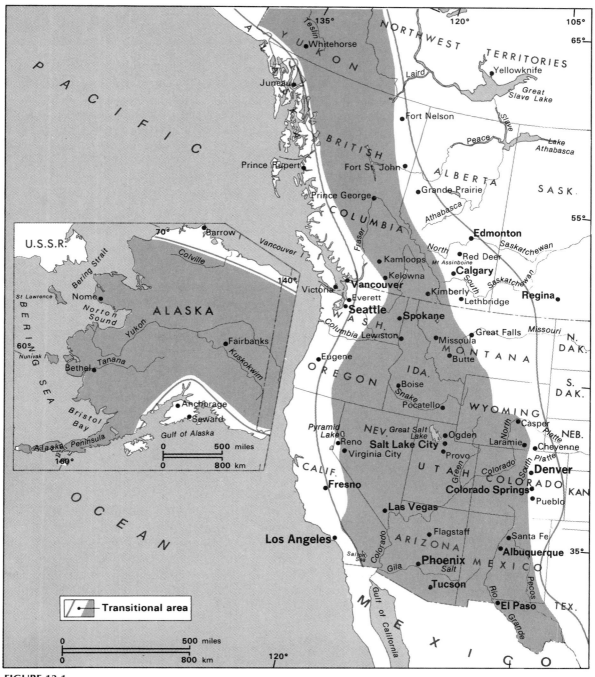

FIGURE 13-1
THE EMPTY INTERIOR.

about a place are heavily influenced by the past experiences of the people involved. Most Americans, products of the eastern United States and northwestern Europe, found the grasslands of the midsection of the continent rather alien. Many of the early evaluations of the Plains are biased by this perception. The reaction of most Americans to the Empty Interior of the West can be viewed in much the same light. Eastern Americans were (and are) products of the woodlands; they are accustomed to an undulating terrain where variations in elevation are seldom dramatic. Where mountains occur, their size, certainly when compared to those of the West, is often moderate. Most of the mountains of eastern North America, with the possible exception of a few in the Presidential Range of New Hampshire, do not contain an elevation change from the mountain's base to its top that exceeds 1000 meters (3000 feet), and even this degree of terrain variation is rare. By comparison, dramatic changes of 1000 meters or more are common in the interior West. The scale of altitudinal variations is vastly different from that found in the East (see Chapters 7 and 8).

Variation in local relief is only one aspect of the difference between the East and the interior West. A second somewhat related element of the physical geography of the region is its ruggedness. Most of the mountains of the eastern United States appear rounded and molded; the ranges of the West present abrupt, almost vertical slopes, and the peaks frequently appear as jagged edges pointing skyward. This difference is due partly to age. Most of the western mountains, although by no means all of them, are substantially younger than the eastern ranges. Thus, erosion has been active for a much shorter time. One result of long-term erosion is an eventual smoothing of the land surface and the elimination of many of the jagged edges that may have been characteristic at an earlier time.

A drive across the mountains on Colorado's Trail Ridge Road, for example, affords a striking scene of the effects of erosion. The road bisects the alpine country in the higher elevations of Rocky Mountain National Park. Much of the region was arched upward during several different geologic periods; interspersed periods of limited uplift allowed erosion to eventually "smooth" the surface. Today a trip along the road begins with dramatic increases in elevation, scenic vistas of deep canyons and jagged ridges, and a steeply rising, almost frightening drive to the crest. Then, for nearly 18 kilometers (12 miles) along the highest parts of the road, with much of the elevation above 4000 meters (12,500 feet), the terrain is gently rolling. If you can ignore the rugged mountains that are visible in almost every direction, the alpine vegetation, and the low-lying clouds below you, the road is similar to many country drives in the East. You have passed through areas of erosion on the flanks of the uplifted area and into its core. Here, far away from the base of the mountains, is the last area to receive the substantial effects of the geologically new cycle of erosion.

Climatic conditions, both historical and current, have been important contributors to the rugged character of these western mountains. During the most recent period of geologic history, the Pleistocene, substantial parts of this region were affected by the actions of ice. The massive ice sheets that constituted continental glaciation were confined in this region to portions of British Columbia and northward into Alaska, but smaller areas of ice, commonly called alpine glaciation, existed as far southward as the mountains of northern New Mexico. The carving done by these mountain glaciers did much to form the topography of the interior West. Remnants of the glaciers can still be found in parts of the region. Most widespread in the Pacific mountains of southern Alaska, smaller glaciers are found as far south as the central Rockies in Colorado and the Sierra Nevada of California.

Alpine glaciers form in higher elevations and gradually "flow" downhill as the volume of ice increases. The moving ice is a powerful agent for erosion. Where this erosion pattern continues for a sufficiently long time, a deep U-shaped

Trail Ridge Road in Rocky Mountain National Park, Colorado. The elevation here is 4000 meters (12,500 feet), and glaciers have carved deep valleys into this rolling upland, creating a pattern of gentle slopes dramatically bisected by steep cliffs. (National Park Service.)

valley is created with almost vertical sides and a relatively flat bottom. If two glaciers flow side by side, a narrow ridge line called an *arête* is formed, characterized by jagged small peaks. Where several glaciers formed in the same general locale and flowed outward in different directions, a sharp-faced pyramidal peak resulted. The Matterhorn (in Europe) is certainly the best known such mountain, but Mount Assiniboine in the Canadian Rockies is also a fine example of this feature.

Within North America, much nomenclature for various physical features is borrowed from sections of Europe where the feature is widespread. Thus, most features of alpine glaciation have French names, reflective of the French Alps. Many of the names for arid landscape features, such as canyon and *arroyo*, are from the Spanish language and reflect the earlier Spanish and Mexican presence in the arid lands of the U.S. Southwest.

The general result of alpine glaciation, then,

is a rugged landscape. Yosemite Valley in the Sierra Nevada in California, an almost classic glacially carved valley nearly a mile deep, is perhaps the region's most photographed example of alpine glaciation. The consequences of such glaciation are visible in the higher ranges throughout the entire region.

The Plateau Country

Most of the Empty Interior is occupied by plateaus rather than mountains. With some exceptions, these plateaus are the basic feature of all the area between the Rocky Mountains and the Sierra Nevada-Cascades-Coast Ranges-Coast Mountains complex from Mexico through Alaska. The term "plateau" is, however, somewhat misleading because parts of this section have great local relief.

Probably the most scenically dramatic portion of this section is the Colorado Plateau along the middle Colorado River. Although there are some large structural changes in relief, most of the area is underlain by gently dipping sedimentary rocks. The major landscape features are the result of erosion by the *exotic streams* that cross the plateau, most notably the Colorado River and its tributaries. In this arid environment, streams are easily the predominant erosive influence. Thus, when accompanied by recent substantial geologic uplift over much of the plateau, great downward erosion has resulted and been con-

centrated primarily in the immediate vicinity of the streams. The canyonlands that have been produced, again particularly along the Colorado, are some of the best known examples of America's natural scenic resources. One recent major travel survey conducted in the United States identified the Grand Canyon of the Colorado River as the most widely recognized natural scenic attraction in the country. River erosion deeper than the Grand Canyon occurs within parts of the Snake River Canyon. The greater attractiveness of the Colorado canyon, however, particularly in its more fully developed form in Grand Canyon National Park, is partly a result of its many deep tributary canyons. A system has been created that is at places more than 16 kilometers (10 miles) wide. In addition, the variable resistance of strong and weak rocks in these sedimentary formations has created an angular pattern of scarps and benches that is especially characteristic of the area.

Filling the plateau country from the Colorado Plateau as far north as Oregon and Idaho is the Basin and Range region. This wide area is composed of a series of north-south trending linear mountain ranges of generally moderate elevation that are usually no more than 120 kilometers (75 miles) long within a broad, flat basin. North and west of the Colorado River basin, most of the Basin and Range area has interior drainage; that is, streams begin and end within the region, with no outlet to the sea. One result is that much of this area has received vast quantities of alluvia eroded from the surrounding mountains.

During the late Pleistocene, substantial parts of this section of the region were covered by lakes that resulted from the melt of alpine glaciers (Figure 13-2). The largest, Lake Bonneville, covered 25,000 square kilometers (9500 square miles) in northern Utah. Most of these lakes are now gone or greatly diminished in size because stream flow is now dependent on annual precipitation; many of the lakes that remain, such as Pyramid Lake in Nevada or Utah's Great Salt Lake, are heavily saline. Flowing water always

MAJOR METROPOLITAN AREA POPULATIONS IN THE EMPTY INTERIOR, 1980	
Phoenix, Ariz.	1,509,032
Salt Lake City-Ogden, Utah	936,255
Tucson, Ariz.	531,443
Las Vegas, Nev.	463,087
Albuquerque, N.Mex.	454,499
Spokane, Wash.	341,835
Provo-Orem, Utah	218,107
Reno, Nev.	193,623
Boise, Idaho	173,036

Monument Valley, in northern Arizona, was made famous by countless "cowboy and indian" movies of decades past. These narrow buttes and broader mesas are erosional remnants of a sedimentary rock bed now largely removed by water and wind action. (Santa Fe Railroad photo by Don Erb.)

picks up small quantities of dissolvable salts; normally, these salts make a minor contribution to the salinity of the world's oceans. But because they lack an outlet to the ocean, lakes in the Basin and Range area have increased their salt concentration. The famous speed-racing course at Bonneville Salt Flats is on the former lakebed of Lake Bonneville. The Great Salt Lake, covering about 5000 square kilometers (1900 square miles), is the remnant of Lake Bonneville and today has

a salt content much higher than that of the oceans.

North of the Basin and Range region, the Columbia Plateau is the result of a gradual buildup of lava flows. Contained by the surrounding mountains, these repeated flows, each averaging 3 to 6 meters (10 to 20 feet) thick, have accumulated to a depth of 650 meters (2000 feet) in some areas. A few small volcanoes or cinder cones dot the area, but the prime features of volcanic

FIGURE 13-2

PLEISTOCENE LAKES. Large sections of the Empty Interior were filled by lakes during the Ice Age. The Great Salt Lake is a remnant of the largest of these lakes, and the even surface of Bonneville Salt Flats in northwestern Utah is former lakebed.

activity here are the vast flows of formerly molten material. Here, too, streams have eroded deep, steep-sided canyons. The canyons lack the stepped appearance of the Grand Canyon, however, because the lava beds present a more homogeneous erosive material. With some gaps, the pattern of eroded plateaus continues northward into the Yukon Territory in the area between the Rocky Mountains and the Pacific Ranges. In central Alaska, the drainage basin of the Yukon River occupies the territory from the Alaska Range to the Brooks Range. The surface materials are mostly sedimentary rocks. Large portions of the middle and lower Yukon valley are quite flat, and, thus, marshy in summer.

Climate and Vegetation

Much of the Empty Interior has a precipitation regime that is, at best, semiarid. This is increasingly true toward the region's southern zones. In the north, especially in northern British Columbia, the Yukon Territory, and Alaska, the long cold season is the dominant climatic theme. Nearly all of the area in the United States that can be classified as desert is found in this region or in the southwestern borderlands. This aridity is of fundamental importance to the region; the water deficiency has played a greater role in the limited population growth of the area than has the ruggedness of the terrain.

There is a strong pattern of greater precipitation at higher elevations throughout the interior West. In fact, a precipitation map of the region would also serve as a reasonably accurate topographic map. The entire region is almost totally dependent for surface water on the streams that flow down and outward from the mountains. These are *exotic* streams, so called because they carry water—something otherwise unknown (exotic)—into the arid plateau environment.

The association between topography and precipitation results in a marked altitudinal differentiation of vegetation throughout the Empty Interior. The lowest elevations are generally covered with desert shrub vegetation. In the far south, there is a modest late summer increase in precipitation that allows a sagebrush/grasslands combination. Elsewhere, this combination is found at elevations above the desert shrub. Upslope from the sagegrass is a tree line. The forests are at first a transitional mix of grass and small tress, like piñon pine and juniper. At higher elevations these blend into more extensive forests of larger trees, such as ponderosa pine or Douglas fir. If the mountains are high enough, smaller tree varieties and then a second tree line is encountered. Above this upper tree line, a

295

Much of the eastern two-thirds of Oregon is arid or semiarid. Here the ruts of the Oregon Trail are still clearly visible in the sage and short-grass country more than a century after they were created. (The Bettman Archive.)

combination of high winds and a short growing season render tree growth impossible, and the trees are replaced by tundra. The actual elevation at which these vegetation changes occur varies by latitude. Both tree lines are found at substantially lower elevations in the north. The boundary between forest and tundra is above 3700 meters (11,500 feet) in New Mexico, whereas in southern British Columbia, it occurs at an elevation of less than 2000 meters (6000 feet).

THE HUMAN IMPRINT

Public Land Ownership

The human use of the interior West is strongly influenced by the federal and provincial governments' extensive land control in the region. Most of the land in the Empty Interior in both the United States and Canada remains in govern-

mental hands (Figure 13-3). In Nevada, for example, the various branches of the U.S. government control almost 90 percent of all land. Although the percentages are lower elsewhere, the basic pattern of governmental predominance is universal in the region.

It is not particularly surprising that so much of the land in the interior West remains in governmental hands. In both countries, this area and the Far North were the final regions to be occupied by any substantial number of people. And in both cases, federal land programs of land distribution, designed to encourage agricultural use, were not relevant because little of the region held any real agricultural promise. In the United States, the Bureau of the Census proclaimed the end of the settlement frontier in the country in 1890, a time when much of the Empty Interior still remained unsettled. Although significant settlement expansion continued on the prairies

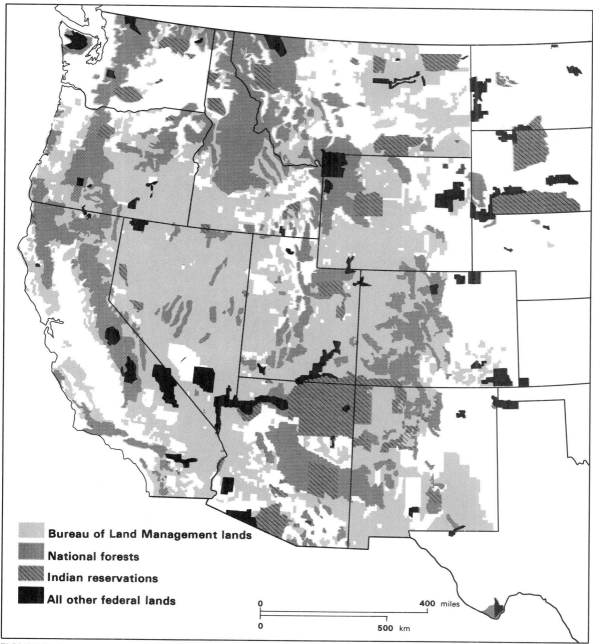

FIGURE 13-3
FEDERAL LANDS. Many Westerners question the wisdom of such widespread federal land control. They argue that it lessens the tax base, damages opportunities, and often does not even result in environmentally desirable land use practices.

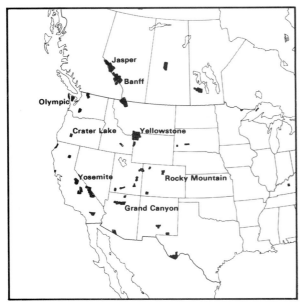

FIGURE 13-4

MAJOR NATIONAL PARKS. The interior West, with its often dramatically rugged topography, is the location of many national parks in Canada and the United States.

of Canada for several more decades, there, too, most of the interior West was ignored. Also, by the time other interests, such as lumbering or mining, began to push for greater private land ownership, the federal governments were seriously reevaluating their earlier programs in which they distributed land almost for free. Again, focusing on the United States, Theodore Roosevelt's creation of the National Forest Service in the first years of this century was largely meant to counter a threatened transfer of much of this land into private ownership. This act was a landmark in the beginnings of an effective conservation movement in the United States. Because Canadian settlement of the interior West tended to lag somewhat behind that of the United States, an even larger portion of the nonurban, nonagricultural land in Canada remains in the hands of the federal and provincial governments.

These public lands have been put to a wide variety of uses in both countries. In each, but especially in the United States, a substantial part of the total national park system is found in the interior West (Figure 13-4). The U.S. and Canadian national park systems have as their basic task the preservation of the unique (or, at least, highly unusual) sections of their respective country's natural environment, and much of the physical landscape that fits this requirement is found in the Empty Interior. Some of the continent's most famous national parks, such as Yellowstone, Glacier, Banff, and Grand Canyon, are found there.

The national parks represent the best known use of this vast resource of governmental lands, but they are only a small portion of the total public land area in the region. The largest share of these lands in the United States is held by the Bureau of Reclamation, a part of the Department of Interior, which puts this land to many uses, grazing being easily the most important. The bureau has also been the main agent in the construction of irrigation and hydroelectric dams in the area. The National Forest Service, based in the Department of Agriculture, is the second largest of the U.S. federal landholders. Much of the Forest Service's lands were set aside during a few frantic days of the administration of Theodore Roosevelt when the president feared that impending congressional action would release to private interests most federal lands that had not been specifically set aside for other purposes. The Forest Service today controls a major portion of the forested lands in the interior West. The service has traditionally emphasized logging and grazing under its multiple-use charge. In recent years, it has also increased the quality and quantity of recreational uses of the land. This has been, at least partly, in response to increased pressures on the limited facilities available in the national parks as tourist numbers have boomed.

In the Canadian portion of the region, most governmental land not in federal national parks is controlled by the provincial governments. Most of the land is forested. Alberta and especially British Columbia have used this land resource to control and expand lumber production, and the latter has also established a number of provincial parks.

Many people question this pattern of extensive governmental land ownership, particularly in the American West. There are a variety of reasons for this concern. Some believe that governmental ownership has a negative impact on regional economic growth. Governmental decisions on land use often do not satisfy local interests. No property tax is paid on the governmental land, and many feel that federal payments fail to cover that tax loss. Other critics are concerned about what they see as frequently poor management of governmental lands. A recent study indicates that misuse has damaged much of the federally owned grasslands in the West, damage that is difficult to reverse in arid and semiarid regions. The development of U.S. National Parks plus a growing number of areas set aside under the Wilderness Act of 1964 (which basically states that humans must honor a hands-off approach to all areas designated as wilderness) has eliminated most kinds of economically productive use. Perhaps the greatest concern of Westerners over this question focuses on their inability to plan and control their own region's future. Long-range economic and environmental plans must depend in large part on decisions made by the federal government, and local and national authorities often disagree on planning goals (see the Alaska discussion in Chapter 16).

Two other uses of parts of the Empty Interior say much about the region's past and about America's attitude toward the land's quality and usefulness. First, some of the largest American Indian reservations are found here, especially in northern Arizona and New Mexico. The social and economic implications of this are discussed in Chapter 14, but the existence of these reservations is partly a measure of the low regard Americans had for much of the region during the settlement era. In general, if territory controlled by the Indians was perceived as valuable, it was seized from them. Also, some of the country's largest bombing and gunnery ranges as well as its only atomic bomb testing facility are found here. The population is sparse, and alternative demands on the land are not great.

The Mormon Influence

The region's paucity of population is a principal feature of the Empty Interior. The combination of rugged terrain and widespread aridity limits the possible scope of agriculture. As the agricultural frontier moved westward in the late decades of the nineteenth century, it largely swept past the interior West. In fact, were it not for minerals, transportation, and the Mormons, few people would have chosen the region until well into this century.

Settlement patterns within the Empty Interior have been strongly affected by members of the Church of Jesus Christ of Latter-Day Saints, or more commonly, the Mormons. Established in upstate New York, the church and its followers were attacked repeatedly, both verbally and physically, for what were considered their "unusual" beliefs. The Mormons moved several times, each time searching for a place to practice their religion in peace. In 1844, a mob killed the church founder, Joseph Smith, in the Mormon town of Nauvoo, Illinois (which at the time was vying with Chicago for the honor of being the largest city in the state). Over the next few years, many Mormons, often on foot, pushed into the West to settle far from American settlement where they hoped to create an independent Mormon state.

The locale they selected for their initial western settlements was the Wasatch Valley, tucked between the Wasatch Mountains and the Great Salt Lake in northern Utah. It must have seemed an unlikely spot to begin an agricultural settlement. The climate was dry, the lake saline and useless, and the landscape barren. Nevertheless, the Mormons quickly began their agricultural operations; their settlements expanded as new arrivals came, for the Mormons were, and are, fervent missionaries, and a high birth rate pushed their population numbers upward. They dreamed of founding a country to be called De-

The Great Salt Lake Basin, Utah, along the western edge of the Wasatch Mountains, is a very dry area. Careful use of the meltwater from the mountains has enabled the Basin's farmers to create a major mixed-farming agricultural region. (Georg Gerster/Rapho—Photo Researchers.)

seret, stretching northward into what is now Oregon and Idaho and southward to Los Angeles, and Mormon communities were established at greater and greater distances from Salt Lake City.

The Mormons ultimately failed in their hopes for creating Deseret. American expansion moved through and beyond the Mormon area, and the Mormons again found themselves under the will of the United States. Interestingly, the initial site that the Mormons had found so satisfying in its isolation lay directly astride what would become the California Trail, and the discovery of gold in California initiated a large movement of non-Mormons through the Mormon settlements. Deseret was divided eventually among a half-dozen different states.

This failure did little to lessen the impact of the Mormons on the settlement of the region. They were the first Americans to face the problems of life in the interior West, and they solved the majority of them. Many of their solutions were later adopted freely by non-Mormons. None was more important than their innovations in dryland irrigation techniques. Americans had previously had no need for extensive irrigation. The techniques and careful central control necessary to collect and move water to a large number of agricultural users were almost unknown in North America. The Mormons, with their theocratic government providing the strong central organization, constructed a large number of storage dams on the western slopes of the Wasatch Range; many miles of canals moved the water to the users in the valley below. The results of these efforts today cover much of the valley with agricultural crops, trees, and green lawns. Even a casual observer flying over the area in early summer is struck by the greens of the irrigated fields contrasting dramatically against the browns of unirrigated portions of the valley. These early efforts at large-scale irrigation were the beginning of an irrigation boom in the Empty Interior, and today most of the agricultural crops here are produced on irrigated lands.

Mormons continue to have a substantial impact on the interior West. Of the roughly 10 million persons found in this region, over 1.5 million are Mormons. They comprise three-fourths of the population of Utah, a majority of that in south-central Idaho, and substantial populations in Nevada, northern Arizona, and western Wyoming. Mormon beliefs are definitely procreative, and Mormons have by far the highest birth rate of any major religious denomination in the United States. Their strong belief in cooperation and support within the church has made this religion a central influence in business as well as family life in its region. For example, one of the two principal department stores in Salt Lake City is owned by the church. The Mormons have created a dynamic, optimistic society, although one that some argue is closed to most non-Mormons.

The regional boundaries and variable nature of this distinctive culture region have been defined by Donald Meinig (Figure 13-5).[1] He defines the core, with its focus on the populated Wasatch Valley, as the center of Mormondom. The domain, covering most of the rest of Utah plus southern Idaho, is dominated by Mormons, but it lacks the complexity of the core. The sphere is that area where substantial numbers of Mormons live in a largely non-Mormon society.

FIGURE 13-5

THE DISTRIBUTION OF MORMONDOM (reproduced by permission from the *Annals of the Association of American Geographers*, **55** (1965), Donald J. Meinig). More than 90 percent of the population of some rural counties of southern Utah are Mormon. By comparison, Mormondom's core in Salt Lake City has a large non-Mormon population.

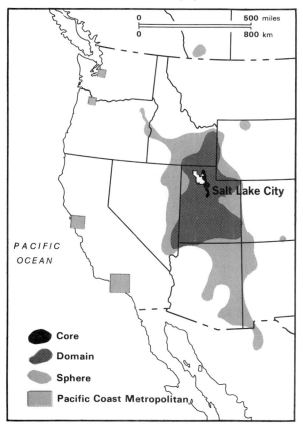

Core
Domain
Sphere
Pacific Coast Metropolitan

A DISPERSED ECONOMIC STRUCTURE

Irrigation and Agriculture

As mentioned, irrigated crops dominate the overall agricultural production of the Empty Interior (Figure 13-6). Without irrigation, no stable agricultural economy would be possible in this dry region. Today much of the flow of several of the more important rivers is diverted for various uses, with irrigation claiming the lion's

[1] Donald W. Meinig, "The Morman Culture Region," *Annals of the Association of American Geographers*, 55 (1965), pp. 191–220.

301

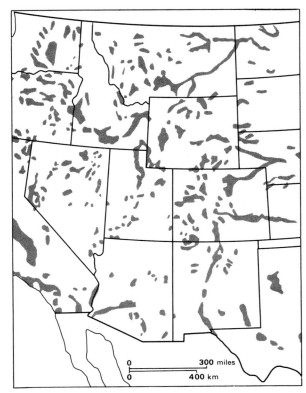

FIGURE 13-6
IRRIGATED AREAS. These areas are important in
the interior West as settlement foci in a region of
generally sparse population as well as the location
of most of the region's agricultural production.

share. This massive diversion would be impossible under the water laws in effect in most parts
of the United States, which usually require that
usage not noticeably diminish stream flow. The
unique water problems of the Empty Interior
resulted in the development of a new set of standards, the Doctrine of Appropriation, that permits total diversion of a stream.

The Reclamation Act of 1902, passed with
strong Western backing, provided for federal
support for the construction of dams, canals, and
eventually hydroelectric systems for the 17 states
(excluding Alaska and Hawaii) west of the 100th
meridian. Today over 80 percent of the water

from these federally supported projects is used
for irrigation, much to the consternation of many
urban water users. Over 4 million hectares (10
million acres) is irrigated with this water. Agricultural water users have been required to pay
only a small portion of the real cost of supplying
irrigation water. A 1981 report by the San Francisco Federal Reserve Bank suggested that
farmers in California's Central Valley between
1948 and 1981 had paid somewhere between 2
and 20 percent of real costs for federal water,
depending on the accounting system used.

The use (some would argue misuse) of the
largest river in the region, the Colorado, offers
an interesting example of several aspects of the
region's attitude toward water ownership. By
1915, nearly the entire flow of the river had been
allocated, mostly at downstream locations in Arizona and California. This meant that the upper
basin states of Wyoming, Colorado, Utah, and
New Mexico had little right to the stream flow
that originated largely in their mountains. In
1922, the Colorado River Compact reallocated
these water rights, giving half of the river's total
allocated flow of 15 million acre-feet to the upper basin and half to the lower basin states of
Nevada, Arizona, and California.

Most streamflow in the region results from the
spring and summer melt of winter snows. The
amount of winter snow accumulation is, thus,
quite important, and quite variable. Several winters in the late 1970s were dry in the Sierra Nevada and central Rocky Mountains, with resultant water deficiencies the following summers.
The winters of 1982–83 and 1983–84 were two
of the snowiest in history. Water spilled over the
top of Boulder Dam, Salt Lake City turned one
of its main streets into a temporary river, and
the Great Salt Lake, historically shrinking because of irrigation, reached its largest areal extent in over a century. The lake's water, six times
saltier than the ocean in the summer of 1982,
had only three times the ocean's salt content by
the summer of 1984.

Heavy irrigation, especially in the Imperial
Valley in southwestern California, greatly in-

Hoover Dam and Lake Mead on the Colorado River. The canyonlands of the Colorado River are an efficient area for water storage—the lakes can be very deep, minimizing water loss due to evaporation. The several lakes along the Colorado and its tributaries have created a major power-boat recreation industry in the arid Southwest. (Bureau of Reclamation/D.O.E.)

creases the salt content of the river by the time it flows past the U.S.-Mexico border. This salt buildup occurs as farmers use part of their water allocation to "flush" away the salts that have been deposited as the irrigation waters evaporate rap-idly in the hot, dry environment. Without this flushing procedure the salt would eventually render the land unusable, a fate that befell large sections of the ancient irrigated lands of North Africa and South Asia. Mexicans argue that this

303

water, directed back to the river by return canals and, hence, to Mexico as part of its allocation, creates a river flow that is so salty that it has a strong negative impact on the quality of irrigated agricultural land in Mexico. Studies have indicated that these complaints are justified, and the United States has promised to reduce this practice. The state of California, fearful of the possible loss of agricultural lands in the Imperial Valley because of salt buildup (which some experts suggest could happen within a half-century, even with moderate flushing), has recently drilled a series of deep wells. It is hoped the wells will provide a supplemental supply of almost salt-free water to the Imperial Valley.

Competition among the states for the Colorado River's water is intense. The states of the upper Colorado have, until recently, used relatively little of their allotted share. The lower states, California and Arizona, responded by increasing their withdrawals until they had included most of the water "surrendered" by their northern neighbors. Arizona, confronted with water supplies inadequate to support its continuing population and economic booms, argues that California's share is too large and Arizona's too small. The northern states, especially Colorado, have no desire to relinquish permanently any of their allotted share. The Big Thompson River project, discussed in the previous chapter, represents one part of Colorado's move to increase its usage.

The original allotments of Colorado River water have been modified in response to these conflicting interests. The total amount of water distributed will be reduced. Mexico's share will be increased substantially, water available to the upper basin states will be reduced, whereas that for the lower basin states (primarily California and Arizona) will be increased somewhat. No decision can possibly satisfy everyone, and additional reallocations are likely in the future.

The Colorado River projects are simply the most notable of a number of examples of intensive irrigation development in the intermountain region. Diversion by ranchers of water from

the principal stream flowing into Pyramid Lake in western Nevada has led residents of the Indian reservation to demand a halt to the practice before their lake disappears. Irrigation projects along the Gila and Salt rivers in south-central Arizona were using virtually the entire flow of the Gila by the early 1930s. Farmers and the growing urban populations in the area then turned to wells for additional supplies. Today Arizona depends on these receding subsurface waters to supply several times as much water as the Gila-Salt river system. In short, the pressures on a deficient water supply are growing more acute throughout the drier southern half of the region.

Future water resources are a matter of pressing concern throughout the dry west. Politics and economics have discouraged the federal government from proposing major new impoundment and diversion programs. Urban and industrial water users demand a larger share of the water. (By some estimates, Arizona could support 20 million people with its current water supplies instead of its present 3 million people if none of its water went to agriculture.) In 1982, Arizona was overdrawing its subsurface water by an estimated 2 million acre-feet annually and groundwater levels had dropped by over 30 meters (100 feet) in much of the Gila and Salt River valleys.

As an example of an innovative response to these problems, Arizona in 1980 passed a comprehensive water-planning law. All large wells are now metered and withdrawal fees charged. Water rights can now be sold, with the expectation that they will be obtained by urban and industrial users. If that does not happen, the state will eventually begin buying farmland with water rights and retiring it from agriculture. The goal is to reduce the state's 526,000 hectares (1.3 million acres) of irrigated land to 324,000 hectares (800,000 acres) by 2020. Finally, the state is paying most of the cost of the Central Arizona Project, the goal of which is to build an aqueduct to carry water from the Colorado River to the Salt and Gila valleys.

Although irrigated lands are scattered across much of the Empty Interior, a few areas are sufficiently large to merit special mention. The 1 million hectares (2.5 million acres) irrigated in the Snake River Plain makes Idaho the region's leader in terms of the number of acres irrigated. This enables the state to be among the country's leaders in sugar beet and potato production; alfalfa and cattle are also important. The Columbia Valley Reclamation project, supplied by the bountiful waters of the Columbia River impounded behind Grand Coulee Dam in western Washington, contains well over 400,000 hectares (1 million acres), producing such crops as alfalfa, sugar beets, and potatoes. Irrigation along the Wasatch Valley has expanded little since the first decade of Mormon settlement. About 400,000 irrigated hectares there are devoted primarily to sugar beets and alfalfa. The Grand Valley, along the Colorado River in west-central Colorado, produces alfalfa, sugar beets, and potatoes as principal crops, although tree fruit, especially peaches, are also important. In Washington, a series of tributaries of the Columbia River flowing eastward out of the Cascades, notably the Yakima and Wenatchee in Washington and the Okanagan in British Columbia, supply water for both countries' most famous apple-producing regions. The Okanagan Valley (spelled Okanogan in Washington) has been irrigated since the 1860s. It was one of the first major irrigated areas in the Interior West. A combination of an attractive product and extremely effective marketing have enabled this dry area with an average annual precipitation of less than 25 centimeters (10 inches) to grow a third of each nation's apple crop.

It should be noted that each of these areas produces a limited set of crops. The short growing season precludes production of most long-season crops. Local demand is limited, minimizing the need for dairying or many fresh vegetables. Livestock, sugar from beets, apples, and potatoes, however, can be grown locally and shipped profitably to distant markets.

The more southerly irrigated districts, al-though particularly affected by irrigation-water shortages, nevertheless, have one major advantage over their northern counterparts—a far longer growing season. The Imperial Valley, with a frost-free period in excess of 300 days, is currently one of the premier truck-crop producing areas in the country. Much of the winter head-lettuce supply in the country comes from here, as do grapes, cotton, and alfalfa for fattening beef. The growing season is long enough to support double-cropping (growing two crops annually on the same area of land), and of course this increases overall productivity. The Coachella Valley north of the Salton Sea produces such crops as dates, grapes, and grapefruit. The Yuma Valley along the lower Colorado River supplies principal crops of cotton, sugar beets, and oranges. In the Salt River Valley, winter lettuce, oranges, and cotton are the major crops. These southern crops, unlike those grown farther north, face little direct competition from agricultural centers in the major market areas of the eastern United States.

Transportation Services

Transportation has also played a major role in the population size and economic development of the Empty Interior. With the growth of population along the West Coast, the interior West has constituted a broad barrier to transportation between more densely populated areas in both countries. Because little traffic is generated within the area, a prime goal of transport developers has been to permit movement across the region as speedily and inexpensively as possible. Consequently, most major highway and rail routes pass through the region east-west, from the urban centers of the Midwest to those of the West Coast, as directly as the topography will allow.

Despite these requirements of transport design, the great width of the region demands development of many service facilities for the traffic passing through. Many of the smaller towns of the region as well as a fair number of the larger ones began as rail centers established to service and administer the railroads. These centers were founded wherever railroad personnel were needed, whether the region was populated or unpopulated. The arrival of automobile traffic in this century provided a demand for gas stations, car-repair facilities, motels, and restaurants for travelers. Although fewer local railroad workers were needed as technological innovations were implemented after World War II, this population reduction was more than compensated for by the growing need for people to serve truck and automobile traffic. Few regions today have as large a share of the total work force involved in transportation services as the Empty Interior.

Although the provision of transport services was the principal early influence on the growth of urban centers in the region, cities that have become the largest have usually been aided by some additional attribute. Spokane, Washington, with a population of 340,000, has, for example, become the principal center for the "Inland Empire" of Washington. That area, geographically defined and half encircled by the sweep of the Columbia River across central Washington state, has long been a region with substantial agricultural production. Albuquerque, New Mexico, with a Standard Metropolitan Statistical Area (SMSA) population of 455,000, has gained a role similar to Spokane's through its centrality and accessibility in that state. Phoenix, Arizona, grew initially as an agricultural center and then boomed as Americans flocked to its warm, dry environment. Its 1.5 million people rank it 26th among the country's urban regions. It has become a retirement center as well as a focus of manufacturing activities. Industries that produce a small, high-value product, such as the electronics industry, have been particularly important in the city's growth. The higher transportation costs engendered by locating a great distance away from most of their markets are not of major significance to these light industries. The amenity factor of outdoor recreation opportunities, which attracts both labor and management, far outweighs the slight economic locational disadvantage. Locational considerations that largely ignore a traditional factor, such as accessibility to either resources or a major market, have grown more common as technological advances place a greater share of all manufacturing in such industries (Chapter 5).

Ogden, Utah, is one community that exists as a major rail center and was early among the most important such places in the region, but it has not become a major urban place. It is about 55 kilometers (35 miles) north of Salt Lake City, a city whose continued dominance derived from its early development and its key functions as the capital of Mormondom.

Tourism

Any mention of automobiles and travelers in the Empty Interior brings to mind the great importance of tourism to the regional economy. The great variety and appeal of the region's scenic wonders attracts millions annually. Most of the major attractions are controlled by the federal government, but this has not slowed the growth of tourist-oriented industries. Visitors to most of the major parks must first pass a long, garish strip of motels, snack bars, gift shops, and other sources of ersatz local color. In addition, distances between attractions are usually great, and services are, thus, needed in countless locations. When Nevada's gambling operations are included as part of the area's tourist industry, the overall regional impact of tourism becomes even greater.

The total volume of tourism has grown so much in this thinly populated region that many of the tourist attractions have become overburdened. The U.S. National Park Service, faced with substantial local environmental damage from overuse of its facilities, has chosen to cut facility availability in some parks. In Yellowstone, where the summer morning lineup to obtain emptying campsites is as much a part of the landscape as the roadside bears once were, the service has reduced the number of campsites. Part of the road along the south rim of Grand Canyon National Park has been closed to private vehicles, as have sections of the road system in Yosemite Valley

Yosemite Valley in Yosemite National Park, California. Eroded primarily by a series of alpine glaciers, this is one of America's premier natural features. The valley is so heavily used in summer that the National Park Service has been forced to limit camping and to close a portion of the valley to private automobiles. (National Park Service.)

in Yosemite National Park in California. One must have a reservation to stay in many of the more populated national park campgrounds during the busy summer season. This tremendous tourist growth has exerted pressure on the National Forest Service to provide recreational facilities on its lands to decrease the pressure on the national parks, a program encouraged by the Federal Lands Policy and Management Act of 1976, which fosters multiple use of National Forest Service and Bureau of Land Management lands. This has been done in some areas. Many also hope that substantial portions of these forest lands will be selected under the Wilderness Act of 1964, which would restrict human use of areas designated as wilderness.

In spite of a trend toward excess demand, tourism can fluctuate, with consequent effects on the local economy. The summer of 1979 was particularly difficult for the tourist industry in much of this region of the United States. The sudden dramatic increase in gasoline prices plus concern over its availability encouraged many would-be travelers either to remain home or visit less distant locations. The result was a business decline that averaged about 30 percent that year. Although gasoline prices later declined, that summer's experience demonstrated the fragile nature of the region's tourist economy. If changing energy prices produce a significant alteration in American driving habits, the interior West could be especially affected because of its great distance from the major concentrations of American population.

Lumbering and Ranching

Ranching, lumbering, and (to a lesser extent) mining depend on governmental land for many of their basic materials; the holdings of the National Forest Service and the Bureau of Land Management are open to grazing and most lumbering here is carried out on Forest Service lands in the United States and Canada. Levels of production per acre for both ranching and lumbering products are relatively low, particularly in the United States. This is especially true when a comparison is made with land held in private hands.

There are several reasons for this apparent inefficiency. A greater interest in conservation on the part of governmental agencies is probably not one of these reasons, for there is little evidence to support that conclusion. A more likely explanation is the limited quality of the land, particularly that controlled by the Bureau of Land Management. In many drier areas, 40 hectares (100 acres) per head is needed for satisfactory cattle grazing. Many who oppose the government's land-use policies contend that poor management has resulted in widespread overgrazing of federal lands and that the land's carrying capacity has declined as a result.

The great seasonal climatic variations found in much of the region make this one of the few areas in North America where *transhumance* is practiced. This is a seasonal movement of flocks by those who tend them from the lowlands in winter to mountain pastures and meadows in summer. It is especially important in the sheep-

Overgrazing in the late 1800s destroyed much of the sparse grassland vegetation that formerly covered large sections of the interior West. This area in Arizona has been protected from grazing for 50 years; still, little grass has returned because of competition from mesquite trees. (U.S. Forest Service.)

ranching economy. Many Basques, expert shepherds from the Pyrenees of Spain and France, came to this area as contract laborers to manage the herds. Today descendents of the Basques are a substantial part of the population of several states, especially Nevada.

The wood products industry in British Columbia remains the cornerstone of the provincial economy. Wood products and paper and allied products are its two most important manufactured goods. Over 90 percent of the forest is Crown Land—held largely by the provincial government. The government has actively encouraged both cutting and reforestation.

Mining

The other element of the region's economic resource base that has supported the growth of substantial population nodes is mining. The Kondike gold discovery in 1898 was important in the early settlement of northern British Columbia, the Yukon Territory, and the Yukon Valley of Alaska. In the United States, miners followed shortly on the heels of the Mormons to become the second largest group of settlers in the region. The discovery of the Comstock Lode in Nevada gave rise to Virginia City, which grew into a city of 20,000 during its heyday before nearly disappearing with the decline of high-quality ore. The boom in gold and silver mining in the years immediately following the Comstock discovery created rapid population growth in Nevada. This culminated in the admission of the state into the union in 1864, long before most of its neighbors. The depletion of much of this mineral resource by the late nineteenth century resulted in a widespread population decline in Nevada, one from which it did not fully recover until well into the twentieth century. Today the mining economy is of little importance to the state, although some of the abandoned mining centers, notably Virginia City, are important tourist attractions. In fact, gold, a foundation of the mining boom over much of the region during the last century, is today of little economic importance in any part of the Empty Interior.

Continued high gold and silver prices have encouraged some renewed mining, but most of this new interest remains at a small scale of operation.

Leading today's list of mineral contributions to the region's economy is copper, with production concentrated in Arizona, Utah, and Montana. The vast open pit of the Bingham mine outside Salt Lake City, the largest human-made excavation in the world, has yielded some 8 million tons of copper. Among the several score major and minor copper-mining centers in Arizona, the most important is at Morenci in the eastern part of the state. Other important mines are at San Manuel, Globe, and Bisbee, all in southern Arizona.

Most of the copper ore mined in the Empty Interior is low grade, with a metal content of under 5 percent. Consequently, most mines have a smelting or concentrating facility located nearby to lessen shipment costs. Refining is, thus, a major manufacturing industry in the region, employing, for example, 14 percent of the total manufacturing labor force in Arizona.

Lead and zinc follow copper in regional importance in the United States but are of greater importance in the Canadian portion of the region. The vast mine near Kimberley, British Columbia, dominates Canadian production of both metals. Production is more widespread in the United States, with the two often joined by several other metals mined at the same location. The Butte Hill mine, for example, was long a significant producer of lead and zinc as well as copper. The Coeur d'Alene district in northern Idaho produces gold, silver, lead, and zinc; the Leadville district in Colorado has those four plus molybdenum. The latter metal now ranks above silver in regional importance owing primarily to work at the Climax mine north of Leadville. Some three-quarters of the world's supply of molybdenum comes from the Leadville district. Uranium exploration has also been widespread in the region, and today Utah and Colorado are the principal producing states. The northern interior of British Columbia is one of Canada's main

Butte, Montana, is built atop a mountain of ore rich in a number of minerals. Expansion of the Berkeley Pit, the large mining excavation in the foreground, nearly pushed the city off its site. (Craig Aurness/Woodfin Camp.)

FIGURE 13-7

OIL SHALE DEPOSITS. Few people presently live in this isolated area. How much local population growth would a major development of this resource create?

resource frontiers. Large coal deposits are being developed for export to Japan.

The economic recession of the early 1980s was accompanied by increasingly effective foreign competition and, in some cases, by concern over the environmental and health consequences of mining and smelting. This was devastating to the interior West's mineral economy. The massive Butte Hill mine in Montana is closed, its underground mines flooded because the constant pumping needed to keep the mines dry has been stopped. The economies of Butte and the nearby smelting center of Anaconda have been severely affected. The large zinc smelter in Kellogg, Idaho, is closed. Many other mines have cut production.

One potential resource of the region must be emphasized. Spread across thousands of square miles of the area where Utah, Colorado, and Wyoming meet are the vast oil shale deposits of the Green River Formation (Figure 13-7). Locked in these rocks are as much as a trillion barrels of oil, vastly more than the entire proven alternative oil reserves of the world, including those of the Athabasca tar sands of Alberta (see Chapter 12). The technology needed for economically feasible extraction, however, is just developing. Processing the shale requires crushing and heating the material, so that the oil turns to gas and separates from the rest of the rock. The gas is then cooled, forming a liquid oil-like substance. Unfortunately, much of the shale is covered by overburden that is several hundred feet thick in many places. It has been suggested that the shale-converting process may leave behind enough waste to level the topography of the entire production area if most of the petroleum is removed. Also, most of the extraction process requires large amounts of water, a problem in this dry area where most available water is already used for agriculture. These major environmental conflicts are likely to become more serious as the country's demand for energy grows.

There has been little sustained or substantial urban growth based on these mineral resources. As with all such resources, they are ephemeral, with depletion a standard segment of the extrac-tion process. Cities, other than temporary boom towns, seldom develop on such a basis. In this region, copper provides the only real exception. Even in this, the results are modest. Butte, with a population of 37,000 in 1980, is perhaps the region's largest city developed with mining as the main base of the economy, yet it is an important processing center for agricultural products as well.

In summary, this is a region of sparse settlement, vast open spaces, and a quality of ruggedness and a sense of the outdoors that greatly appeals to its residents. The population is concentrated largely in isolated nodes. They know theirs is a region to which many travel thousands of miles for vacations, and they are generally firm in their belief that life in their region is superior. The Empty Interior's isolation and its lack of a broad resource base are clearly important here, making it for many, then, a delightful place to live, but a difficult one in which to earn a living.

Lake City, New Mexico, once a thriving mining community, is now nearly abandoned. "Ghost" towns are common in the Empty Interior. Developed to serve mining, they lost their economic function entirely when the mineral resource was gone. New Mexico alone has 2000 such ghost towns. (Courtesy New Mexico Tourism and Travel Div. Commerce and Industry Department.)

Additional Readings

Ballard, Stephen C. *Water Policy and Western Energy.* Boulder, Colo.: Westview Press, 1982.

Brown, Robert H. *Wyoming: A Geography.* Boulder, Colo.: Westview Press, 1980.

Clawson, Marion (ed.). *The Federal Lands Revisited.* Baltimore: Johns Hopkins University Press, 1983.

Cole, David B., and John L. Dietz. "The Changing Rocky Mountain Region," *Focus,* 34 (1984).

Federation of Rocky Mountain States. *The Rocky Mountain Region: Unity of Interests.* Denver: Federation of Rocky Mountain States, 1975.

Francaviglia, Richard V. *The Mormon Landscape.* New York: AMS Press, 1978.

Franklin, Philip L. *A River No More: The Colorado River and the West.* New York: Alfred A. Knopf, 1981.

Griffiths, Mel, and Lynnell Rubright. *Colorado: A Geography.* Boulder, Colo.: Westview Press, 1983.

McPhee, John. *Basin and Range.* New York: Farrar, Straus & Giroux, 1981.

Rogers, Garry F. *Then and Now: A Photographic History of Vegetation Change in the Central Great Basin Desert.* Salt Lake City: University of Utah Press, 1982.

Stegner, Wallace. *The Gathering of Zion: The Story of the Mormon Trail.* New York: McGraw-Hill, 1964.

Stegner, Wallace, and Page Stegner. "Rocky Mountain Country," *Atlantic Monthly,* 241 (1978), pp. 44–91.

CHAPTER FOURTEEN

THE SOUTHWEST BORDER AREA: TRICULTURAL DEVELOPMENT

The Southwest is a distinctive place to the American mind but a somewhat blurred place on American maps which is to say that there is a Southwest but there is little agreement to just where it is.

DONALD W. MEINIG

The area known generally as the Southwest is one of the most widely recognized yet also one of the most transitionary of American regions (Figure 14–1). Many Americans, when asked to regionalize the United States, will create some place that they call the Southwest. Its extremities may be vague, but its existence is clear. However, many geographers, when asked to do the same task, create no place that even vaguely resembles the Southwest. Of the existing textbooks on the geography of the United States and Canada, probably no more than half choose to identify the Southwest as a separate place. We do feel that it should be treated as an identifiable region, but, at the same time, we include almost all of its territory in four other regions. No other region shares portions of its territory with as many other regions as the tricultural border area.

This combination of a clear regional identity for most Americans but a rather ambiguous one for geographers is unusual, although its roots are understandable. As mentioned in Chapter 1, regions of the kind used in most textbooks are mental constructs, somewhat dependent on the whims of their creators. Even so, identifiable regionalizations follow some logic. For geographers, this logic frequently depends heavily on a distinctive combination of physical and economic conditions. The Southwest, however, is a region of substantial economic diversity. Moreover, its apparent physical uniformity can be attributed primarily to its aridity but it is little different than the Empty Interior in this regard. Topographically, the region includes the broad flatlands of the lower Rio Grande Valley, the plateaus of New Mexico, the dramatic mesas and buttes of Monument Valley in Arizona, and the Sangre de Cristo Mountains of New Mexico.

Most Americans, we feel, would call the Southwest a region because of its clear, dry climate. To us, it is a culture region; it is made distinctive by the coexistence of Spanish-American, American Indian, and Northwest European (Anglo) cultures. Its special character results from this cultural complexity; although

313

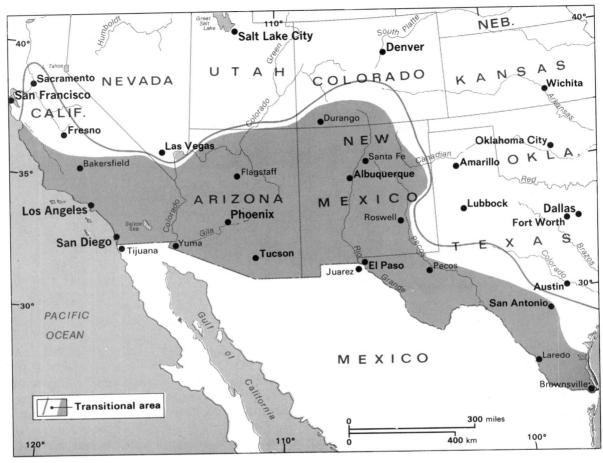

FIGURE 14-1
THE SOUTHWEST BORDER AREA.

somewhat unusual, the physical environment is almost like a stage that serves to emphasize aspects of each culture and further sets apart the region from the rest of the country.

The tricultural border region is preponderantly non-Spanish and non-Indian. Perhaps 1 person in 5 has a Spanish surname, and little more than 1 in 100 is American Indian. The expectation, then, might be that these minority populations would be engulfed by the much larger and relatively homogeneous Anglo population. This has not been the case. Both minority groups have had a major sustained impact on the region. No other region in the United States

is as clearly tied to ethnic patterns. The region's outer limits are directly defined by the distribution of the two minority groups.

The canyonlands of northern Arizona and southern Utah provided an effective barrier to Spanish expansion northward from Mexico. The Spanish moved up the Rio Grande to the broad expanse of the Rocky Mountains, north of which little Hispanic settlement developed. In Texas, most settlement remained concentrated along the Rio Grande and the Nueces River. The extensive cattle-grazing industry that the Spanish introduced into south Texas was not well suited to the moist forested lands of the eastern portion

The adobe brick house of a Mexican-American family in southwest Texas. The Spanish/Mexican cultural influence and dry climate of the region has encouraged the use of easily eroded adobe brick in construction. (Baldwin—Watriss/Woodfin Camp.)

of the state. That area, left as a frontier, was largely unsettled by the Spaniards. Most migration by Spanish-Americans beyond this original settlement area has been to urban places, where their cultural impact has been far more limited. Thus, the region was most clearly established by the extent of Spanish settlement, and its basic outline was manifested well before the arrival of the westward migrating American population.

The aridity of Arizona, New Mexico, and bordering areas in Utah and Colorado discouraged large-scale Anglo agricultural settlement in the nineteenth century. The slow pace of Anglo penetration ensured that substantial numbers of American Indians remained in those states, although Indian strength was as important as An-

glo hesitancy. The Pueblo of the upper Rio Grande Valley had developed the technologically most advanced pre-European Indian civilization in what was to become the United States; they remain important in New Mexico. The Navajo, Hopi, and Apache, all primarily in Arizona, also survived the European wave better than had most eastern tribes.

ETHNIC DIVERSITY

Most Americans take pride in the concept of "America, the melting pot," in which numerous and diverse peoples have formed a population now united by common goals and sharing basically a common culture. There is much truth in

315

MAJOR METROPOLITAN AREA POPULATIONS IN THE SOUTHWEST BORDER REGION, 1980	
Los Angeles-Long Beach, Calif.	7,477,503
Anaheim-Santa Ana-Garden Grove, Calif.	1,932,709
San Diego, Calif.	1,861,846
Riverside-San Bernardino-Ontario, Calif.	1,558,182
Phoenix, Ariz.	1,509,052
San Antonio, Tex.	1,071,954
Tucson, Ariz.	531,443
Oxnard-Simi Valley-Ventura, Calif.	529,174
El Paso, Tex.	479,899
Las Vegas, Nev.	463,087
Albuquerque, N.Mex.	454,499
Bakersfield, Calif.	403,089

the melting pot idea: 45 million Europeans moved to the United States, and many of their progeny continue to hold dear the customs of their European homelands. In such divergent characteristics as eating and drinking habits, religious preferences, and musical tastes, Americans demonstrate their heterogeneous ethnic heritage. As we have seen in Chapter 5, they sometimes choose to live with others of the same background in closely knit neighborhoods. Yet nearly all of these people are Americans who have far more in common with other Americans than with the present residents of their European homelands.

American Indians

It is somehow ironic that the residents of the United States who have been least well integrated into this American cultural fabric are those who have had the longest contact with American soil. Of all the country's major ethnic groups, American Indians are the least accepted in American society. They, in turn, have shown considerable reluctance to become members of that society. Of course, American Indians are not all members of a single ethnic group; the variety and complexity of American Indian cultures is great. But because most other Americans are ignorant of American Indian culture, this has created a bond among the Indians, bred of adversity. This association is by no means simply a creation of the contemporary Red Power movement. On a fateful day just over a century ago, General Custer faced a proud alliance of Plains Indians at the Battle of Little Big Horn. Many in that alliance were former enemies; they had joined forces only in response to the engulfing tide of U.S. settlement.

Although there are American Indians in every part of the United States, today most are found in areas that white settlers had deemed undesirable or in areas that were not part of the earlier settlement frontiers (Figure 14-2). The tribes of the East have largely disappeared, succumbing more often to disease than to the bullet; others were moved westward, out of the way of European settlement. The Sioux and other Plains tribes and the tribes of the interior West (particularly in the Southwest) fared far better numerically. Their lands were not easily suited to crop agriculture and, within a few decades after their first substantial contact with the United States, the wanton governmental disregard of Indian rights was replaced with a somewhat more benign approach. As a result, the great majority of all American Indian reservations as well as most Indians are in the Plains and Western states (Figure 14-3).

The American Indian population is by no measure evenly distributed across the region. Most American Indians are found in the major reservation areas, especially those centered on the Four Corners area where the states of Colorado, Utah, Arizona, and New Mexico meet and in the cities of California. Arizona, with over 150,000 Indians in 1980, ranked a close second to Oklahoma in the size of its American Indian population. California was third; New Mexico was fourth. The Navajo reservation had 10 times the population of any other reservation. The Los Angeles-LongBeach SMSA had far more Indians than any other urban area.

Taos Pueblo, New Mexico, is a frequently visited tourist attraction. It is also an Indian village, one that has been inhabited for hundreds of years, and is the best known and perhaps most picturesque of several score pueblos that continue to exist in the state. (U.S.D.A.)

Hispano-Americans

What is in a name? Sometimes a great deal. For the Spanish-surnamed population of the Southwest, the issue of what to call themselves has been important but difficult to agree on.[1] Many younger militants, angered over injustices suf-

fered at the hands of Anglo society, call themselves *la raza* (the race) or *Chicano*. Recent migrants from Mexico also often prefer the latter term. Traditional Spanish-surnamed people from the Southwest prefer *Hispano*. Many in Texas call themselves *Texanos*. *Mexican* is usually reserved for people born in Mexico, although some would also use the term when referring to the older, rural, conservative population. Somewhat arbitrarily, we will use *Hispano*.

History texts in the United States commonly

[1] For an interesting discussion of this problem of nomenclature, see Arthur L. Campa, *Hispanic Culture in the Southwest*. Norman: University of Oklahoma Press, 1979, pp. 6–10.

317

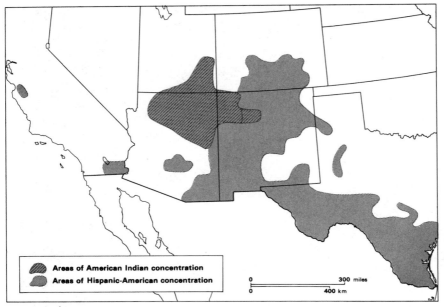

FIGURE 14-2

AREAS OF HISPANIC AND AMERICAN INDIAN POPULATION. Hispanic
or American Indian populations numerically dominate much of rural New Mexico, south Texas, and northern Arizona.

FIGURE 14-3

AMERICAN INDIAN RESERVATIONS. Ignored by Anglo settlers in the nineteenth century as of little economic worth, some reservation lands are now valued for their resources, notably the coal deposits that underlie portions of the
Navajo Reservation. Most, however, remain isolated and outside the main pattern of U.S. economic growth.

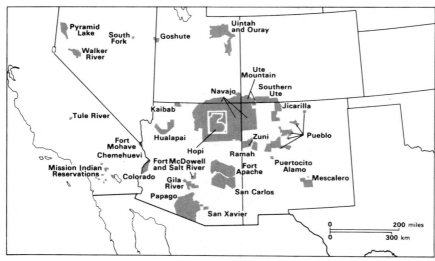

cite the early coastal settlements, such as the Raleigh colony in North Carolina or the Pilgrim landing in Massachusetts, as the start of European occupation of the country. In fact, Spanish settlements in Florida and the Southwest easily predated the English arrivals. St. Augustine, founded in 1565, validly claims to be the oldest European city in the United States. Some southwestern Hispanic communities were founded more than 200 years before Anglo settlement of the region began during the early nineteenth century.

All of what is today the southwestern United States was incorporated into the Spanish Empire during the early years of the sixteenth century. By 1550, the Spanish had explored widely across the region. They found little of the gold and silver they so eagerly sought, but their travels did serve to establish a firm Spanish claim to the territory. The lack of any identified, easily extractable riches coupled with the great distance to the core of Spanish development in Mexico (near present-day Mexico City) minimized Spanish concern for their northern territory. Before 1700, the only permanent Spanish settlements north of the present U.S.-Mexico border were along the valley of the upper Rio Grande in New Mexico. Santa Fe was founded in 1610 and other *pueblos* (civilian communities that could loosely be viewed as small towns), notably Taos and Albuquerque, soon followed. This area was to remain the most important part of northern Mexico.

A tentative Spanish occupation of Arizona began in 1700. The Apache Indians were a constant threat, repeatedly raiding Spanish settlements in the area; Arizona, consequently, remained a small settlement outlier. Colonization of Texas began at about the same time, with long-range results that were considerably more successful. Nacogdoches was founded in 1716, followed two years later by San Antonio. During the middle 1700s, the lower Rio Grande valley was settled. Still, by the early nineteenth century these and other centers of Hispanic settlement were viewed by the authorities as a small, inadequate occupation when compared with the large

number of Americans then pushing westward toward Texas. Thus, foreigners, mostly Americans, were allowed to develop settlements there during the 1820s and 1830s; it was hoped that these settlers would be loyal to Mexico (independent from Spain after 1817) and form an effective bulwark to American expansionist designs on Texas. History was to prove this hope a false one.

California, the most distant of Spain's northern territories, was the last to be settled. A mission and *presidio* (military post) were not established at San Diego until 1769. During the next two decades, a string of missions, with a few presidios and pueblos as well, was established along the coast as far as Sonoma, north of San Francisco. This thin band of coastal occupation was encouraged partly in response to a growing British and Russian interest in the west coast of the continent.

After U.S. acquisition of Texas in 1845 and the Mexican Cession, which ended the Mexican-American War in 1848, the estimated Mexican population of this broad territory was 82,000. Of this number, roughly 60,000 were in New Mexico, 14,000 in Texas, 7500 in California, and 1000 in Arizona.[2]

By 1850, the Mexican population of Texas and California represented less than 10 percent of the two states' total populations. There were good reasons for the rapid increase in the non-Spanish population. Texas was the new western frontier for Southern settlements, and the discovery of gold in California in 1848 was largely responsible for the influx of non-Spanish peoples in that area. Only in New Mexico did the Hispanic population continue to dominate for a few more decades.

This original Hispanic population of the Southwest has been greatly increased by substantial immigration, especially during the twen-

---

[2] For a presentation of the settlement geography of the Hispano to 1960 see Richard L. Nostrand, "The Hispanic-American Borderland: Delimitation of an American Culture Region," *Annals of the Association of American Geographers*, 60 (December 1970), pp. 638–661.

The village of Chimayo, north of Santa Fe in New Mexico, is one of a small number of examples of a Spanish influence on town form in the region. Spanish towns were usually small and often unplanned, and their pattern has usually been lost with the urban growth of the last 100 years. (Dick Kent/FPG.)

tieth century. However difficult the existence of many Hispano-Americans may seem, present living conditions for these people are seen, by them, as a substantial improvement over the quality of life experienced in northern Mexico. Between 1900 and 1980, there were roughly 2.1 million legal Mexican immigrants to the United States; California, Texas, and Arizona easily ranked as the leading destinations for these immigrants. Although an increasing number of Mexican people are moving into areas quite distant from the Southwest, most remain in the border states of the region.

Because of a substantial illegal immigration, it is difficult to estimate exactly how many Mexicans enter the United States each year. Although legal migration into the United States has been considerable, general quota restrictions and numerous denials of individual Mexicans' requests to immigrate have meant that far more have wanted to enter than have been admitted. The illegal "wetback" movement (so named because many entered the United States by swimming or wading across the Rio Grande into Texas) has existed along the Mexican-American border for decades. Many of these illegal migrants are caught by immigration authorities and returned to Mexico. Many simply renew their efforts until they are ultimately successful.

Although an accurate figure for such illegal migration is impossible, a major congressional report in 1979 placed the number living in the United States at 5 million. Recent estimates have ranged from 2 to 12 million, a variation that suggests the lack of reliable data. Perhaps half of the illegal migrant population is from Mexico. In 1982, 800,000 illegal aliens were arrested in the states from Texas to California. In 1983, the number reached nearly 1.1 million. Most were citizens of Mexico, and many had been arrested multiple times. A high rate of population growth in Mexico coupled with widespread unemployment and poverty in the north provide strong reasons for this migration. The international

This mechanical tomato harvester reduces greatly the need for migrant labor. The tomatoes it harvests are bred to survive rough treatment and so that all of the fruits ripen at the same time. (John Colwell/Grant Heilman.)

border, although more closely regulated than that between the United States and Canada, is permeable and can be crossed with relative ease. Despite increased efforts by border officials to control access and despite attempts by Mexican and U.S. officials to negotiate a migration agreement that is enforceable, there is little likelihood of a significant decline in this flow of people in the near future.

Much of this illegal movement is a temporary labor migration; workers who enter the United States usually work in the agricultural harvest in the Southwest for a few months, then return to Mexico with the profits of their labor. Until the early 1960s, the United States admitted large numbers of such temporary workers into the country legally. Then, the federal government decided that this influx of low-wage labor had a detrimental impact on living and working conditions of migrant farm laborers. Organized labor, attempting to unionize the migrant agricultural labor force, also opposed a large temporary labor migration. Labor leaders argued that it held down wages by providing a plentiful and inexpensive labor force. The number of such temporary workers admitted into the United States was, therefore, greatly restricted. It is debatable whether this decision had a positive effect on the working conditions of U.S. migrant workers. It certainly has been a factor in the rapid mechanization of the harvesting of crops that had formerly been picked by hand. Tomatoes, iceberg lettuce, peaches, asparagus—these and other types of farm produce can now be picked by machine. What is known is that many workers like those who had formerly entered the country legally decided to enter illegally.

The impact of illegal entry may be great on both the individual and on the area into which he or she moves. The individual, existing outside the law, loses nearly all the protection offered other residents of the country. Access is denied to police protection, a guaranteed minimum wage, welfare services, and nearly all jobs for which identification is required. Thus, the person is relegated to temporary jobs, often at

wages well below minimum rates; becomes an inviting target for crimes; lives in substandard housing; and is far more likely to be in ill health than the average American. Employers often realize they are hiring an illegal migrant, but the temptation to acquire a docile, low-wage worker is too great to ignore. For the community, negative effects can include depressed wages for low-skilled entry-level jobs, higher crime and disease rates, and a pattern of general social disorganization.

The 1979 congressional survey portrayed a contrasting image of illegal immigrants. It suggested that most illegal immigrants had become well integrated into American society. Most live reasonably normal lives, with little evident social disorganization. Although most do not have high-paying jobs, they are, nevertheless, regularly employed at above minimum-wage levels. The study concluded that illegal immigrants have a basically positive impact on American society and the U.S. economy.

In 1980, persons of Spanish surname represented 16.2 percent of the population of Arizona; in California, the proportion was 12.9 percent; in Colorado, 11.6 percent; in New Mexico, 36.5 percent; and in Texas, 14.0 percent. These percentages represented gains over 1970 for all but Texas, ranging from an increased share of 1.0 percent in Arizona and California to 6.2 percent in New Mexico. (However, the method of identifying someone as of Spanish origin had changed somewhat, making such comparisons rather unreliable.) In short, the proportion of the Spanish-surname population in the states that lie wholly or partly in this tricultural border region increased, even though all of these states, except New Mexico, had a substantial immigration of non-Spanish-surname people as well. This Spanish-surname population increase is due both to immigration and to a higher fertility rate as compared to that for the total population. For example, in 1980, the average woman in Arizona with a Spanish surname, aged 40 to 44, who was or had been married, was a mother of 4.2 children. The average for the total population of women in Arizona in the same age range was 3.1. For Texas, comparable statistics were 4.3 and 3.1, respectively; for California, 3.9 and 2.8, respectively. Completed family size for Spanish-origin families, then, averaged 1.1 more children than for the overall population of those states. Today there is a larger percentage of Spanish-surname people in California and Texas than there was in 1850.

The present Hispanic population is concentrated in those areas where most were found in 1850 (Figure 14-2). Nearly every county in the Rio Grande Valley is populated predominately by people with Spanish surnames. Major clusters exist along the lower Rio Grande in south Texas and in northern New Mexico and southern Colorado. In some south Texas counties, 90 percent of the population is Hispano. The Hispanic population is also overwhelmingly urban. More than 2 million Spanish-origin people reside in the Los Angeles-Long Beach SMSA, a population that represents roughly 1 out of every 7 Hispanos in the country. San Antonio has about 480,000 Hispanos, El Paso, 300,000. Of the 10 SMSAs with the largest Hispanic populations, 7 are in the Southwest. The popular image of the Hispano as a stoop-shouldered farm worker stoically following the harvest is no longer any more valid than that of a Southern black picking cotton or a midwestern white riding atop an expensive corn-picking machine. Today all are far more likely to be urban residents with urban jobs and urban attitudes.

Socioeconomic Disparities

The cultural and economic gap between the Anglo population, on the one hand, and the Hispanic and American Indian groups on the other hand, is immense in many instances. Most of the cities of the region are marked by patterns of segregation similar to those found in other American cities shared by blacks and whites (Figure 14-4). A tremendous variety of statistics exists that illustrates the economic disparities in the region. For example, the 1979 median family income for whites in California was $22,748; for persons of Spanish heritage, it was $16,081; for American Indians, $16,548. For New Mexico, the comparable statistics were $15,713, $12,062, and $9,893, respectively. One result is that in most of the region's cities, Hispanic neighborhoods are characterized by a low-income population with large families living in poor-quality housing (Figure 14-4). Across the region there is a close association between a high percentage of the population living below the poverty level and a high percentage of the population being Hispanic or American Indian (Figure 14-5). The average white family in California in 1980 had 3.11 members; in New Mexico, 3.25. Averages (California and New Mexico) for Spanish-origin people were 4.00 and 3.79 family members, respectively; and for American Indians, 3.46 and 4.59, respectively. In other words, much smaller incomes were serving to support the larger Hispanic and American Indian families. Finally, 33 percent of white Californians, aged 30 to 34, had completed four years of college; for Spanish-origin people this proportion was 13 percent; for American Indians 10 percent. Comparable New Mexico statistics were 28, 11, and 7 percent, respectively.

These data could easily overwhelm their message—the gap between Anglos and the other two cultures in the region is considerable, with American Indians falling statistically somewhat below their Hispanic neighbors. California and New Mexico were selected to illustrate trends in both an urban and a more rural part of the region. Differences in urbanization account for some of the variation that exists between these three groups; in the Southwest, Anglos are the most urbanized; American Indians, the least. Urban Americans tend to have higher incomes, more education, and fewer children than rural Americans.

The cultural differences among Anglos, Hispanos, and American Indians are important over much of the Southwest, especially in rural areas. Here the American melting pot has worked to only a limited extent, and a wide cultural gap

FIGURE 14-4
SOCIAL AND POPULATION PATTERNS IN SAN ANTONIO, TEXAS,
1980. Hispanic neighborhoods, like black residential areas in most large American cities, are characterized by low income, limited education, and high fertility.

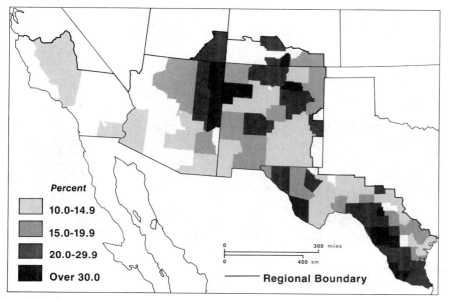

Percent

10.0-14.9

15.0-19.9

20.0-29.9

Over 30.0

| 0 | 300 miles |
| 0 | 400 km |

Regional Boundary

FIGURE 14-5

PERCENTAGE BELOW THE POVERTY LEVEL, 1979. In general, the higher
the percentage Hispanic and American Indian, the greater the portion of the
population living below the poverty level.

remains between these various groups. The Hispanos, residents of the region for nearly two
centuries before the Southwest was annexed by
the United States, suddenly became part of a
country with a different language, a different
legal system, and a different economic structure. The Anglos, quickly and clearly in control
of the politics and economics of the entire region, also grew rapidly in numbers to predominate over most of the region. As a group, Anglos viewed the Hispanos with a feeling of
superiority; equality of opportunity seldom existed. Perhaps if their numbers had not been as
large, if they had been more scattered, or if they
had not had such a long and close tie with the
region, the Hispanos might have merged with
the Anglo culture. Although the situations are
different, clear similarities exist between this
continued cultural separation and that which has
persisted between the English and French populations of Canada.

Placing most American Indians on reservations effectively maintained the distinctions between Indian and other cultures. Reservations
were located almost by definition in out-of-the-
way areas that offered little obvious economic
opportunity and little contact with Anglos, let
alone integration into Anglo culture. Until recently, reservation educational opportunities were
minimal, and local opportunities to use an education were almost nonexistent. In spite of this,
most Indians did not choose to leave. Ties to the
land and to family, plus a legitimate fear of the
problems of coping successfully with an urban,
Anglo culture, lessened what might otherwise
have been a wholesale movement off the reservations. The large reservations of the Southwest, such as the 62,000-square-kilometer
(24,000-square-mile) Navajo reservation centered on northern Arizona or the Hopi or Pa-
pago or Apache reservations, provided with their
size and isolation a separate, although poverty-

325

This occupied Navajo house, or hogan, is on the Navajo reservation in Arizona next to the Black Mesa Peabody Power Company plant (note top of crane operating beyond the hill). (Ilka Hartmann/Jeroboam.)

stricken existence.

In many small but, nevertheless, meaningful ways, conditions for the two minorities have changed for the better in recent years. Developments on the Navajo reservation, although not entirely typical, are indicative of altered reservation conditions. Final authority remains, much to many Navajo's dislike, with the Bureau of Indian Affairs, but an elected tribal council now makes most economic decisions for the reservation. Since the passage of the Rehabilitation Act of 1950, appropriations to the reservation have increased dramatically. All-weather roads now cross the reservation, greatly reducing isolation. Health and educational facilities have been improved. Huge reserves of fossil fuel, particularly coal, have been found on the Navajo land. Several large power plants located on the reservation serve southern California, which has banned such plants from its own area because of the atmospheric pollution they produce. Additional plants are under construction or planned. Many Navajo oppose this disruption of their tribal lands; others feel they are receiving far too little monetary return for the disruption. The large amounts of water used by power plants are also a concern because readily available water is in

limited supply and its diversion to power production may restrict the development of irrigated agriculture or other forms of industrialization on the reservation. This controversy has split the Navajo leadership at times. Nevertheless, the power companies annually move millions of dollars into the reservation economy. The reservation also has expanded greatly its tourist industry and has attracted a number of new industries with its large, available, and now better-educated labor force.

Still, family incomes on the reservation remain less than half of that for white rural Americans, and the reservation fertility rate is twice the national average. Few Navajos enter college; few of those who do, graduate. Even so, the general situation has improved.

ECONOMIC CONDITIONS TODAY

Along with its distinctive cultural diversity, the principal identifying features of the Southwest in the minds of most Americans are sunshine and aridity. This is a reasonably accurate mental image. This is the sunniest and driest of all the regions considered in this textbook. Throughout the area, the characteristic vegetation is bunch grass, mesquite, and—although far less common than most believe—cactus. Temperature conditions across the region vary widely. The southern Pacific Coast and south Texas normally have hot summers and only a short mild winter; the hot summers of the Rio Grande valley are balanced by winter conditions that can drive temperatures far below freezing. The common image of the Southwest as a warm area will be discounted easily by anyone who has been in northern New Mexico or Arizona—where elevations are generally much higher than in the south—during a midwinter storm, when cold, wind, and snow can combine in chilling intensity.

Despite this variety, the sunny climate of the Southwest, especially in mild winter areas, has proven to be a powerful attraction for many Americans. The spectacular growth of southern California's population during the late nineteenth century and the emergence of its vast metropolitan areas in the twentieth century will be discussed in the next chapter. This population and economic growth has spread well beyond California. Arizona was the second most rapidly growing state on a percentage basis between 1970 and 1980. Only three states—California, Florida, and Texas—added more to their populations between 1970 and 1980 than Arizona, which grew by an estimated 943,000 new citizens. In fact, all of the region's states during

The street system of a large land development project near Albuquerque, New Mexico, cuts across the landscape. Plans such as this one have placed several hundred thousand potential second home and retirement lots in rural parts of the state. This boom, a criterion of the American movement toward the "Sun Belt," has resulted in a supply of building lots that currently far exceeds the demand for such lots. (Georg Gerster/Rapho—Photo Researchers.)

this period grew at a rate well above the national average. The city of Phoenix doubled in size between 1950 and 1960, added an additional 50 percent to its population between 1960 and 1970, and then added yet another 60 percent between 1970 and 1980. It is now a booming urban area with a frequent cover of atmospheric pollution to attest to its status. Greater Tucson grew from 266,000 in 1960 to 530,000 in 1980. These low-density urban areas now roll for miles across large expanses of former desert.

Some of the early attraction of the Southwest stemmed from the healthful effects of the dry environment for people with respiratory ailments. The warmer parts of the region today attract many thousands of retired Americans. Large retirement communities have grown in the desert. One, Lake Havasu City, Arizona, with a population of some 16,000, gained national recognition when its developer purchased London Bridge and carefully transported it to Arizona, where it now arches over a newly created and artificially maintained lake.

Still, Arizona's growth cannot be attributed only to retirement migration. Many industries and corporate offices have been attracted. An aircraft industry developed in Phoenix during World War II, taking advantage of its proximity to the large aircraft complex in southern California plus the promise of good flying weather. Many employers have located in southern Arizona because the environment has a strong appeal for a work force (and for the employers themselves). The relative isolation of the state from most of the major national markets, once a significant constraint on Arizona's growth, has lost much of its impact with the emergence of high-value, low-weight manufactured goods, especially electronics.

Other SMSAs, such as El Paso, Texas, and Albuquerque, New Mexico, roughly doubled in size between 1950 and 1970 and since then have continued to grow rapidly. Both cities, and San Antonio as well, have benefited from the presence of large military bases, although they also share in the diversified growth of light industry common to the entire region's cities.

Elsewhere, in New Mexico and in that part of Texas included in this region, growth has been far more spotty. Many rural counties of the lower Rio Grande valley and most in southern Colorado and eastern New Mexico have lost population during the last few decades, sharing the fate of most other strongly rural areas in America. Today, New Mexico remains one of the least industrialized states in the country.

PERSISTENCE OF A PLURAL SOCIETY

The Hispanic and American Indian cultures give portions of this region a unique place in the American landscape. The rural highlands of central and northern New Mexico, the principal core of Spanish settlement in the United States, continue to show characteristics remarkably unaffected by the Anglo tide that has engulfed Albuquerque and southern Arizona. Hispanos make up, perhaps, 70 percent of the highland population of northern New Mexico and comprise the entire population of many small towns. American Indians, mostly one of the Pueblo peoples, are a much smaller but highly visible element of the region's rural non-Anglo culture.

Traveling the back roads north of Santa Fe, one can get the impression of leaving entirely the twentieth-century Anglo society. Old adobe villages, poverty-stricken agriculture, and public signs in Spanish dominate the cultural landscape. Along the highway near the intensely Anglo city of Alburquerque, as throughout the north-central part of the state, are several centuries-old apartmentlike Indian villages; their ancient appearance is in striking contrast to the low, sprawling modern city. These villages, like the Hispanic settlements which followed, are called *pueblos*. The pueblos each control substantial areas of surrounding lands that insulate them from the Anglo community. Pueblo society and traditions are not merely surviving remnants of a past period, they are vibrant and thriving. New Mexico's city of Sante Fe, in the heart of Hispanic country, retains a Spanish flavor that gives it a pleasant distinctiveness, despite its rapid

This 1970s street scene is from a small town in New Mexico. The Spanish influence, and the poverty, are obvious. (Danny Lyon/Magnum.)

growth and increasing Anglo influence. Hispanos, in many cases poor and cut off from the economic advancement offered by the Anglo economy, have become an important political force in New Mexico and frequently hold statewide elective offices.

The Winter Garden area of the lower Rio Grande valley in Texas, also overwhelmingly Hispano, is a major area of irrigated agriculture. The average growing season is longer than 280 days and supports such crops as oranges, grapefruit, and winter lettuce and tomatoes. The Hispanic population has long provided stoop labor for this agriculture. Here, too, politics have become an important Hispanic activity.

In Los Angeles, Hispanic enclaves may contain hundreds of thousands of inhabitants. Despite far more acculturation into Anglo society than has occurred in the upper or lower Rio Grande valley, Hispanic traditions remain important; Spanish-language radio stations and newspapers abound, and major Mexican-American festive occasions attract huge throngs.

In the Southwest border area, then, the impact of the Hispanic and the American Indian cultures remains strong. The abundant evidence of a flourishing plural society continues to distinguish this region from others in the western United States.

Cultural Geography Case Study: The Hopi Reservation

Land, and their attachment to it, is of central importance to many people in the world. Some have argued that the European population of the United States and Canada, with its great willingness to relocate and leave the land, does not share fully in this attachment. That is not the case with many American Indian tribes in both countries.

The Hopi, a tribe of perhaps 8500, are direct descendents of the Anasazi, the people who built the magnificent cliff dwellings of such places as Mesa Verde in Colorado and Hovenweep in Utah plus equally impressive pueblos like Chaco, New Mexico. The Anasazi dominated a large area of the southwest for centuries. A succession of droughts in the 1200s apparently forced the abandonment of their great urban centers and irrigated agriculture. Some moved southeastward to contribute to the emerging pueblo culture along the Rio Grande. After a period of wandering, another group settled around a series of three mesas in what is now northeastern Arizona. These mesas form the core of the current Hopi reservation (Figure 14-6).

The Hopi lived peacefully around their mesas for several centuries. They settled into a series of compact villages in the Anasazi fashion. They continued (and still continue) an agricultural economy focused on such crops as corn, squash, and beans. In the late 1600s, they removed their pueblos to the mesa tops to avoid anticipated Spanish reprisals after the Pueblo revolt against those Europeans settling along the Rio Grande. At about the same time, the Navajo began raiding and then settling in Hopi land from their center in current northwestern New Mexico. The aggressive, expansive Navajo soon surrounded the Hopi, forcing them from much of their ancestral land. The Hopi even invited Tewa Indians from the pueblo country of New Mexico to join them in their battles. The Tewa still live in a village on the Hopi reservation and maintain their cultural identity.

The U.S. government ignored this ongoing conflict in the decades after the Mexican War of 1848. The Hopi were finally granted a reservation in 1882. A portion of the order establishing the reservation said it might also be used by "such other Indians as the Secretary of the Interior might see fit to settle thereon." The relatively small Hopi population used only a small part of the reservation for their agriculture. The numerically dominant Navajo had adopted a pastoral economy based on sheep. Their need for grazing land for their animals was great in the dry environment of northern Arizona. Thus, the Navajo had soon moved onto the Hopi reservation so that only a relatively small portion remained exclusively Hopi. A federal panel recognized this situation by proclaiming all of the reservation except the Hopi-dominated section a Joint-Use Area. The Navajo were the larger group, however. In practice, the Navajo occupied virtually the entire Joint-Use Area. Hopi attempts to claim what they considered to be their share of the area were repelled.

In 1974, Congress mandated a division of the Joint-Use Area. The two tribes attempted and failed to negotiate a boundary. Finally, in 1977, a federal district court divided the area equally between the two tribes. The decision requires the relocation of 60 Hopi and over 3000 Navajo, a migration that will not be complete until the end of the 1980s, if ever. Despite government promises of substantial funding of all moves, many Navajo bitterly oppose being forced from their homes.

The Hopi are a conservative people. They cling to their traditional ways of life. The land is important to them, and they feel a strong sense of responsibility to it. The Hopi were disturbed at the persistent overgrazing of what they considered Hopi land by the sheep herds of the pas-

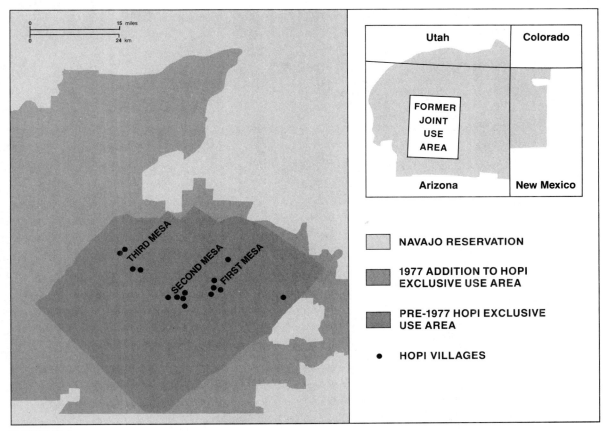

FIGURE 14-6
THE HOPI RESERVATION. Hopi communities are clustered on the tops of
the reservation's mesas. Navajo settlement is generally more widely distributed.

toralist Navajo. They see the 1977 settlement as just. The Navajo, for their part, have lived on what is now Hopi land for a century and now also feel an attachment to the land. They consider their overgrazing problem a result of too little land for a large and rapidly increasing population. The Hopi, they believe, do not use Hopi land effectively. To the Navajo, the settlement is an unfair limitation of their right to earn a living by traditional pastoral means.

Additional Readings

Acuna, Rodolfo. *Occupied America: The Chicano's Struggle Toward Liberation.* New York: Harper & Row, 1971.

Comeaux, Malcolm R. *Arizona: A Geography.* Boulder, Colo.: Westview Press, 1982.

Davis, Cary, et al. "U.S. Hispanics: Changing the Face of America." (Population Bulletin No. 38.) Washington, D.C.: Population Reference Bureau, 1983.

Goodman, James O. *The Navajo Atlas: Environments, Resources, People and the History of the "Dine Bikeyah."* Norman: University of Oklahoma Press, 1981.

Hansen, Niles. *The Border Economy: Regional Development in the Southwest.* Austin: University of Texas Press, 1981.

Hecht, Melvin, and Richard Reeves. *The Arizona Atlas.* Tucson: University of Arizona Press, 1981.

Horgan, Paul. *The Historic Triad: Essays in the Social Energies of Three Southwestern Cultures.* New York: Holt, Rinehart & Winston, 1970.

Meinig, Donald W. *Imperial Texas: An Interpretive Essay in Cultural Geography.* Austin: University of Texas Press, 1969.

Meinig, Donald W. *Southwest: Three Peoples in Geographic Change, 1600–1970.* New York: Oxford University Press, 1971.

Nichols, John. *The Milargo Beanfield War.* New York: Random House, 1974.

Nostrand, Richard L. "The Hispano Heartland in 1900," *Annals of the Association of American Geographers,* 70 (1980), pp. 382–396.

Romo, Ricardo. *East Los Angeles: History of a Barrio.* Austin: University of Texas Press, 1983.

Stoddard, Ellwyn L., et al. (eds.). *Borderlands Sourcebook: A Guide to the Literature on Northern Mexico and the American Southwest.* Norman: University of Oklahoma Press, 1983.

CHAPTER FIFTEEN

CALIFORNIA

Perhaps no American region excites such a diversity of perception among people living outside the region as California. Many see it as the desirable ideal of the modern, outdoor-oriented American life-style. At the opposite extreme, others see in it a decadent example of what can go wrong with the country. Whatever viewpoint one chooses, California, home to more than 10 percent of all Americans, is a central element in the American cultural fabric.

Our identification of California as a single region is based on its political unity, its national image as a single place, and its growing role in influencing national lifestyles (Figure 15-1). By most of the criteria used in the definition of regions, California is not a single unit. Culturally, the agricultural population of the Imperial Valley in the southeast is quite different from the urban population of San Francisco. The striking flatness of the San Joaquin Valley is in sharp contrast to the ruggedness of the Sierra Nevada. In the same region are found broad areas of desert in the southern interior and heavily

forested slopes along the coastal north, where average annual precipitation exceeds 200 centimeters (80 inches). The lowest and highest elevations in the conterminous United States, Death Valley and Mount Whitney, respectively, are almost within sight of each other.

California's dramatic and, certainly, varied physical environment has played a strong role in the state's human settlement. The variety and value of agricultural crops grown there is unsurpassed by any other state. Most of the state's population today is crowded into a small part of its territory, constricted by expanses of rugged topography and a widespread lack of water. Californians have invested untold billions of dollars in attempts to overcome these two problems. Their success is partial, at best, and in many areas it is fragile. It is a surprising contradiction that this mecca for America's worship of the outdoor life, this migration destination for hundreds of thousands seeking to free themselves from the indoor enslavement of the colder, physically less attractive midsection of the country, should also

FIGURE 15-1
CALIFORNIA.

surpass every other state in its level of urbanization. Yet it does. No state has a higher percentage of its population classified as urban by the Bureau of the Census. Several factors account for this, but the restrictive aspects of the physical environment are certainly an important element.

THE PHYSICAL ENVIRONMENT

Topography
California is associated with what geologists have

called the Ring of Fire, a belt of intense earthquake and volcanic activity that encircles much of the Pacific Ocean. In what almost amounts to a landslide of intellectual activity during the past 20 years, earth scientists have improved their understanding of the reasons for this kind of concentrated, violent earth movement. Their complex findings extend far beyond this book's needs, but at least some discussion of this research is needed here. It is important to both the physical and human geography of California.

Most geologists now agree that the crust of the earth is composed of a number of large zones called plates that move slowly on the molten rock which lies beneath them. These plates, perhaps 85 to 100 kilometers (50 to 60 miles) thick, may pull apart forming a deep trench bordered by ridges. A mid-Atlantic ridge, stretching nearly the entire north-south extent of that ocean, is a prominent example of such a feature. Elsewhere, plates may be plunging beneath the crustal surface as they move toward one another, such as at the major ocean trenches. Mountain-building activity is often found where two or more plates collide. The general theory given to explain these associated activities is called *plate tectonics*.

The coast of California is lined by a series of long, linear mountain ranges that trend in a generally northwesterly direction. They are collectively called the Coast Ranges. Most are not particularly high—summits are between 1000 and 1600 (3000 and 5000 feet). They are heavily folded and faulted as a result of the pressures of plate contact just to the west. Faults are fractures in the earth's crust where relative displacement of the rocks has occurred. The California faults follow the same northwesterly trend as the Coast Ranges (Figure 15-2). The most famous of these faults, the San Andreas, extends from the Gulf of California through the Imperial Valley to Point Arena well north of San Francisco, where it extends into the Pacific Ocean. Lateral earth movement (i.e., horizontal shifting as opposed to vertical displacement) was as much as 6 meters (20 feet) along the fault at the time of the 1906 San Francisco earthquake.

Small earthquakes are common across large sections of the region, especially from the San Francisco Bay area southward to near Bakersfield and from the Los Angeles area southeastward through the Imperial Valley. Seismologists (scientists who study earthquake activity) believe the regions of major continuous stress, such as the Coast Ranges area, will periodically experience large earthquakes as a result of this stress as it gradually accumulates during periods of less

violent activity. Many believe that this fault zone, which has had no major earthquake since 1906, is due for a large one. The area near Los Angeles is considered especially vulnerable because it has not experienced a major earthquake from movement along the San Andreas fault for 120 years.

Earthquake activity represents a substantial hazard for many Californians. The two major foci of population growth in the state, the San Francisco Bay Area and the Los Angeles Basin, are also principal centers of seismic activity. Thus, the majority of all Californians must live with the threat of an earthquake. Major highways, commercial and industrial concerns, and thousands of homes are located on or near known fault zones. For many, earthquake insurance is prohibitively expensive; insurers understandably have fears of suffering catastrophic financial losses in the event of a major earthquake.

The actual impact of this ever-present environmental hazard on the lives of Californians is minimal. Most have adopted an it-can't-happen-to-me attitude, a standard psychological reaction to the persistence of imminent danger. Still, they cannot avoid recognizing that earthquakes are very much a part of life in their state. Small quakes occur regularly. Half-joking references to "The Big One" that is predicted and local tales and earthquake stories including a novel that describes the sinking of coastal California into the ocean, suggest Californians' recognition of the situation. Not surprisingly, California is also a leading center for earthquake research in the United States.

Perhaps the most visible impact has been on the skylines of the state's larger cities. Until recently, local "earthquake laws" limited the height to which buildings could be constructed as a protection against great loss of life during an earthquake. The result for many years was a low, even skyline, with no skyscrapers in San Francisco and only City Hall rising more than a few stories in Los Angeles. Their pleasantly homogeneous skylines (especially that of San Francisco) set these cities apart visually from their

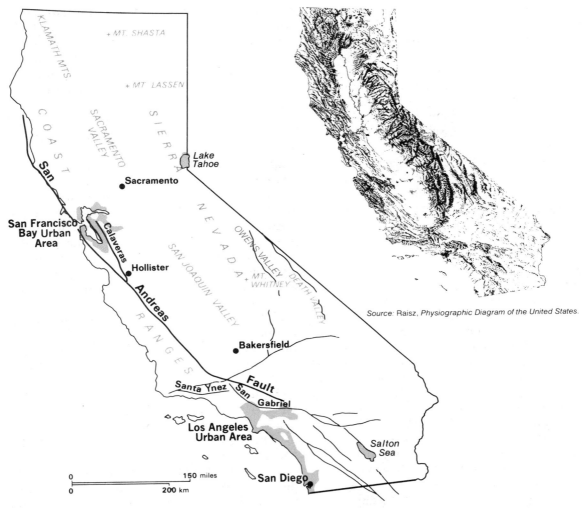

FIGURE 15-2

EARTHQUAKE FAULT ZONES. The topographic linearity of coastal California is the result of land movement along these fault zones.

counterparts to the east. Substantial innovations in structural materials and construction techniques have led to modification of many of these laws; now a small but growing number of tall buildings punctuate the skylines in both cities. Many native Californians find these changes in the state's urban landscapes visually jarring, especially in the case of San Francisco.

To the east of the Coast Ranges for much of their distance lies the Central Valley. This valley is extremely flat, extends 650 kilometers (400 miles) from north to south, and is nearly 150 kilometers (95 miles) wide in places. The Central Valley was originally a massive extension of the Pacific Ocean, open to the sea only at the Golden Gate of San Francisco Bay. A recipient of the erosion material carried off the western slope of the Sierra Nevada, this former section of the sea has been filled in with sediments. The result is a low-relief landscape rich in potential for large-scale agricultural pursuits.

To the east of the Central Valley, the Sierra Nevada rise gradually and have been heavily eroded. In contrast, the eastern face of the mountains offers a dramatic change in elevation. These are *fault-block mountains*, large rock masses that rose as a whole unit, and the eastern side was lifted far more sharply than the western face. Because they reach higher elevations and contain few passes, the Sierra Nevada have proven a major barrier to movement between the northern half of California and areas to the east. Today the tragic tale of the Donner Party or the exploits of Chinese laborers during construction of the first transcontinental railroad are partly matched by the problems, if not the drama, of highway construction in the "Sierras."

Two other landscape areas, less indentifiably Californian, complete the state's topography. In the north, the mountain-valley-mountain pattern breaks down, and much of the northern tier of the state is generally mountainous. The central plateau portion, directly north of the Central Valley, contains two of the state's major vol-

The modern skyline of San Francisco, built after the repeal of the "earthquake laws" that formerly limited building height in the city, towers over the older buildings that extend westward toward the Golden Gate. (San Francisco Examiner.)

MAJOR METROPOLITAN AREA POPULATIONS IN CALIFORNIA, 1980	
Los Angeles-Long Beach	7,477,503
San Francisco-Oakland	3,250,630
Anaheim-Santa Ana- Garden Grove	1,932,709
San Diego	1,861,846
Riverside-San Bernardino- Ontario	1,558,182
San Jose	1,295,071
Sacramento	1,014,002
Oxnard-Simi Valley-Ventura	529,174
Fresno	514,621
Bakersfield	403,089
Santa Barbara-Santa Maria- Lompoc	398,694
Stockton	347,342

canic peaks, Mount Lassen and Mount Shasta. This is the southern extension of the Cascades of Oregon and Washington. To the southeast of the Central Valley, the Great Basin of the interior extends well into California; topography there is composed of low-lying mountains (at least by comparison with the Sierra Nevada) interspersed with large areas of fairly flat land.

Climate and Vegetation

Variations in the state's climate and vegetation are nearly as great as the diversity of its topography. Much of the weather that spreads over the state is initiated offshore, along the eastern margins of the Pacific Ocean. Of greatest importance for precipitation is the frequent movement of moisture-laden air southeastward out of the northeastern Pacific. This is a large source area of moist maritime air. The distance southward that storm systems from this air mass are able to move is influenced substantially by the occurrence of a stable high-pressure center usually found somewhere off the west coast of Mexico. Such a stationary center blocks the southward movement of storms emerging from the maritime air mass and forces the moisture-

producing system eastward onto shore. This blocking high tends to drift northward during summer and southward during winter in conjunction with the latitudinal movements of most climate zones during the year.

A definite north-south gradient in average annual precipitation occurs in California as a result of the interaction of these two major air masses; the north is much more moist on the average than the south. Furthermore, summers are characteristically drier than winters, especially in the south. In summer, southern California frequently experiences long periods during which there is no rainfall. Consequently, the wooded mountain slopes grow dry as tinder. Forest fires, another of the region's recurring environmental problems, are most frequent in late summer and fall, toward the end of the long dry period. A climate with moderate precipitation concentrated almost entirely in the winter months and that has mild winters and hot summers, is called a Mediterranean climate, so named because it is found in much of the Mediterranean Basin lands of Europe, Africa, and Asia. In California, the entire coast from San Diego northward past San Francisco Bay, all of the northern Central Valley, and the western margins of the southern valley represent the only zone of Mediterranean climate found in North America (see Figure 2-3).

Northward from San Francisco along the coast, average annual precipitation increases greatly, and its seasonality partially disappears. The northern coasts of the state have a climate similar to that of the Pacific Northwest—mild temperatures with relatively little seasonal variation, plentiful year-round precipitation, and frequent periods of overcast skies. Such a climate is called Marine West Coast, an indication of its customary location along the western margins of continents in the middle latitudes (see Chapter 16).

The Central Valley is much drier than the coastal margins of the state. As air masses pass over the Coast Ranges they rise, cool, and lose much of their moisture. When those air masses descend into the valley, they warm. Little moisture is added to the air and its relative humidity (the percentage of the *absolute humidity* represented by the moisture actually in the air) declines. The result is a much lower average annual precipitation. Annual precipitation in the Central Valley is usually less than half that found at a similar latitude on the western slopes of the Coast Ranges. This dryness is especially pronounced along the west side of the valley.

To the interior of the Coast Ranges and the Sierra Nevada, in California's southeast, is a broad region of dry steppe or desert environment. The limited precipitation that reaches southern California is reduced further as the air masses move eastward over the coastal mountains. In addition, the entire area frequently comes under the influence of very dry air generated from the south and east. During the summer months, these dry winds occasionally press westward to the coast, bringing extremely low humidity and temperatures that may register more than 40° C (100° F), even along the coast. The southeastern interior area of California receives on the average less than 20 centimeters (8 inches) of precipitation annually.

Vegetation patterns closely parallel variations in climate in the state. Nearly all of lowland southern California and the area east of the Sierra Nevada-Cascade ranges is covered with sage, creosote bush, chaparral, and other characteristic desert and semidesert growth. The Central Valley and the valleys of the southern Coast Range are somewhat better watered than areas farther south; they are steppe grasslands. Wrapping around the Central Valley and following the coast from Santa Barbara to Monterey Bay are mixed open forests of live oaks and pines. The coast from Monterey Bay northward is the home of the redwoods, the world's tallest trees. At higher elevations in the Coast Ranges and Sierra Nevada are mixed forests of pine and fir, and high in the Sierra Nevada are found subalpine hemlock-fir forests, including those of sequoia.

CLIMATE AND CALIFORNIA'S GROWTH

California's climate, which provides much of the unquestionable attraction of the state, is also at the basis of some of its greatest problems. Viewed from a purely economic standpoint, it is almost inconceivable that fully 1 out of every 10 Americans should live in California. Geographers, and many others, have long viewed the population growth of a place as a result of several factors; two of the most important are the natural resources of that place and the strategic advantages of its location. [For a discussion of relative location (situation), see Chapters 3 and 4.]

Perhaps California's greatest disadvantage, at least until recently, is its location at the far western periphery of the country. The United States was settled primarily from east to west. The major population concentrations and most areas of principal economic importance remain near or east of the Mississippi River. Thus, California is located as much as 3500 kilometers (2200 miles) from the most important areas of economic demand and supply in the country. This relative isolation is compounded by the nature of most of the land that lies between the Sierra Nevada and the South and Midwest, a broad section of the country that generates little local freight (except at a few sites). Therefore, nearly the entire cost of movement between California and the cities east of the Mississippi River must be absorbed by through traffic. Normal economic considerations would suggest that high transportation costs would limit economic growth in California.

California is certainly not without its natural resources; San Francisco Bay, for example, is one of the world's premier harbors. Still, these are not nearly sufficient to account for the state's population size. More than anything else, climate has been the key in overcoming the state's apparent locational disadvantage; climate has been important to both the state's settlement history and its agricultural development.

Settlement History

Although Spanish explorers brushed the edge of California in the mid-1500s and claimed it as a part of Spain's large North American holdings, they basically ignored California for the next two centuries. The region was at the extremity of Spain's North American empire (see Chapter 14). Not until concern arose over British and Russian expansion in western North America, just before the American Revolution, did Spanish missionaries establish a string of missions from San Diego to Sonoma. These mission settlements (now some of the most sentimentally revered artifacts of the state's past) were joined by presidios (forts) and a few pueblos (towns). And during the next few decades, the Spanish and Mexican governments granted a series of large landholdings (ranchos) to immigrants. Still, the area remained peripheral; the towns were small and ramshackle, and hides and tallow were the ranchos' most important exports.

Following on the heels of the U.S. seizure of California in 1846, the great gold strike in the foothills of the central Sierra Nevada in 1848 brought the first significant change in the region's settlement fortunes. Within a year, 40,000 people had come to the gold fields by sea, passing through San Francisco harbor. Perhaps as many more came overland. By 1850, California was a state. The frantic period of the gold rush lasted only a few years, but it succeeded in breaking the state's massive isolation from the rest of the country. It also stimulated the growth of San Francisco, a growth that easily weathered the end of the gold boom. San Francisco remained the largest U.S. city on the West Coast until World War I.

Southern California, the center of Spanish occupation in the state, did not share in the early population expansion. The completion of the Southern Pacific Railroad to Los Angeles in 1881 and soon thereafter the arrival of the Atchison, Topeka, and Santa Fe Railroad to the same destination brought a sudden end to the area's quiet existence. In an effort to create a demand for their facilities, the railroads advertised widely for

A San Francisco street scene. Chinese have been important in the life of the city since the gold rush days of the 1850s. (San Francisco Convention Visitors Bureau.)

settlers, aided new arrivals in finding housing and jobs, and lowered fares. For one day in 1887, rail fares on the Southern Pacific from Kansas City to Los Angeles were $1. During the first southern California land boom, between 1881 and 1887, the population of Los Angeles grew from 10,000 to 70,000. Land speculators offered literally hundreds of thousands of town lots for sale in communities throughout the Los Angeles Basin, and many paper millionaires were created almost overnight.

Although this first land investment boom burst in 1888, the growth of Los Angeles, like that of San Francisco after the gold rush, continued. Promotional literature propagandized the delightful and healthful climate, appealing to sufferers from tuberculosis and asthma. As this propaganda continued to spread, thousands of the ailing flowed into the area.

A large number of crops were also introduced into southern California during this period, including the navel orange (1873), lemon (1874), Valencia orange (1880), avocado (1910), and date (1912). They were in demand in eastern markets and, at that time, only southern California could provide them in large quantities. Several technical improvements in food handling, notably artificial dehydration (1870) and refrigerated freight cars (1880), meant that these crops could reach eastern markets with only a minimal loss owing to spoilage. Agriculture was to remain the backbone of southern California's economy until after World War I.

California's Agriculture Today

California, by some measures the country's most urbanized state, is at the same time its most agricultural state in terms of total farm income. In 1982, the total market value of agricultural crops sold in California was $14.2 billion. Its closest competitors were Iowa ($10.3 billion) and Texas ($9.7 billion). The state produces more than half of all the country's lettuce, artichokes, broccoli, Brussels sprouts, cauliflower, garlic, honeydew melons, navel oranges, lemons, grapes, peaches, plums, apricots, avocados, dates, figs, and walnuts. In addition, it is the leader in the sale of tomatoes, asparagus, cantaloupes, celery, spinach, strawberries, honey, and pears. It is also third nationally in cotton, sheep, and dairy sales; fourth in rice; fifth in beef cattle; and eighth in wheat. In short, California's agriculture, although typified by its many specialty products, is, in fact, broadly based. The wide variety of climatic regions and the market demand of the state's own large population are the principal contributors to this diversity.

The importance of the many specialty crops in California does much to explain why the state's farmers have had such success in penetrating the markets of the distant East. These are crops that can be grown, or at least grown on a large scale, in only a few parts of the country. Most require long growing seasons. Thus, there is no local competition in the demand areas. Southern California, especially the Imperial Valley, is also able to provide vegetable crops, such as tomatoes or lettuce, during the winter season when competition is minimal and sale prices are at a maximum. Cooperatives such as Sunkist have been organized to facilitate marketing and to guarantee low shipping costs.

Although the state as a whole produces many agricultural products, local areas within the state tend to specialize in the growth of one or a few local crops. There are several reasons for this pattern of local specialization. One factor is reflected in the general trend toward agricultural specialization that exists across the country. The image of a typical farm as producing a variety of agricultural goods and being at least partly self-sufficient in terms of food is no longer accurate in American agriculture. Often, the local market is structured in a way that makes it difficult to sell all but a few products. And as the required technology and knowledge for production has increased, many farmers have found it too difficult or too expensive to handle more than a few basic products. For all these reasons, the national trend has definitely been toward increased specialization on both a regional and an individual farm basis (see Chapter 11).

A second important factor is the major role played by large agricultural operations in the state. Some of the specialty crops produced in California are grown by only a handful of farmers. In the San Joaquin Valley (the southern half of the Central Valley), some landholdings extend over many thousands of acres. Such large-scale growers generally consider carefully the profitability of their operations and have been leaders in the move toward further specialization.

Finally, the diversity of climatic and physiographic nitches within the state is a major influence for regional specialization. Many of the specialty crops are particularly sensitive to what may seem minor variations in climate or soil type. The Coastal Range valleys that open onto the Pacific are frequently foggy and have moderate temperatures. Vegetables, such as artichokes,

California's orange groves, squeezed out of the Los Angeles basin by urban expansion, are concentrated in irrigated areas of the southern San Joaquin valley. Wind machines and smudge pots are needed to protect trees from frost damage, especially during the early spring blossoming period. (Courtesy Sunkist).

lettuce, broccoli, or Brussels sprouts, grow well under such conditions. The varietal grapes that people often feel produce the best of American wines need a mild, sunny climate, such as that found in the inland Coastal Range valleys around San Francisco Bay. The grapes of the San Joaquin Valley or of southern California, where summer temperatures are much higher, are used

for table grapes, raisins, or for what most (although by no means all) wine drinkers consider less distinguished wines. Most flowers grown for seed in the country are planted in the Lompoc Valley west of Santa Barbara. Navel oranges and particularly lemons are grown almost exclusively along the coast and the interior surrounding the Los Angeles Basin. It is unlikely that early spring frosts would occur there to damage the emerging flowers or that heavy winter frosts would kill the trees themselves.

Such local concentrations, although economically advantageous, mean that moderate environmental changes in a relatively small area of California may have a major impact on the overall national production of specialty crops if they are grown in only a few places across the country. Because of this, the rapid expansion of California's urban areas represents an acute problem for many agricultural areas in the state. For example, the Napa Valley, containing many of the finest vineyards in the state, is being "invaded" by waves of residential construction. Although most vintners are relocating elsewhere (total wine grape acreage in the state has increased greatly in recent years), these new locations are less suitable for growth of quality wine grapes. South of San Francisco, some of the coastal valleys, such as the Santa Clara, are urbanizing rapidly. Production of such crops as apricots, prune plums, and pears are threatened by this expansion. California's orange and lemon groves were concentrated in the Los Angeles Basin through the 1940s. Urbanization and smog damage (and competition from the Florida citrus industry, see Chapter 10) have forced almost all production from the Basin.

Whenever land-use competition develops between urban and agricultural users, it is the latter who almost always lose. Urban users, both commercial and residential, are willing and able to pay far more for land than their agricultural counterparts. In addition, rural land has traditionally been taxed in the United States on the basis of its appraisal sale value and not on its value as a strictly agricultural unit. Thus, the twin pressures of increasing taxes and higher and higher offers from people who wish to buy land, both normal consequences of urban expansion, may convince even the most dedicated farmers to sell (see the discussion of this point in Chapters 4 and 6).

The consequences of this changing land use, estimated to result in the annual changeover of 800,000 hectares (2 million acres) from rural to urban uses in the United States, have been of little national concern in the past. It appeared to many that enough rural land remained to provide adequate agricultural production and to offer the aesthetic reward many find in a trip to the country. But this lack of concern has begun to change. Many cities are located in the midst of good farmland, and because urban land users tend to prefer the same flat land that is most valuable to farmers, the farmland lost to urban expansion tends to be better than average. Also, in an increasing number of areas, residents have come to feel that in losing local farmland, they are losing a valued part of the landscape.

Attempts to preserve farmland in an urbanizing environment have followed several tactics. Most common is a decision to grant special tax breaks to farmers; this is usually accomplished through the valuation of their lands for tax purposes as farmland rather than as potential urban land. One problem is that urban expansion, already quite jumbled in its spatial pattern, may become even more so as some farmers choose not to sell their land until the surrounding land is urbanized. Another, more negative problem is that urban services (water, sewer, streets, etc.) tend to become less orderly; cost increases are often the result. Within the Los Angeles region, several dairying areas attempted to protect themselves by incorporating, thereby creating their own tax structure. This system broke down when individual farmers within the "town" decided to sell to residential developers. These new residents soon outnumbered the farmers, and the urbanites' demands for improved services began to push taxes upward. Other areas have reacted with substantial zoning restrictions, such as a 2- or 4-hectare (5- or 10-acre) minimum for residential plots. Often used to keep "undesirable elements" out of exclusive areas, it may also keep land in farms. However, the future of such restrictions are in doubt because some courts have found this type of zoning to be illegal.

Today nearly every state has laws that provide some property tax relief to farmers. California

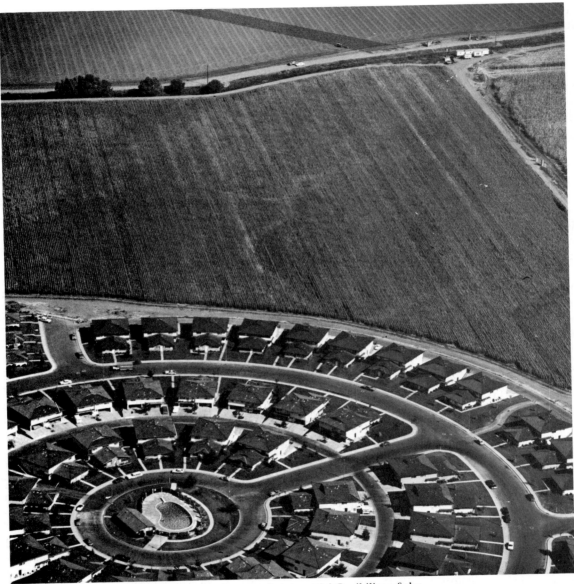

The family automobile, which greatly increased the spatial flexibility of the
American population, encouraged a very rapid expansion of urban space. America's urban areas extend onto many hundreds of thousands of acres of agricultural land every year. Much of what several decades ago was the best agricultural
land in southern California is now covered with a vast suburban sprawl.
(U.S.D.A.)

requires owners to enter into a long-term agreement not to develop their land as a condition for eligibility for reduced property taxes. Nearly 300 counties nationally have some type of agricultural zoning to preserve farmland. A few locales have programs to purchase development rights to ensure that the land remains in farming.

Water Supply

California's agriculture, more than its manufacturing or urbanization, has created a massive demand for water across much of the state. Some 70 percent of the state's precipitation falls in the northern mountains and valleys and in the Sierra Nevada where few farms or urban places exist, whereas 80 percent is used in the drier south. Of the state's major farming regions, only areas north of San Francisco and a few coastal valleys to the south receive as much as 50 centimeters (20 inches) of precipitation in an average year; two of the most important farming areas—the southern end of the Central Valley and the Imperial-Coachella area in the southeast—average annually less than 25 centimeters (10 inches). Consequently, California has more farm land irrigated than any other state, about 3.5 million hectares (8.5 million acres) in 1982. Only Texas has as much as half as many irrigated acres as California. Nearly all of California's cotton, sugar beets, vegetables, rice, fruits, and nuts are grown on irrigated farmland. California farmers use more than one-fourth of all the irrigation water used in the United States.

Farms are the principal users of water, but it was the cities that initiated the development of the state's tremendous water-movement complex. At the beginning of this century, Los Angeles outgrew its local groundwater supplies and looked for a supplementary source in the Owens Valley, east of the Sierra Nevada and some 300 kilometers (185 miles) north of the city. By 1913, the Los Angeles Aqueduct was carrying water to Los Angeles, much to the dismay of Owens Valley farmers who lost virtually their entire water supply (Figure 15-3). This aqueduct still provides half of the city's needs. In 1928,

Los Angeles and 10 other southern California cities formed the Metropolitan Water District to develop an adequate water supply for their entire area. By 1939, the group had completed the Colorado River Aqueduct, which carried water from Parker Dam on the Colorado to coastal cities from San Diego to the Los Angeles Basin. Today the Metropolitan Water District, which serves 6 counties, over 130 cities, and half of California's population, is one of the state's most powerful political bodies.

Farmers were by no means ignored, however. Perhaps the most spectacular episode in the state's water history occurred in 1905 in the Imperial Valley. In 1901, private groups constructed canals to carry water from the Colorado River into the Imperial Valley; the result was an immediate agricultural land boom. Then, in February 1905, the Colorado River flooded, broke out of its channels, and flowed into the irrigation ditches. Before a massive effort returned the river to its channel in the fall of 1906, 1100 square kilometers (400 square miles) of the valley had been filled by the Salton Sea, a body that still exists, its size maintained by ongoing drainage from irrigated farmland in the valley. Most of the valley's irrigation water is provided today from the Colorado River by the All-America Canal, built by the federal government in the 1930s.

In the 1940s, the federal Bureau of Reclamation began the Central Valley project. The aim of the project was simple—to improve the local availability of irrigation waters in the Central Valley. Most of the best potential irrigation land was in the southern San Joaquin Valley, but most of the water was in the northern Sacramento River. Today the project is fully functional. Water, which is removed from the Sacramento River by the Delta-Mendota Canal, flows southward along the west side of the San Joaquin Valley to Mendota, where it is put into the San Joaquin River. This transferred water meets the irrigation needs of the San Joaquin Valley from Mendota northward. Thus, most of the normal flow of the San Joaquin River can be used for

FIGURE 15-3
WATER MOVEMENT SYSTEMS. During the very dry summer of 1979, the residents of the San Francisco Bay area had to adopt severe water conservation measures to conserve dwindling supplies while urban southern California, benefiting from this massive water movement system, had no supply problem.

irrigation in the southern part of the valley. The southern valley is the leading agricultural region in the state, supplying nearly 40 percent of the state's total agricultural output. Of the country's 10 leading agricultural counties, measured in terms of value of farm products sold, 5 are in the San Joaquin Valley. Only 7 states, other than California, exceed the combined value of agri-

cultural output for these 5 counties. The value of farm products sold in Fresno County alone is greater than that for 17 states.

In 1957, California brought all of the existing water projects and all schemes for new ones under the California Water Plan, which was administered by the state. The plan developed the California Water Project, funded initially in 1960 through the passage of a $1.75 billion bond issue. The project, easily the largest water movement program in American, perhaps in world, history, focuses on a massive system to move water from the north to the south in the state. The California Aqueduct, also starting from the Delta area, carries water southward; this water is used partly in the western San Joaquin Valley and partly to supply the ever-growing water needs of the southern coast. Most of the cost of the southern extension is borne by users in the Metropolitan Water District. The total cost of the system continues to increase, with the estimates now exceeding $20 billion.

It is predicted that even this monumental system will satisfy the state's increased demands for only a few decades, with critical water shortages perhaps by the year 2020. Some hope that desalination of seawater will provide a major new source of fresh water by then. However, existing processes are expensive and are heavy users of energy. Others look northward to the Columbia River, whose flow dwarfs that of the Colorado, for a future supply. As early as 1949, the federal Bureau of Reclamation suggested the possibility of transferring Columbia River water to California. Apart from the huge costs of such an undertaking, people outside of California in the Northwest show no inclination to accept such a suggestion because they believe they will eventually need all of their own water supplies.

Several central points emerge from a study of water use in California. One is fairly obvious—given adequate financing, we have a tremendous technical ability to move water from one place to another. This, in turn, has led to a rapid increase in the demand for water. In recent decades, California's consumption of water has increased even more rapidly than has its population size. Suburban swimming pools dot southern California and a growing industrial demand has also been important, but farm irrigation has been the major user. Between 1964 and 1982, the total acreage of California's farms dropped by more than 20 percent, but both the number of acres irrigated and the volume of water used per acre

increased. The main reason is that irrigated lands in this generally dry environment produce very high yields. California's cotton production per acre, for example, is double the national average. As mentioned earlier, governmental agencies tend to absorb most of the costs involved in irrigation projects, thus providing water at an artificially low price to farmers, and this encourages expanded use. Finally, there is a clear tendency to reach across ever-increasing distances to meet existing needs. The water supply of the state and of the entire Southwest is limited. The region is quite close to using all there is.

Three problems in connection with the water supply in California are worth mentioning. First, much of the Imperial Valley and Coachella Valley face an impending problem because of a gradual buildup of salts in the soil. Once these salts, carried in large concentrations in the Colorado River water, are in the soil they can be removed only through the application of large quantities of water to flush them out. If left, the salts will eventually destroy the soil's productivity. As discussed earlier (Chapter 13), much damage is already apparent in the irrigated lands just south of the California border in Mexico. California hopes to find a solution through the addition of nearly salt-free well water to the river water supply in order to provide adequate water for flushing.

Second, many northern Californians are not happy about the transfer southward of much of what they view as their water supply. They too feel that their part of the state will eventually need all the water it can get. California's center of population is in the mountains just north of Los Angeles. Hence, in any statewide political decision, the south has far more potential voting power than the north. Northerners see the water decision as one proof of that. Antagonisms have long existed between the two sections of the state. Angelenos denigrate what San Franciscans like to view as "The City" by calling it "Frisco." For their part, San Franciscans feel a certain condescending pity for those who they feel are the uncultured people of the south. Mention is even

The California Aquaduct, and the Delta-Mendota Canal in the background, carry water southward to irrigate the San Joaquin Valley and for metropolitan Los Angeles. Note the intensity of land use in the irrigated area. (U.S. Department of Water Resources.)

made occasionally, usually by northerners, of splitting the state in half. A bill proposing such a split reached Congress in 1940. Although a separation is more an interesting speculation than a likelihood, the differences between the two areas are real; political conflict generated by the state water-movement program has been a significant reflection of these differences.

Third, many Californians question the economic and environmental costs of water-transfer projects. In 1980, the state legislature proposed construction of a Peripheral Canal to carry water from the Sacramento River southward east of the Delta to increase the volume of flow to the dry south. The state estimated that the project would cost $5.4 billion; opponents argued that $20 billion was closer to correct. Environmentalists were concerned that the loss of freshwater would damage the rich aquatic and waterfowl populations located in the intricate lacework of waterways of the delta. Delta farmers feared loss of their rich vegetable producing lands. San Franciscans again worried over a water loss to the south. A coalition of these diverse interest groups defeated the canal in a 1982 referendum.

URBAN CALIFORNIA

Despite the lengthy treatment of California's agriculture—necessary in the light of the state's agricultural importance nationally, its unique character, and because agriculture's influence on the state far outstrips its share of the total pop-

ulation—its population is overwhelmingly urban and still increasing. Over 90 percent of the people live in one or another of the state's 14 SMSAs. In 1982, California had 25 of the country's 176 cities of over 100,000. Los Angeles had passed Chicago to become the nation's second largest city. Most of the state's population live in one of its two major urban regions, one centered on Los Angeles, the other on San Francisco.

The Southern Metropolis

"A city in search of a center!" This is the title that many give to the loose assemblage of cities normally lumped under the general name, Los Angeles (Figure 15-4). The land boom of the

FIGURE 15-4
INCORPORATED CITIES IN THE LOS ANGELES AREA. Los Angeles annexed large portions of the San Fernando Valley northwest of the early focus of the city to avoid being entirely surrounded by other independent urban places.

1880s led to the establishment of several score of cities scattered across the Los Angeles Basin and southern California coastlands. As their populations increased, these communities squeezed out the intervening rural lands that formerly separated them spatially. At the same time, the communities maintained their jurisdictional independence. The result is a jumbled complex of politically independent but economically intertwined units. The cities have surrendered some of their individual decision making and taxing powers in recognition of a common regional need that can only be met cooperatively. The Metropolitan Water District is the best example of such a regional agreement. Still, the political complexity results in many problems in taxation and planning (see Chapter 4).

Most of the 300-kilometer (185-mile) stretch of coastline from Santa Barbara to San Diego is now occupied by one long megalopolis. The home of about 13.5 million Californians, the vast spread of this urban landscape is a reflection of a growth process shared by other cities in North America and, at the same time, a strikingly unique metropolis.

The entire complex is basically a creation of the twentieth century. In 1900, its population was less than one-tenth of its present size, and, as mentioned earlier, it was not until after World War I that the area began to shift from an essentially agricultural to an urban region. Thus, many of the elements of eastern cities, elements placed on the landscape largely during the nineteenth century and the first two decades of the twentieth century, are not present in Los Angeles. Among these missing features we might list the four- or five-story walk-up apartment buildings, warehouses of about the same elevation, fixed-rail elevated (or underground for that matter) public transportation lines, and a strong nodality centered on the central business district (CBD). Knowledgeable Angelenos could undoubtedly expand this list greatly, but it is representative.

The southern metropolis's major identifying characteristics are obviously of the twentieth century. The most important stimulus has been the family automobile. Without it, the region would surely strangle. Automobiles offered a degree of spatial flexibility unknown previously. People could live, work, shop, and recreate where they chose, and the bonds of structured travel patterns were removed from urban residents. If we measure distance by time rather than by miles (a far more relevant method of measurement to most travelers), the maximum one hour's distance that a worker might be willing to travel between residence and place of work increased during the present century from a 5-kilometer (3-mile) walk to a 90-kilometer (55-mile) drive. Furthermore, the linear inflexibility of public transport was no longer an important factor in Los Angeles; this made it possible to live far from work and not be bound to the limited residential opportunities located along streetcar lines.

Today over half of the central portion of Los Angeles has been surrendered to the automobile, either for roadways or for parking. The urban area's dense system of freeways—often spoken of in a disparaging manner but, in fact, quite effective—makes possible fairly high-speed movement across much of the metropolitan region. The Los Angeles area has more cars per capita than any other part of the country. It also has only a minimal public transportation system. Although neither fact should be surprising given the diffuse pattern of housing, employment, and commercial facilities, nevertheless, it means that the private car carries the overwhelming percentage of the region's daily population circulation.

Much about southern California's urban landscape reflects the impact of the automobile. The population density of Los Angeles is 17,000 per square kilometer (6100 per square mile); in San Diego it is about half that. By comparison, Philadelphia has an average density of 38,000 per square kilometer (14,200 per square mile). Housing units throughout the area are dominantly of the detached, single-family type. Even low-income areas tend to be composed of such housing, a feature that often leads the casual observer to conclude mistakenly that slum or near-

The Los Angeles River, usually a trickle but a major stream after winter rains, is paved for much of its distance across the Los Angeles basin. The basin's freeway system connects its many cities. (Georg Gerster/Photo Researchers.)

slum conditions do not exist in Los Angeles.

Finally, as suggested earlier, Los Angeles is a city without a center. The traditional singular CBD, as a focus of urban activity, barely exists here. Los Angeles is really many cities that have grown together as they increased in size; 14 of these cities currently have populations of more than 100,000. The absence of a single dominant CBD results, in part, from the continued existence of independent centers for each of these communities. Many of these cities have grown recently, and the predominant use of cars and trucks for daily movement has led to the growth of shopping centers and suburban industrial parks. Thus, employment and shopping opportunities are almost ubiquitous throughout much of southern California.

The polynucleated nature of the urban area has taken on a new dimension in recent years. Influences, such as rising housing costs (a typical single-family detached house in California costs 50 percent more than the national average) and population densities, have encouraged an increased focus on multifamily housing and business around the larger urban cores. The result is an urban population density gradient and built-up environment increasingly similar to that of Eastern and Middle Western cities.

The automobile is also partly responsible for one of Los Angeles's most infamous features—smog. Ringed by mountains to the north and east with hot, dry deserts beyond and bordered by the cool waters of the Pacific to the south and west, the Los Angeles Basin frequently is subjected to *temperature inversions*. Such a situation occurs when a body of warm air lies above cooler air. This warmer air forms an atmospheric "lid," and pollutants in the cooler air are unable to rise into the higher, warmer air. Onshore winds push the pollutants inland, but a combination of the inversion lid and the surrounding mountains keep much of this pollution within the basin. Palm Springs, just east of the basin, has grown up as a high-status community because it is beyond the smog's reach. In reality, Los Angeles produces only moderate amounts of pollution, for it has

little heavy industry, burning is carefully controlled, and there are strict controls on automobile emissions. The city has made a continuous effort to lessen smog problems since World War II. At first rather simplistic "keep your eyes open for polluters" kinds of efforts were used, then they changed to one of the most comprehensive and effective atmospheric pollution-control programs in the country. The area's peculiar physical geography, a pattern that led early Spanish explorers to comment on the buildup of smoke from Indian fires, is the remaining, probably unbeatable, problem. Some have suggested, semiseriously, that a large cut be carved through the mountains and fans set up to blow the smog out. People living in areas beyond the mountains, already hit by some effects of the smog that drifts over the mountains, are understandably opposed to such suggestions.

If it is difficult to explain the existence of more than 25 million Americans in California using the normal logic of location theory, it is nearly impossible to explain the presence of over 13 million Americans in southern California. Although the area is not without resources, their overall importance is not overwhelming. We have already looked at agriculture. Petroleum production is also important; three of the country's major fields are in southern California. In addition, offshore development began in 1965. Although the Santa Barbara oil spill was a sad indication of the environmental tradeoffs involved in offshore oil drilling, Long Beach has already reaped a large financial harvest from the sale of leases and from a tariff on oil pumped. Many find the offshore pumping stations, built as islands complete with trees and waterfalls, visually less offensive than they had expected. The heavy demand for petroleum products, especially gasoline, results in the local consumption of virtually all of southern California's production.

Southern California is known worldwide as the location of Hollywood, long the center of the American motion picture industry. In the early days of filmmaking, outdoor settings and natu-

MacArthur Park, Los Angeles, around 1950. A mild climate, swaying palms, relaxed living—these were at the core of the American image of southern California in the first half of this century. Does that image still exist? (David M. Mills/American Stock Photos.)

roads emphasized the area's fine climate and scenery for recreation late in the nineteenth century. Pasadena's Rose Parade was partly developed as an advertisement for the area's mild winters. Today, these natural advantages have been supplemented by some of the country's largest and best-developed recreation facilities. Balboa Park in San Diego, with its excellent zoo, Knott's Berry Farm, and Marineland are major attractions. Disneyland has become an American phenomenon, and the main destination of countless tourists.

Still, all of this does not add up to resource support for 13 million people. Heavy resources, such as coal or iron ore, are virtually nonexistent. San Diego has a good harbor, but Los Angeles's harbor, entirely human-made, is only average. Neither is Los Angeles a major break-in-bulk center for goods moving into or out of the United States. As a West Coast financial center, it is surpassed by San Francisco. To better understand this population concentration, three additional elements must be included—general national affluence, the growth of a postindustrial economy, and the role of federal government spending.

Southern California has profited from government spending far more than most other areas of the country. California now receives about 20 percent of all Department of Defense spending and nearly half of that of the National Aeronautics and Space Administration (NASA). San Diego is the West Coast home of the U.S. Navy, and the navy is easily the city's principal employer. San Diego has relatively little manufacturing employment for an urbanized area of nearly 2 million, surely an important part of its justifiable claim to be one of the country's most livable cities. Southern California has been the country's leading aircraft-manufacturing center for nearly 40 years, with much of the business generated by the purchase of military aircraft. More than one-third of all manufacturing employees and one-tenth of the entire civilian labor force is employed in the aerospace industry. This means that the regional economy is cyclical, with the fluc-

ral light were the norm. The area's cloudless skies and short cold temperature periods made its streets and fields a fine home for countless motion pictures. Los Angeles remains one of the centers of American filmmaking and television, both of which have provided invaluable national and international advertisement for California as well. The silent movies introduced California's palm-lined streets, and contemporary television continues to glamorize the region.

The Mediterranean climate has also supported the recreation industry. That, and the varied scenery, especially along the coast, early made southern California one of the country's centers of outdoor recreation. Towns and railtuations dependent on government defense spending.

The declining share of total employment held by manufacturing and the corresponding growth of the tertiary and quaternary sectors has tended to free American workers from even the tenuous locational impact that the physical resource base has on manufacturing. This shift is essentially a transition into a postindustrial economy. The nature of manufacturing itself has also shifted; an increasing share of all workers are in light industries, where the value added by manufacture is high and worker skills particularly important. Electronics is such an industry, and it is important to southern California manufacturing. In this postindustrial period, new factors, such as the availability of an adequate secretarial pool or the presence of major universities with their diverse intellectual and research resources, become more meaningful locational considerations. The question of where the workers and more particularly, perhaps, the management want to live becomes highly significant (see Chapter 10). Southern California, with its mountains, seashore, its mild climate, and its outdoor image, has proved to be a most attractive location for many. As never before, we have the ability to choose where to live; studies show that southern California offers the kind of environment that many Americans find especially desirable.

Southern California is also the major destination for Latin American and Asian migrants entering the United States. Today only a minority of Los Angeles county's 7 million people are Anglos. Found there are 2.1 million people of Mexican ancestry; 200,000 Iranians and Salvadorans; 175,000 Japanese and Armenians; 150,000 Chinese, Koreans, and Filipinos; and 130,000 Arab-Americans. And the list goes on. Nearly 1 schoolchild in 4 in the Los Angeles school district speaks one of 104 different languages better than English. Los Angeles is close to Asia and Latin America, its climate is moderate, and in their numbers, its diverse ethnic population offers some security to new arrivals.

The cityscape of Los Angeles reflects this ethnic diversity. Recent migrants especially tend to settle in ethnic neighborhoods. Little Tokyo, long a part of the city, has a renewed vibrance. One can select from dozens of Chinese language schools. A rich diversity of ethnic restaurants, many found within ethnic enclaves, dot the city.

The large size of southern California's population, and its great distance from traditional sources of manufactured products in the East and Midwest, has resulted in the development of broad-based manufacture of consumer goods: automobile assembly, food processing, mobile home manufacture—the list is long. These industries are present because of local demand.

The Bay Area Metropolis

By comparison with their upstart competitors in the south, San Franciscans choose to view their city as old, cosmopolitan, and civilized. There is something to be said for this attitude. The city was the northern core of Spanish and Mexican interest in California. It served as the supply center for the California gold rush. By 1850, it was the largest city on the Pacific Coast, a ranking it maintained until 1920. The completion of the first transcontinental railroad in 1869 coupled with the city's size and its excellent harbor made it not only the focus of western U.S. growth, but also the key location for U.S. commerce with the Pacific as well. Into the city came large numbers of immigrants from Asia, especially Chinese, plus substantial numbers of other foreigners. They created a cosmopolitan ethnic mixture that today remains a readily apparent aspect of the city's character. Much of the early history of the city was more bombastic than civilized, but San Francisco is certainly the oldest West Coast city.

The romantic flair of this early history is one piece of the mosaic that makes San Francisco among the most popular of American cities. Its physical geography provides a splendid setting for a city, steep slopes that offer dramatic views of the Pacific Ocean or of the Bay coupled with a mild climate that escapes the sometimes staggering summer heat of southern California. Re-

The San Francisco Bay area is unusually hilly for a major urban center. In newer sections the housing often appears like ropes coiled over the hills. (Peter Menzel.)

cent quantitative evaluations of the quality of American cities, which tend to emphasize readily available socioeconomic statistics, have generally given the city a good ranking. Most high rankings tend to go to the newer cities of the West; hence, San Francisco's position is neither surprising nor exceptional. Even so, it is another component of the pattern that places San Francisco at, or near, the top of a list of "desirable" places in studies of urban perception. One consequence of this perception has been a continued migration of newcomers into the city. This, in turn, has led to unemployment levels generally well above the national average.

In some ways, this attitude is based on a false perception of the area. The city of San Francisco is currently home for fewer than one-eighth of the Bay region's 5.4 million people. Hemmed into its small peninsula, the city is actually losing population while the entire urban area grows. Most of this growth has been diffused, random, and very much like that found in most large metropolitan regions (Figure 15-5). The exotic appearance of Chinatown, or the vaguely romantic seediness of the Mission District, both in San Francisco, today must be matched against the stark ghettos of Oakland, the refineries of Richmond, or the almost endless sprawl of San Jose. A folk song lamenting "houses made of ticky-tacky," where they "all look just the same," refers to the San Francisco SMSA, not Chicago or New York. The sweep of brown hills and rich vineyards around the region, so long a part of its special nature, pass daily under the bulldoz-

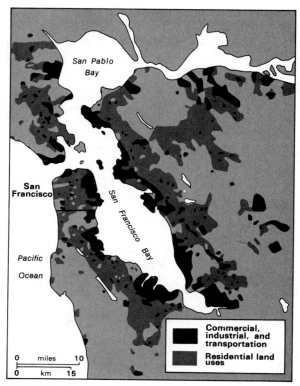

FIGURE 15-5

LAND USE IN THE SAN FRANCISCO BAY AREA. Industrial and large-scale transportation activities hug the margins of the bay. Accessibility provided by this supurb site has been critical to the area's economic growth.

ers of the metropolitan area's leapfrogging urban growth.

The Bay Area today is really composed of several different areas, each with its own special character. The East Bay is the most varied, with a mix of blacks, Berkeley students, large tracts occupied by middle-class residents, and most of the port facilities and heavy industry of the region. Some have suggested that it is not very different from northern New Jersey in its landscape character. The San Jose-South Bay area is upper middle class, white, with new houses, fine

yards, and major regional shopping centers. Along the Bay north of San Jose is Silicon Valley, so named because of its concentration of businesses engaged in chemical and electronics research associated with production of computer components. North of the Golden Gate Bridge, the cities are smaller, there is little manufacturing, and the conflict between agricultural and urban land use is sometimes obvious. Here are found the well-to-do urbanites searching for a place in the country. The city of San Francisco itself, with its grid pattern of streets incongruously placed on a hilly terrain, its closely spaced late nineteenth- or early twentieth-century housing, and its ethnic diversity maintains a special appeal, despite the recent appearance of skyscrapers along the skyline. San Francisco has adopted zoning ordinances aimed at maintaining the special character of the city. Buildings over 18 meters (60 feet) high are banned from many areas, architectural appropriateness is encouraged in public housing, private redevelopment plans are encouraged, and the city, in general, takes an uncommon interest in preserving what it sees as its special character.

San Francisco is far less the focus of the area than it once was. Oakland is the principal center of the East Bay; San Jose is the hub of Santa Clara county and is now almost as populous as San Francisco. The bay, so important to the quality of San Francisco's site, is in some ways also a problem. Movement between sections of the settled area has been difficult and expensive, with few bridges crossing the bay. As the urbanized area grew, water barriers made a continued regional focus on San Francisco difficult. The Bay Area Rapid Transit (BART) system, one of the few new mass transit systems in the country, was designed in an attempt to alleviate part of this problem. Only three of the area's nine counties chose to join the system however. It serves only the East Bay and San Francisco, thus doing little to ease circulation in the entire region. Still, and despite a few widely publicized early breakdowns in its computerized system,

This air photo of a portion of San Francisco Bay is a dramatic indication of human encroachment into the Bay. The irregularly shaped areas that surround much of the Bay are areas for salt collection. They are repeatedly flooded, then allowed to dry. The salt that remains can then be scooped up and sold. (NASA.)

BART is an innovative attempt to meet the difficulties of urban movement.

Unlike Los Angeles, the existence of a major urban center in the Bay Area is not surprising. Its excellent harbor and good climate are important site factors. In fact, access to the bay has been so important and sizable tracts of flat land for construction so limited that large portions of the bay have been filled to create new land. Communities around the bay have created a regional government authority to restrict such landfill development, spurred by the fear of huge economic and ecological losses that might result from excessive filling; little reshaping of the bay has been allowed by the authority in recent years.

Furthermore, the relative location, or situation, of the bay region also remains excellent. By volume, it is the major Pacific port in North America. Its rail and highway ties to the East at least equal those for any other West Coast city. Just as Megalopolis is America's hinge with Europe, San Francisco is its hinge with Asia.

Cities of the Valley

Beyond the southern coastline and the San Francisco Bay region, the largest cities in California are in the Central Valley. The most populous is Sacramento, the state capital, with an SMSA population of 1.1 million. In common with all of the valley cities, it is a major agricultural processing center. Much of the processing of fruit and vegetables for canned or frozen products is accomplished near the source. It also has a substantial aerospace industry. Two large nearby air force bases are additional important contributors to the regional economy.

Fresno (SMSA population 515,000) and Stockton (population 347,000) are regional service and agricultural-processing points for the central and northern San Joaquin Valley. Bakersfield (population 403,000) serves a similar role at the southern end of the valley. It has also been influenced by local petroleum extraction. Bakersfield is the home of country-western music in California.

A STATE OF IMMIGRANTS

"California is a state of immigrants, mostly from the Middle West." "A native Californian is a rarity." These are commonly held beliefs about Californians. As with most such statements, they contain a substantial kernel of truth, but they also can be misleading. Although more than two-thirds of all native-born Americans live in their state of birth, it is true that fewer than half of all Californians were born in the state. California has been an important destination for U.S. internal migration in nearly every decade since 1850 and has become the major state of residence, other than the home state, for people born

357

in an area extending from California eastward and northeastward through the Midwest. The political climate, especially in San Diego and Los Angeles, often reflects the conservatism of the area's southern and Midwestern migrants. Between 1940 and 1980, California's share of the total population of the United States grew from 5 percent to slightly over 10 percent. California had, for decades, attracted more migrants than any other state until Florida gained that distinction for the 1970s.

Despite this, nearly half of all Californians were born within the state. Certainly a native Californian is no rarity. Also, migration to California has definitely slowed. Between 1970 and 1980, California's population increased 18.5 percent. Although this was well above the national growth of 11.4 percent, it nevertheless represented a substantial slackening from the rates of the previous three decades. Many still move to California, but their number has decreased and more are now leaving the state than in the past, especially to go to neighboring states in the West. California will surely continue to keep its place solidly as the country's most populous state, but it appears that an era of quite rapid growth has ended.

Interestingly, as the overall growth rate of the state has slowed, the ethnic diversity of the population has increased. Today, blacks represent 10 percent of the state's population, compared with 2 percent in 1940 and 7 percent in 1970. During the same periods, California's Oriental population increased from 2 percent (1940) to 4 percent (1970) to 6 percent (1980) of the total. In addition, nearly 20 percent of the population is now of Spanish heritage (see Chapter 14).

California's Future

It is unlikely that the slowdown in the state's population growth rate foreshadows an end to Americans' migration toward California. The rapid growth of the 1950s and 1960s will surely not return, but Americans remain among the world's most mobile and affluent people, and they continue to place a high value on an attractive physical environment. California's ranking as the most populous state in the nation seems assured, and it should continue to grow at a rate above the national average.

Many have suggested that California has emerged as the country's cultural core, replacing New England and the Middle Atlantic states. Certainly, it represents what many Americans view as the good life. One author has written that southern California may be "the super-American region or the outpost of a rapidly approaching post-industrial future." It is certainly a new kind of America, one that is copied if not totally admired.

California's problems are as great as its possibilities, with growth the main creator of both. The state's physical environment, so admired and desired, is a fragile one; the desert and the alpine ecosystems are easily damaged. Southern California's environmentally destructive cross-country motorcycle races, with more than 1000 participants dashing across the Mohave Desert, have demonstrated this fragile nature to all but the most ardent off-road enthusiasts. The outdoor recreation demands of 25 million Americans are immense and could easily outstrip the state's resources. Only so much seacoast exists, and much of it is already unavailable for public use. Although the state has some of the country's most rigorous, well-enforced environmental laws, the magnitude of the problems makes these laws barely adequate.

Perhaps the two greatest problems California faces are related to urban expansion and water supplies. Although cities are expanding everywhere, the loss of farmland in this Mediterranean climate, a climate found nowhere else on the continent, poses a serious threat to the domestic production of certain crops. The greatest losses of land, centered around Los Angeles and San Francisco, have been offset largely by expansion elsewhere, particularly in the San Joaquin Valley. Most of the easily available new land is now in use. Additional expansion will require irrigation on a massive scale, and this will bring the need for maintaining agricultural produc-

This Barstow to Las Vegas motorcycle race, held in 1970, is an example of the problems involved in meeting the outdoor recreation demands of Californians within the framework of an often fragile natural environment. (U.S. Department of Interior, Bureau of Land Management.)

tion levels into conflict with the state's limited water supplies; if present trends continue, water demands will outstrip California's available supply early in the next century. The chances of obtaining water from the Columbia River are, at best, problematic.

It may well be that, in the long run, irrigation is not the most economical use for the state's water. The monetary return for water invested in crop growth is relatively small when compared with alternative industrial uses. The repeated development of new sources of supply has thus far allowed the state to sidestep this difficult question. Soon it may no longer be able to do so.

The problem of water sufficiency makes the future of California's agriculture somewhat difficult to predict. Although California does lead the nation in the value of its agricultural pro-

duction, the relative importance of agriculture within the state's total economy is much less than for states like Iowa or Nebraska. Also, many of the crops grown are not a basic part of the American diet. Demand for such crops may fall if their prices are pushed up by more expensive irrigation water. Finally, these expensive irrigation projects help relatively few farm operators who are often corporations rather than individual farmers. Thus, despite the considerable political clout these few operators may have, the decision might eventually be made to shift water from agricultural land to urban areas where the vast majority of Californians live and where most of the state's income is generated; this might be done in spite of the loss it would represent to corporate farmers, to the state, and to the nation.

Additional Readings

Donley, Michael M., et al. (eds.). *Atlas of California.* Culver City, Calif.: Pacific Book Center, 1979.

Ford, Larry, and Ernst Griffin. "The Ghettoization of Paradise," *Geographical Review,* 69 (1979), pp. 140–158.

Furuseth, Owen J., and John F. Pierce. *Agricultural Land in an Urban Society.* Washington, D.C.: Association of American Geographers, 1982.

Kahrl, William L., et al. *The California Water Atlas.* Sacramento: Governor's Office of Planning and Research and California Department of Water Resources, 1979.

Lantis, David W. *California: Land of Contrast.* Dubuque, Iowa: Kendall/Hunt, 1977.

Liebman, Ellen (ed.). *California Farmland: A History of Large Agricultural Holdings.* Totowa, N.J.: Rowman & Allanheld, 1983.

McWilliams, Carey. *Southern California: An Island on the Land.* Santa Barbara, Calif.: Peregrine Smith, 1973.

Miller, Crane S., and Richard S. Hyslop. *California: The Geography of Diversity.* Palo Alto, Calif.: Mayfield Publishing, 1983.

Nelson, Howard J. *The Los Angeles Metropolis.* Dubuque, Iowa: Kendall/Hunt, 1983.

Rosenberg, M., and C. Crane. *San Francisco Today.* Washington, D.C.: Urban Land Institute, 1979.

Steiner, Rodney. "Large Private Landholdings in California," *Geographical Review,* 72 (1982), pp. 315–326.

Walker, Richard A., and Matthew J. Williams. "Water from Power: Water Supply and Regional Growth in the Santa Clara Valley," *Economic Geography,* 58 (1982), pp. 95–119.

West, Nathaniel. *The Day of the Locust.* New York: Random House, 1939.

CHAPTER SIXTEEN

THE NORTH PACIFIC COAST

Come to visit us again and again. . . . But for Heaven's sake, don't come here to live.

TOM MCCALL, former governor of Oregon

C old, clear mountain streams tumble down rock-strewn courses. The destination: a rugged, unused coastline where precipitous, fog-enshrouded cliffs rise out of pounding surf. In the distance, one sees mountains—lofty, majestic, covered with snow—and tall needleleaf evergreens—pines, spruces, and firs—that cover the land between with a mantle of green. Cities, where they exist, give an impression that they are new. They seem to fit within the general landscape, clean looking and somehow better than most other urban places. The people here are satisfied and friendly. They ask for little but the beauty of their land and a life filled with the virtues of independence and self-reliance. This is North America's North Pacific Coast, or more popularly, the Pacific Northwest (Figure 16-1).

At least that is how many Canadians and Americans (both from within the region and from outside it) think of the coastal zone that stretches from northern California to southern Alaska. The appeal of the region's natural environment and the variety of outdoor activities it supports led one writer to label this region "Ecotopia."[1] It is an image that has induced many to seek their fortunes in this region. A traditional western belief (in both countries) is that progress can be measured in the ability to attract people to fill an empty land. However, a change of attitude occurred in the southern half of the region in the 1960s as residents of Oregon and Washington began to realize that they had something good and the continued influx of newcomers threatened to destroy it. Their fear was given nationwide attention by the plea—quoted at the beginning of this chapter—made on national television by Tom McCall, then (1976) the governor of Oregon. That attitude, too, is now a firm part of the region's image.

Another important element of its regional character is the North Pacific Coast's relative isolation from the rest of North America. The pop-

---

[1] Joel Garreau, *The Nine Nations of North America.* Boston: Houghton Mifflin, 1981.

FIGURE 16-1
THE NORTH PACIFIC COAST.

ulated sections of the region are separated from the other principal population centers of the two countries by substantial distances of arid or mountainous terrain. Residents of the region often view this isolation as a positive element, one that provides a geographic buffer against the less desirable rest of the world. Economically, however, it is a hindrance because high transportation costs inflate the price of Pacific Northwest products in distant eastern markets and also discourage some manufacturers from locating there.

Like many images, that of the Pacific Northwest is made up of equal parts reality and myth. Whether the image of the region is more true or more false depends partly on alternative interpretations of its boundaries. Oregon, Washington, and Idaho are often grouped together with the province of British Columbia, sometimes under the general rubric, "The Northwest." Others would not include Idaho and might find the regional title, "The North Pacific Coast", more acceptable. Regional identification through the utilization of state and provincial boundaries does create a place of great internal diversity. Broad expanses of the eastern two-thirds of Oregon are a desert or semidesert environment, much of it virtually treeless. This comes as a shock to the uninitiated visitor from the East, as an environment is encountered that is certainly unlike the well-watered Pacific margins of the state. Southeastern Washington is a land of large wheat farms that consistently place the state among the country's half-dozen leading producers of that grain.

THE PHYSICAL ENVIRONMENT

As identified here, the North Pacific Coast is, with one or two notable digressions, defined primarily on the basis of its physical environment. Stated simply, it is a region strongly subject to maritime influence and rugged terrain. Precipitation is high and vegetation associated with heavy moisture is located near the coast. There is marked variation between relatively close places because of

the influence of surrounding mountains on the region's climate.

Climate

The mountain is out today.

Any discussion of the nature of the geography of the Pacific Northwest begins with its climate. The heaviest annual precipitation amounts on the continent are found here, averages above 190 centimeters (75 inches) are common, and averages are double that figure on the western slopes of the Olympic Mountains on the Olympic Peninsula in northwestern Washington (Figure 16-2). The 600 centimeters (230 inches) recorded on the northwest end of Vancouver Island easily surpasses even these high amounts and is more than twice the amount received anywhere else in the United States (with the exception of Hawaii) or Canada. During the winter, the cloud cover is often so constant that residents of Seattle will tell each other on those rare days when the slopes of nearby Mount Ranier are visible. The heavy precipitation of the Olympic Peninsula supports rainforests where ferns and mosses grow in profusion and trees (including western hemlock, red cedar, Sitka spruce, and the world's largest Douglas fir) are frequently more than 60 meters (200 feet) tall. The naturalist Roger Torey Peterson suggests that the Olympic rainforest contains more weight of living matter per unit of land than anywhere else in the world.

The North Pacific Basin is a spawning ground for great masses of moisture-laden air. As these air masses move outward, they are pushed south and east by the prevailing winds onto the Pacific shores of North America. The high-pressure system off the coast of California in the summer and off northwestern Mexico in winter prevents many of these maritime air masses from drifting farther southward and ensures that most of their moisture will fall over the Northwest coast. The seasonal movement of weather belts has a noticeable impact on precipitation in the Pacific

363

Average annual precipitation (inches)

☐	Under 15
▨	15-40
▩	40-80
▦	80 or over

FIGURE 16-2

TOPOGRAPHY AND PRECIPITATION (topographic map from Raisz, *Physiographic Diagram of the United States*). Notice the association between areas of rugged topography and higher precipitation.

Northwest. Winter precipitation amounts are everywhere higher than summer levels, but the seasonal difference is more marked on the southern margin of the region. The coast of southern Oregon and northern California receives less than 10 centimeters (4 inches) of precipitation during the summer months of July and August, only a tenth of the amount that falls between December and February.

It is tempting to suggest that this area has generally high precipitation and then leave the subject. That would be an error; considerable portions of the region are semiarid. Parts of the borderlands of Puget Sound in Washington receive only about 60 centimeters (25 inches) annually, no more than the amount that falls on Wichita, Kansas, in an average year. Portland's average is similar to that for Boston or New York and is far less than the precipitation of a southeastern U.S. city, such as Atlanta. Perhaps these areas of the Northwest seem wetter because of the high amount of cloud cover during the midfall to midwinter season. Precipitation seldom falls in the form of heavy thundershowers; instead,

MAJOR METROPOLITAN AREA POPULATIONS IN THE NORTH PACIFIC COAST, 1980	
Seattle-Everett, Wash.	1,607,469
Vancouver, B.C.	1,268,183[a]
Portland, Oreg.-Wash.	1,242,594
Tacoma, Wash.	485,643
Spokane, Wash.	341,835
Salem, Oreg.	249,895
Victoria, B.C.	233,481[a]
Anchorage, Alaska	174,431
Bremerton, Wash.	147,152

[a] Indicates 1981 population.

it is a gentle, light, frequent rain that often feels more like a heavy mist. One consequence is that runoff, so normal in heavy rains, is lessened, and vegetation is able to make maximum use of the moisture.

The region's mountains are the major influence both on the high precipitation along the coast and on the substantial variations that often exist in close proximity. As the eastward and southeastward moving air masses pass over land, they strike the several mountain ranges that line the North Pacific coast and are forced to rise. As they rise, the air cools, the moisture-carrying capacity is reduced, and precipitation results. Such precipitation is called *orographic* (mountain-induced) precipitation, and it is common on the windward side of most extensive mountain chains.

Along a belt extending from south-central Oregon to southwestern British Columbia, the Coast Range in the United States and the Insular Mountains in Canada are backed by a trough of low-lying land, including the Willamette Valley in Oregon, the Puget Sound lowland in Washington, and the Fraser River lowland in British Columbia. As the east-moving air descends into these lowlands it warms, and its moisture-carrying capacity is increased. Because additional moisture is not introduced into the air, little precipitation will occur. Thus, the same air masses that provide abundant coastal moisture and heavy precipitation on the western slopes

of the coastal ranges provide substantially less moisture along the lowland trough.

To the east of the lowland is a second north-south-trending band of mountains. The moving air must again rise, and orographic precipitation, although not quite as heavy as along the coast, results again. These inland mountains, called the Cascades in the United States and the Coast Mountains in British Columbia, are high: Mount Ranier in Washington has an elevation of 4390 meters (14,140 feet), and many peaks are in the range of 2750 to 3750 meters (9000 to 12,000 feet). Winter precipitation falls in the form of snow, making this the snowiest portion of the continent.

Finally, in the eastern extension of the region in interior Washington, air masses again descend and warm. The little moisture that remains in the air is retained, and most of eastern Washington averages less than 30 centimeters (12 inches) of precipitation yearly. It is only through irrigation and dry-farming techniques that most of the area's important crop agriculture is possible.

South and north of this mountain-valley-mountain system, the mountains squeeze together and the separating valley disappears. Heaviest precipitation amounts are concentrated in a single band along the northern coast, including the central and northern coast of British Columbia and Alaska's southern peninsula. This coastline receives substantial moisture throughout the year, and average cloud cover percentages are among the continent's highest. The climate of peninsular Alaska is especially dominated by moisture and by cloudiness.

Average precipitation levels drop sharply along the coast of Alaska north and west of the southern peninsula; most of the southward-facing coast of central Alaska averages 100 to 200 centimeters (40 to 80 inches) annually. As mentioned, many of the weather systems created off southern Alaska's shores move south and east, often only brushing the shoreline of central Alaska, and many residents in Anchorage look with disdain at the damp climate of the southern peninsula, favoring their own clearer, if somewhat colder, weather.

In addition to bringing rain, the region's maritime location provides a moderate temperature regime. Summers are cool. Winters are surprisingly warm, although the dampness creates a rawness in the air that makes one less comfortable outdoors than the thermometer might suggest. Still, Seattle is warmer in winter than St. Louis and Juneau, Alaska, is no colder than New York City.

A third consequence of the seasonal air mass movement is frequent periods of high winds along the region's coastal margin, especially in Oregon, northern California, and southwestern Washington. The large storms built up during passage across the North Pacific can strike the coast with tremendous ferocity. These storms are not hurricanes (which are tropical storms), but it is not uncommon during the long winter months for winds to exceed 125 kilometers (80 miles) per hour during the stormier periods. Although the coastal mountains provide some protection and winds are generally lower in the summer, high winds can reach the more arid eastern portions of the region even in the summer. When they do, the dangers of fire are worsened.

Topography

The Pacific Northwest is a land of mountains; few places in the region do not offer a view of neighboring peaks when the weather is clear. Mount McKinley, at the region's northern extremity is, at 6200 meters (20,300 feet), the highest in North America. The St. Elias Mountains in Canada are the world's highest coastal mountains; Mount Logan is just short of 6000 meters (19,700 feet). Elevations are generally lower farther south, but it remains a land of high mountains and beautifully rugged terrain.

The Coast Ranges of Oregon and Washington are generally lower than other mountains in the region. They are fairly continuous in Oregon, with elevations reaching about 1200 meters (4000 feet). In Washington, they are discontinuous, with several rivers, notably the Columbia and the Chehalis, cutting pathways across them. Coast Range elevations in Washington are seldom above 300 meters (1000 feet).

The Klamath Mountains of northern California and southern Oregon offer a jumbled topography, one that has been heavily eroded by both Pleistocene glaciation and stream erosion. Little pattern is apparent in the terrain, in marked contrast to the parallel ridge and valley structure of the Coast Ranges. This is a wild, rugged, empty area.

The lowlands of western British Columbia and Oregon are part of the structural trough that was created when that area sank at the same time that the Cascades to the east were elevated. This trough extends northward in the form of straits separating Vancouver Island from the rest of British Columbia, then passes through the complex of islands that line the Alaska Peninsula, and provides the Inside Passage north as far as Juneau.

Farther inland, the Cascades extend from the Klamath Mountains northward into southern British Columbia. The southern section of these mountains appears as a high, eroded plateau topped by a line of volcanic peaks. Beginning with Mount Lassen in California (the only volcano in the United States outside Hawaii to have been active in historic time until the 1980 eruption of Mount St. Helens) and passing from Crater Lake to Mount Hood in Oregon, these peaks are especially splendid in their isolation above the surrounding plateaus. This line of peaks marks the zone along which the Pacific plate is passing beneath the North American plate (see Chapter 15). The northern Cascades are more generally rugged and have long proved a difficult barrier to movement eastward from the populous Puget Sound lowland. Here again, extinct volcanos, most notably Mount Ranier, provide the highest elevations and best-defined peaks.

In British Columbia, the Cascades—known there as the Coast Mountains—merge with the Insular Mountains and push against the ocean.

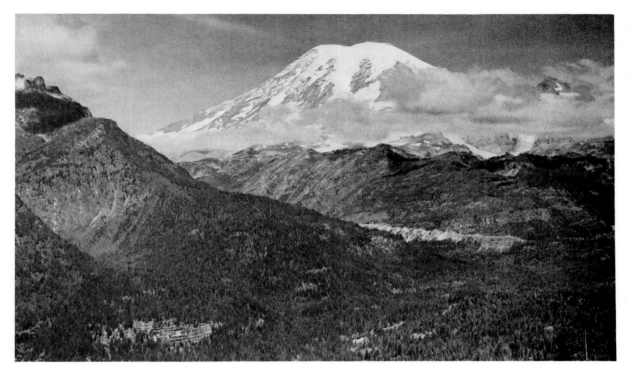

Mt. Rainier, Washington, is one of a series of volcanic peaks that punctuate the southern Cascades. (George Grant/U.S. Department of the Interior.)

This is a land of dramatic coastal mountains cut by many glacially eroded fiords and islands. Land transportation along the coast is nearly impossible; coastal shipping provides the principal transportation connections. From Vancouver northward to beyond the Alaska Peninsula, overland transportation lines connect the coast with the interior at only two points: one at Prince Rupert in British Columbia and the other at Port Chillcoot and Skagway at the northern end of the Inside Passage in Alaska.

Beyond the Alaska Peninsula and the massive, glacier-covered St. Elias Mountains, the mountains again divide in southern Alaska. The Coast Ranges, notably the Chugach and Kenai Mountains, decline in elevation from east to west. The

interior mountains, the Alaska Range, are much higher and more continuous. A large lowland at the head of Cook Inlet is south of a gap through the Alaska Range, and here Anchorage, easily the largest city in Alaska (population 174,431 in 1980), is situated with its good harbor and easy connection into the interior.

Juneau, the capital city of Alaska, is located on a narrow coastal lowland on the peninsula; it is connected to the rest of the state only by air or water. The farthest that one can drive from town is about 15 kilometers (10 miles). This was a reasonable location for the capital when Alaska's wealth was primarily in its peninsular forests and salmon fisheries, and when it also offered access to the Yukon gold fields through Skagway. As

367

the state's economy changed and other resources became more important, the peninsula languished. Both Fairbanks (population 22,645 in 1980)—accessible to central Alaska—and Anchorage—accessible to the southern portion of the state—far outstripped Juneau in growth; the capital city had yet to reach 20,000 in population in 1980. Alaska's voters and political leaders, becoming dissatisfied with the inaccessibility of Juneau, voted in 1974 to move the capital to a point some 80 kilometers (50 miles) north of Anchorage. The cost of constructing a new capital city, now estimated at over $3 billion, has caused many Alaskans to reconsider that decision. In 1978 and again in 1982, the voters refused to allocate money for the move, so that despite its location, Juneau may yet remain the capital.

Trees and More Trees

With magnificent redwood stands in the Klamath Mountains; Douglas fir, hemlock, and red cedar in Washington and Oregon; red cedar, hemlock, and Douglas fir in British Columbia; and Sitka spruce on the Alaska Peninsula, this is a land not just of forest, but of beautiful expanses of tall trees reaching straight for the sky, trees that are among the largest on earth, trees that encourage people to stand and stare in awe or admiration . . . and to cut for profit.

Except for the drier lowlands, where the normal vegetation of the Willamette Valley is prairie grass and that of the land east of the Cascades a mix of grass and desert shrub, and except for the tundra above the tree line, all of the Pacific Northwest is, or was, covered by forest. Tree growth is encouraged by plentiful moisture and moderate winter temperatures. Forest products were long the economic mainstay of the region. Even today, although the Southeast produces more wood for pulp and paper products, no other part of Canada or the United States provides as much lumber as does the North Pacific Coast.

PATTERNS OF HUMAN OCCUPATION

No other coastline, except for the polar areas, was explored by Europeans as late as was the North Pacific Coast. Vitus Bering had claimed the Alaska coast for Russia by 1740, but it was not until 1778 that Captain Cook sailed the coast from Oregon to southeastern Alaska. By that date, several million colonists at the other side of the continent were fighting a war for independence against the very country for which Cook was sailing. By the time Lewis and Clark were stumbling their way across the Cascades to the mouth of the Columbia River in 1805, Philadelphia and New York City, each with about 75,000 people, were vying for the title of the country's largest city. By the mid-1840s, when American settlers began traveling the Oregon Trail to the Willamette Valley in substantial numbers, New York's population was fast approaching 500,000, and its city administration was already wrestling with problems of urban transportation and waste disposal. Since the date of substantial European influence at a place was roughly related to its distance from Europe, this late exploration and settlement is understandable. In the days of sail and slow overland transportation, no nonpolar coast in the world was as far from Europe as this one.

The American Indians

The pre-European population of the region was relatively large. The moderate environment provided a plentiful supply of food throughout the year. Deer, berries, roots, shellfish, and especially salmon represented a natural bonanza of food that seemed without limit. The American Indians responded to this with a simple hunting and gathering economy and no food crop cultivation. Concentrated along the coast, the Indians were divided into many distinct ethnic groups, each occupying its own, often small, coastal valley. They constructed large, impressive houses of red cedar planks and went to sea in dugout canoes made of the same wood. They developed many similarities in culture, notably the *potlatch* (a ritual during which property was given away in a large celebration) and the carving of totem poles (the process of recording on an upright log a series of key incidents in an individual's life).

Along most of this coast, the Indians seemed simply to melt away when Europeans arrived. Because their extreme isolation made organized opposition impossible, each small tribe succumbed quietly, making little impact on European settlement; their numbers dwindled, unnoticed by the incoming European settlers. Today few Indians remain in the south, although one small tribe called the Puyallups recently won in the courts the right (for all Indians) to half of the total salmon catch in Washington. This decision, based on an early agreement in which the Indians surrendered much of their land in return for certain vaguely stated rights, resulted in restrictions on salmon fishing that created a great clamor among the state's many non-Indian sport fishermen. Farther north, the Indian population is proportionately larger and remains a substantial ethnic group in northern British Columbia and peninsular Alaska.

Early Canadian and American Settlement

Much of the early European settlement history of this region reads like a collection of pulp novels, with prospectors, lumbermen, land barons, fur trappers, settlers, sober-minded preachers, railroad magnates, and imported brides trooping across the pages in the best Bret Harte fashion. It is the stuff of good historical fiction, and one wonders why the region's writers have not done better with it. Several events are so important to the region's development that they must be mentioned here.

The Russians were the first Europeans to establish permanent settlements along the coast. They came late in the eighteenth century, initially following the predictable European pattern of searching for easily extracted riches. Here these riches proved to be furs, and the Russians

established a series of trading posts and missions that eventually extended as far south as northern California but were concentrated in southeastern Alaska. The outposts never became self-sufficient in foodstuffs, and the cost of maintaining these scattered, distant sites usually exceeded the income from fur sales. Following several earlier Russian attempts to sell the colony to the United States, a $7.2 million sale price was finally agreed on in 1867. Many Americans considered the price far too high, and the U.S. government was roundly criticized at the time for the purchase. Evidence of the Russian occupation still exists in some parts of Alaska—in the architecture of wooden onion-domed Eastern Orthodox churches, in cemeteries, and in a few pockets of cultural survival from early Russian settlers.

The Hudson's Bay Company moved its fur-trading operations into the Columbia River Basin early in the nineteenth century. It was the dominant influence from northern Oregon northward into British Columbia until the late 1830s when American missionaries and settlers began the long journey across the Oregon Trail from Missouri. Most of these new American settlers moved into the Willamette Valley, but their numbers quickly outweighed the small British population of the entire Northwest. The cry of these American settlers and of Westerners in general, "Fifty-four forty or fight," claimed all of the land as far north as latitude 54° 40′N, and was a significant part of the presidential election of 1846. In that year, the British and American governments agreed to establish the present boundary at latitude 49°N. However acceptable this may have been as a political decision, it disrupted the normal north-south movement pattern along the Puget Sound and Columbia River transportation corridors.

Victoria, on Vancouver Island, was for decades the main urban focus for the British-held portion of the region. As the first permanent European settlement established in 1843, Victoria remained the largest town in British Columbia until 1886. In that year, the settlement of Vancouver, located on the mainland adjacent to the mouth of the Fraser River, became the western terminus of Canada's first transcontinental railroad. Although Victoria continued to be important as the province's administrative and commercial center, Vancouver's eventual position as Canada's main port of entry on the Pacific was ensured once the railroad was completed.

The railroads were of key importance in the eventual growth of Oregon and Washington as well. In 1870, these two states combined had fewer people than did the city of San Francisco. In 1883, the Northern Pacific Railroad was completed to Seattle and was followed a decade later by the Great Northern. This ended the region's overwhelming dependence on oceanic shipment, which sailed via the southern tip of South America to the eastern United States and European markets. The region's bulky resources could now be more easily shipped in volume. For a period after completion of the railroads, twice as many people moved to Washington each year as moved to California.

The immigration that flowed into Washington and Oregon between the 1880s and the beginning of World War I created an interesting difference in the ethnic composition of those states. In 1880, Oregon had 175,000 people; Washington, only 75,000. By 1910, Washington had outpaced Oregon's growth, so that the northern state's population was nearly double, with 1,140,000 persons compared to Oregon's 672,000. Although most of Oregon's population was composed of immigrants from other parts of the United States, booming Washington attracted large numbers of European immigrants as well, especially Scandinavians. Even today ties to that ethnic heritage can be important in Washington, and many of the state's politicians are of Scandinavian extraction. Oregon, by comparison, has a population with a strong New England heritage, the result of early overland settlement.

Population Distribution Today

More than 6 million people live in the Pacific Northwest. Their numbers are increasing at a rate that is more rapid than the national average for either Canada or the United States. Most live in a long cluster of cities and towns along the lowland extending from the Fraser River in

FIGURE 16-3

THE PUGET SOUND—FRASER RIVER LOWLAND. This area, home for a majority of the region's residents, benefits from excellent water transportation, adequate precipitation, and a desirable recreational environment.

southwestern British Columbia through the Willamette Valley of Oregon (Figure 16-3). Three of these cities, Vancouver, Seattle, and Portland, each have metropolitan area populations of more than 1 million people. This land of the great outdoors is, like nearly every other part of the two countries, peopled by an urban population. The cities, in turn, are very much a part of the region.

Vancouver. Now Canada's third largest city, Vancouver is a bustling metropolitan area of more than 1.2 million people. It has grown greatly in size and cosmopolitan character since World War II. Its urban area now extends southward to the U.S. border, and local planners expect the city to double in population during the next 30 years. Vancouver's port has become a true western outlet for interior Canada. The products of the Prairie Provinces, formerly shipped almost entirely to, and through, the cities of eastern Canada, are now to a great extent funneled westward to Vancouver. The development of Asian markets for western Canada's wood products and wheat has encouraged the growth of the city's port facilities. With an annual cargo volume of nearly 50 million metric tons (55 million tons), the city is easily Canada's busiest port. Frequent overcrowding of the port facilities and of the eastern rail lines has prompted Canadian officials to encourage the increased use of the alternative port at Prince Rupert, which the government is considering as the site for construction of a major grain transshipment facility.

Vancouver is the western headquarters for nearly all of Canada's major businesses. Most of British Columbia's manufacturing, with wood processing easily the most important, is in, or near, the city. Only the fact that the provincial capital is across the Strait of Georgia at Victoria on Vancouver Island prevents Vancouver from totally dominating British Columbia.

Seattle. Seattle has been the largest city along the North Pacific Coast since the boom era of

the late nineteenth century, although it is now close to losing that distinction to Vancouver. Founded as a logging center, Seattle began to achieve regional dominance when it was linked to the northern U.S. transcontinental railroads.

The city, since World War I the home of Boeing Aircraft, has been called the world's largest company town. A mainstay of the construction of bomber aircraft in World War II, the company's fame spread after the war as it introduced a series of highly successful passenger jets. Employment at Boeing has fluctuated wildly as airplane orders responded to changes in market demand and the general condition of the world economy. From a high of 103,000 in the late 1960s, employment at Boeing fell to about 50,000 during the 1970s. Consequently, the economy of the city and to an extent that of all Washington suffered; population decline followed the heady growth period of the 1960s, and only nine states grew less rapidly than Washington during the economic slowdown. Several years of comparative employment stability at Boeing and local efforts at economic diversification have enabled the area's economy to recover; today the metropolitan area is again experiencing substantial immigration.

The core of Seattle is tucked onto a narrow isthmus bordered by Puget Sound on the west and Lake Washington on the east. It has, like Vancouver, a beautiful site, with repeated views of mountains and water offered to the residents of its many hills. It is a middle-class city, with pleasant, scattered neighborhoods of tree-lined streets. A subjective evaluation of the 50 largest cities in the United States during the mid-1970s concluded that Seattle was the "best" city in the country in which to live.

Portland. Two other studies of urban quality conducted at about the same time disagreed with this conclusion. A study by the federal Environmental Protection Agency called Portland the "most livable city" in the country. Another, produced by the Midwest Research Institute in

Vancouver, British Columbia. Excellent harbor facilities and easy access into the interior up the Fraser River Valley help make this handsome city Canada's major western urban center. (David Watson/National Film Board—Photothèque.)

Kansas City, evaluated well over 100 variables before agreeing that Portland was indeed the best of the country's 67 largest cities. (Eugene, Oregon, in the Willamette Valley, was judged the best medium-sized city in both studies.) Studies like these have been criticized for overly emphasizing new cities and cities with small ethnic minorities and for discounting less measurable characteristics. Nevertheless, they say much about the general livability of Oregon's principal city.

Residents of Portland greeted both studies with mixed emotions. They were largely inclined to agree, but they feared that such publicity would bring waves of newcomers who would by their

numbers destroy the very qualities that attracted them. An old city by the standards of the region (new by most others), Portland residents consider their city more conservative and more stable than Seattle, and they hope to maintain that atmosphere. Its economy is more diversified than that of Seattle, and its relations with the region's interior are closer because of the lowland routeway to the east provided by the valley of the Columbia River. It is a major transshipment point for the grain from eastern Washington, for example, and wood products and food processing are principal activities of the local manufacturing economy. Portland, inland about 160 kilo-

The Seattle skyline, with its characteristic western U.S. collection of new tall buildings, is seldom blessed with this clear a view of Mt. Rainier. (Seattle-King Convention and Visitors Bureau.)

meters (100 miles) from the coast, nevertheless, rivals Seattle as an ocean port because the lower Columbia River is navigable.

THE REGIONAL ECONOMY

In many ways, the economic structure of the North Pacific Coast region is dominated by the production of staple products (see Chapter 6) and by its isolation from the major markets of both countries. This latter factor is especially applicable for the region's U.S. portion. Less than 3 percent of the American population lives here, although about 10 percent of all Canadians are

in British Columbia. Most major markets of both countries are more than 3500 kilometers (2200 miles) away. British Columbia is the exclusive western terminus for national economic activity in Canada, but goods and people passing in or out of the United States through Pacific coast ports can go through California, bypassing Oregon and Washington completely. Freight-rate structures long placed many of the region's products at a competitive disadvantage in the major markets of the two countries. The region has always contained a number of high-demand products, notably lumber and foodstuffs. However, movement costs have often reduced the

373

ability of producers to get their products to market at reasonable cost; that is, the great distance has limited the *transferability* of these products. Consequently, market areas turned to other sources of supply that were nearer and less expensive to tap. The higher freight rate charged for finished wood products in Canada, for example, has encouraged manufacturing in British Columbia to remain heavily involved in the initial processing stages only.

Agriculture

The region's agriculture is a good example of the impact of transferability problems on the regional economy. The kinds of crops that can be grown in the region are largely duplicated closer to the areas of major consumption in the East (Figure 16-4). Thus, Washington apples are in direct competition in the Eastern markets with those from Wisconsin, Pennsylvania, or Virginia. The same is true for a variety of other fruit and vegetable crops of the region. This is quite different from conditions in central and southern California, where the Mediterranean climate enables growers both to produce many crops that simply cannot be grown in the colder north and to provide more common fruits and vegetables fresh during the winter season. Thus, for decades, California cornered the market for many of its agricultural products (see Chapter 15). *Intervening opportunities* (other sources of supply) were not available, and distance to market had little negative impact on the growth of California's agricultural economy.

The result of all this is that much of the North Pacific Coast region's agricultural products are grown for the local market, not for export. This is evident in Oregon, where the broad Willamette Valley is easily the largest agricultural area near the region's coast. The land has been cultivated for more than a century, and its farms are prosperous and well established. Much of the farmland is in forage crops, and many farmers continue to follow the practice of burning these fields in the fall—with the result that for a period of several weeks, large sections of the valley are covered by a layer of smoke.

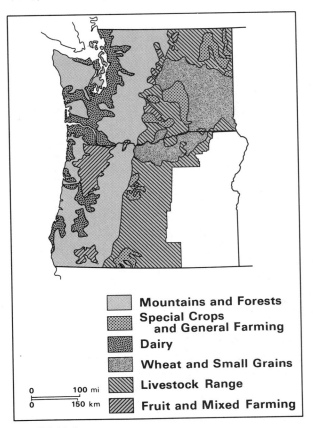

Mountains and Forests

Special Crops and General Farming

Dairy

Wheat and Small Grains

Livestock Range

Fruit and Mixed Farming

0   100 mi
0   150 km

FIGURE 16-4

MAJOR CROP PRODUCING AREAS (from Richard Highsmith, *Atlas of the Pacific Northwest: Resources and Development.* Corvallis: Oregon State University Press, 1968). The Willamette Valley and Puget Sound lowland are clearly identifiable. Areas east of the Cascades emphasize either dry-area crops like wheat or depend on irrigation.

Dairy products, also generated mostly for the local market, are of greatest importance to the agriculture of the Willamette Valley; strawberries are perhaps the most important specialty crop. A variety of other specialty crops also thrive in the valley's climate, including hops, grass for turf seed, cherries, and spearmint. The total value of these crops, however, does not match the area's income from dairy production.

The Puget Sound lowland in Washington is another important dairy area, as is the Fraser

The rolling wheatfields of eastern Washington give the area an appearance that is more like portions of the Great Plains than the usual perception of the Pacific Northwest. (Grant Heilman.)

River Delta in British Columbia. Again, a variety of specialty crops are also grown, with peas leading the pack. Quick frozen and then shipped to markets throughout North America, this cool-weather crop is particularly well adapted to the local climate.

The area east of the Cascades in Washington presents a different kind of agricultural landscape. Most of it is semiarid, and grasses and desert shrubs replace the majestic evergreens of the coast and mountains. Although called the Columbia Plateau, the area has little of the characteristic flatness one expects from the term plateau. In part, it consists of rolling hills. Elsewhere, in central Washington, the landscape has been cut by a series of steep-sided dry canyons, called coulees. That section—properly called "the channeled scablands" because lava pockets dot

the surface with what look to be scablike knolls—is covered by a deep blanket of lava flows eroded by the floods from glacial melt during the Pleistocene ice retreat.

We have placed most of the Columbia Plateau in a different region, the western Empty Interior, because of its aridity, nodal settlement pattern, and economic focus on minerals, grazing, and transportation services (Chapter 13). An exception is found along the Oregon-Washington border and across much of eastern Washington. This is a substantial farming area, easily the most important in the Pacific Northwest. In many ways separate from the coast, residents of the area—who call it the Inland Empire—nevertheless, feel an association with the coast equal to that felt with other parts of the interior. Their crops are primarily sold down the Columbia River; they share a common state government with the coast; and they look to Seattle and Portland for many goods and services. For all of these reasons, the Inland Empire is a part of the Northwest, although an atypical one.

The agriculture of the Inland Empire is quite varied. The hilly country of east-central Washington, called the Palouse, averages between 35 and 65 centimeters (15 and 25 inches) of precipitation annually, somewhat more than other parts of the interior and is today one of the most important dry-farming areas in the country. Dry farming is usually confined to such subhumid climate areas and emphasizes water conservation more than other types of farming. Wheat is the primary crop of the area, with both spring and winter varieties grown. Wheat is normally planted on a field in the Palouse every other year. In alternate years, the land is left fallow; that is, it is plowed, but nothing is planted. This retards evaporation and allows for an increase in soil moisture. The large wheat farms of the Palouse are heavily mechanized and highly productive. Most of the product is exported through Portland to Asia.

Irrigation has played a major role in the Inland Empire's agriculture in recent decades. Two major irrigated areas have been developed. The water from a number of streams flowing eastward out of the Cascades is used to irrigate their relatively narrow valleys. The result is one of the country's most famous apple-producing areas. Effective national advertising has helped convince Americans that these apples from the Yakima, Wenatchee, and neighboring valleys are the best available. More recently, pears from this same area have been given equivalent national marketing attention and distribution.

The damming of the Columbia River at Grand Coulee, northwest of Spokane, was done primarily to provide hydroelectric power. It also made available large amounts of irrigation water to the Big Bend area of south-central Washington. After these waters became available in the late 1950s much growth occurred, and crop agriculture is still expanding in response. Major agricultural products include sugar beets, potatoes, alfalfa, and dry beans.

Forest Products

British Columbia is easily the largest lumber producer in Canada, supplying approximately 45 percent of all timber harvested in that country. In the United States, Washington, California, and Oregon together provide over half of all lumber cut, and Washington and Georgia vie for the lead in pulp and paper production. Although forestry was the first major industry in the region, its rich forest resources did not become nationally important until well into the twentieth century when improved transportation facilities, coupled with the destructive overcutting of many forests farther east, opened the region's woods to lumbering.

Douglas fir (of prime importance as structural supports for houses as well as for flooring, doors, and plywood) is easily the major lumber tree of the region, although each section of the region has its own mix of trees to harvest. In British Columbia, for example, spruces, hemlock, and balsam are most important. In northern California, redwoods remain locally important, although their numbers are dwindling; the western red cedar is also cut widely in the area from Oregon northward.

The large size of many of the trees plus the distances to market tend to encourage large-scale logging operations. Weyerhaeuser Lumber Company, for example, owns some 690,000 hectares (1.7 million acres) of land in Washington and is easily the state's largest private landowner. A substantial part of Washington and a majority of all land in British Columbia, Oregon, and northern California is government owned. Private logging on governmental lands plays a large role in overall production. Effective marketing systems have enabled the lumber products of the region to penetrate all the market areas of both countries, even including the southeastern United States, which is America's other major forestry region. In recent years, the Japanese have become major buyers, especially from British Columbia and Alaska; most of Alaska's lumber is now shipped to Japan.

The changed practices of the forest-products industry are clearly seen in this region. Loggers once cut everything in their path and then moved on, always sure that more trees could be found elsewhere. However, for every tree cut today, a new one is planted. Governments of both countries as well as private companies carry on extensive research programs to develop stronger, straighter, healthier, faster-growing trees; young trees are even fertilized from airplanes and helicopters. Computers now control the process by which the trees are cut into lumber, with each tree cut so that the maximum amount of lumber is obtained. New stands now reach maturity in 40 to 75 years, and yields of most varieties of trees can be sustained indefinitely under these methods of operation.

Because lumber and pulp production are so important to the regional economy, any issue that might affect the availability of trees for cutting is strongly felt and widely discussed. In recent years two major controversies have been subject to debate, especially in the United States. The first involves the redwoods of northern California, the world's tallest trees and among its oldest. The majority of redwoods, the most important lumber tree in California, are on private land. A major environmental drive has developed to keep these trees from being cut. This effort led to the creation of Redwood National Park in the late 1960s. Conservationists contend that the park is too small and that some lumber companies are hastily cutting trees on land that may soon be set aside as parks. Loggers respond that some people would have every tree saved, creating ruin for the lumber industry and large increases in the price of forest products.

The second controversy involves the practice of clear-cutting, a process in which all trees in a given area are cut at one time. Loggers argue that this is the most efficient way to remove trees and to ease reforestation. Many conservationist groups consider clear-cutting to be a major contributor to soil erosion and stream pollution, de-structive of forest ecosystems, and an ugly blotch on the landscape. In 1973, the Izaak Walton League sued to halt clear-cutting in the Monongahela National Forest in West Virginia, contending that the Organic Act of 1897 provided only for the cutting of "dead, matured, or large growth" trees. They won. In 1975, cutting was halted in Alaska's Tongas National Forest on the same basis. A wave of concern swept the lumber industry of the North Pacific Coast. Although the eventual solution to the problem will come from the U.S. Congress, the conflict is indicative of the continuing disagreements between many conservation groups and the forest-products industry.

Power and Dams

The plentiful precipitation and rugged topography of this region provide a hydroelectric potential unmatched anywhere else on the continent—40 percent of the entire U.S. potential is contained in Oregon and Washington alone. The Columbia River, in particular, with a flow volume larger than that of the Mississippi River and a drop of nearly 300 meters (1000 feet) during its 1200-kilometer (750-mile) course from the Canadian–U.S. border to the sea, is a power developer's delight.

Grand Coulee Dam, begun in 1933 and still the region's largest, was the first of many dams constructed on the Columbia River. It was followed by no fewer than 10 dams downstream. The Dalles, between Oregon and Washington, where Lewis and Clark portaged around dangerous falls, is now a lake. British Columbia and the United States agreed to the construction of three additional dams in Canada, which would store water during heavy flow periods and then release the water when flow was low to guarantee adequate levels for power development. The government of British Columbia was given a check for $273 million as part of the agreement. This guaranteed flow has resulted in the addition of new generators at Grand Coulee Dam that will triple its capacity, making it the world's largest single power producer. The treaty also

Clear, or block, cutting, such as is seen here in Washington, is a controversial approach today. Loggers argue that it is the most efficient method, but some environmentalist groups fear that it may increase the ecological damage. (U.S. Forest Service.)

led to expansion of power production elsewhere in British Columbia for provincial use, most notably in the Peace River area.

These developments have provided inexpensive power for the North Pacific Coast. That, in turn, has attracted a number of manufacturers who are heavy power consumers; most notable is the aluminum-smelting industry. Kitimat, isolated on the northern coast of British Columbia, is the site of one of the world's largest aluminum refineries. It is located there because a nearby dam provides the large quantities of cheap electricity needed for aluminum production and the coastal location permits cheap transportation of bauxite, a bulky ore.

Hell's Canyon—the world's deepest gorge—is cut by the Snake River along the Oregon-Idaho border and is unusual in the region because of the absence of power dams. For decades, it has been a center of controversy between power producers and many environmentalist groups. In 1975, the U.S. Congress named much of this stretch a "wild and scenic river," presumably stating that no dams will ever be allowed there.

Fishing

Forestry and fishing at one time formed the backbone of the region's economy. The cold waters of the North Pacific were, and to an extent still are, rich fishing grounds. They attracted large numbers of whaling vessels during the late eighteenth century and the first half of the nineteenth century. Heavy overharvesting has reduced the North Pacific's whale population to a small fraction of former levels, and the few remaining whalers have turned to the waters off Antarctica for most of their catch.

Salmon has long been the most important fish of the North Pacific Coast. It contributed a major part of the foodstuffs of the coastal tribes before the arrival of Europeans and remains the principal fish caught throughout the region. Salmon are anadromous; that is, they migrate upstream from the ocean to spawn in freshwater. Their spawning runs once filled the rivers, and massive catches were easily available to those on the banks. The Salmon Chief of the Colvilles, near Grand Coulee, reportedly could once catch an average of 400 fish a day, each weighing perhaps 14 kilograms (30 pounds), simply by scooping them up and throwing them onto shore. Most of the tribes revered the salmon, making it an important part in their religious rituals.

The size of the salmon catch has declined greatly over the past four decades; today it is less than half its former level. Most salmon are caught off the Alaska coast, with British Columbia following in second place. When the region's streams were dammed, access to many traditional spawning grounds was blocked, especially on the upper Columbia and its tributaries. *Fish ladders*—a series of gradual water-carrying steps that allow fish to jump from level to level and, thus, bypass a dam—have been constructed around some of the smaller dams, but they do not work on the larger ones. Nearly all of the Snake River and its tributaries plus all of the branches of the Columbia above Grand Coulee, are, consequently, closed to salmon. This has led to continued controversy between hydroelectric power developers and fishermen; the two seem totally incompatible. The salmon and those who depend on them have been the chief losers in dam construction.

Overfishing has also had a substantial impact on the North Pacific's fish population. As they fish these waters, Canadian and American commercial fishermen are joined by many others, most notably Japanese. Because of this foreign competition, many American fishermen in the Northwest supported the federal government's 1976 decision to extend exclusive offshore fishing rights to 320 kilometers (200 miles) for a wide variety of fish, including the salmon. The decision on whether or not to favor this extension has continued to be debated vigorously by West Coast fishermen. It cuts foreign fish harvest in these extended American waters and, thus, hopefully provides larger catches for Americans, but it also encourages other countries either to enforce similar limits already established in their areas or to enact new 320-kilometer limits and force American fishermen to observe them. This is especially bothersome to tuna fishermen who are based primarily in California. Much of their tuna catch originates off the shores of Ecuador and Peru, whose 320-kilometer limit these fishermen have largely ignored, claiming it was illegal.

In addition, controls on fishing exerted by Canadian governmental bodies have not always been matched by equivalent conservation measures on the U.S. side of the border. One result has been occasional contention between U.S. and British Columbian fishing interests, which have been similar in nature to the disagreements between the two countries in the Atlantic fishing areas.

ALASKA—A POLITICAL ISLAND

Coastal southern Alaska is clearly a part of the North Pacific Coast, but it must be viewed as somewhat separate from the rest of the region. No railroad connects it with the more populated parts of the continent, and only a single long and partly unpaved highway connects it through interior Canada to the rest of the United States.

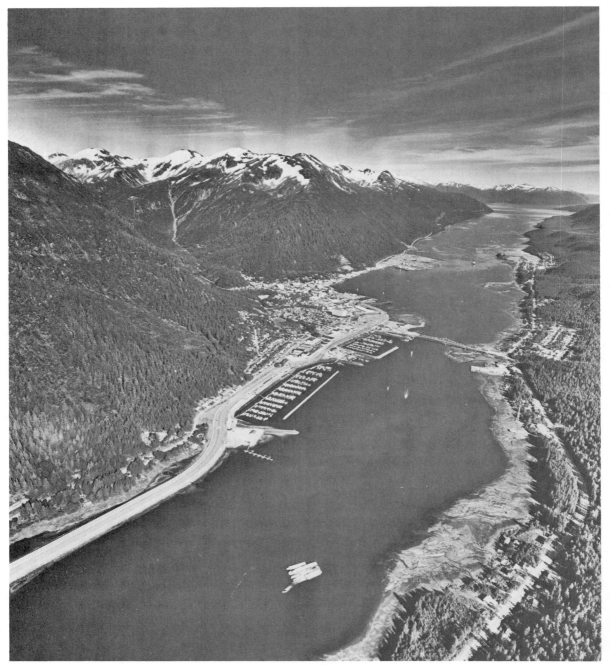

Juneau, the capital of Alaska, is cut off from land communication with the rest of the state by the sea and the encircling mountains. (Air Photo Tech, Inc., Anchorage.)

People in peninsular southeastern Alaska are crowded by coastal mountains onto a narrow shoreline seldom more than a few hundred meters wide. All parts of the area look to air and sea transportation for connection with the rest of the world. This leads to a sense of detachment, of separation, from the activities of the "lower forty-eight" and to an economy of high prices resulting from scarcity and high transportation costs. Anchorage is the most expensive city in the United States in which to live.

The traditional belief is that the Alaskan economy is heavily based on minerals, lumbering, and fishing. In fact, the federal government, primarily the Department of Defense, is the dominant employer in the state. Even the petroleum development boom on the state's North Slope has only modified, not eliminated, this orientation. Anchorage, home of one out of three Alaskans, is dependent on the military as the cornerstone of its economy.

Alaska is almost as close to Japan as to the conterminous United States, and Japan is the major market for many Alaskan products. As indicated, most lumber cut here is destined for Japan. Although the federal government defended its strong support of the construction of the Trans-Alaska Pipeline system as necessary to bring North Slope petroleum to the rest of the United States to meet growing energy demands and to cut dependence on foreign suppliers, some people have suggested that a bigger profit could be gained if a substantial portion of the output was sold to energy-hungry Japan.

LAND OF CONFLICTING OPPORTUNITIES

The Pacific Northwest, where the unspoiled is not yet spoiled, attracts a different kind of person. The weather is too cool, or at times too uncomfortable, to attract the sun-seeking elderly. Economic opportunities rarely provide a quick kill. The Northwest is for people more interested in being than achieving. . . . And in the Northwest

it is not necessary to own a great deal to share in a great wealth, the outdoors all around.[2]

This self-serving statement, written by a lifelong resident of the Northwest, says much about both the region's people and its environment. It is a place where many want to be, and most who are there seem relatively satisfied with their lot. Yet it is not the home of a booming economy. Unemployment is frequently high, particularly when there is a decline in demand for one of its few staple products or as happened with Boeing Aircraft in Seattle in the early 1970s. Although the growth of a diversified manufacturing economy in the northern Willamette Valley is lessening the role of the lumber industry, statewide unemployment in Oregon approached 13 percent in early 1982 as a national decline in home building suppressed the local lumber market. Limited economic opportunity is surely the major reason why far more people have not moved here.

Nearly 200,000 pleasure boats crowd Puget Sound every year. Washington has more than 500,000 licensed fishermen. Literally dozens of magnificent state parks and campgrounds dot Oregon's majestic 650-kilometer (400-mile) coast. To reduce litter, Oregon was the first state to pass a law requiring that beer and soda be sold only in returnable containers. Much of British Columbia remains true mountain wilderness, in spite of development projects undertaken in the last several decades across the province. This is a land of the great outdoors. The environment seems to call people to go out and enjoy it. Most do. That so much of the land is publicly owned simplifies their search. The land is the people's, and they use it. All of this sounds a little mushy, almost quaint. Still, as Thomas Griffith puts it, "Northwesterners are a people by nature possessed."[3]

---

[2] Thomas Griffith, "The Pacific Northwest," *The Atlantic Monthly*, 237 (April 1976), p. 51.

[3] Ibid., p. 47.

A Land-Use Case Study: Alaskan Land—Who Is It for?

When the United States acquired Alaska in 1867, nearly all of the territory's 150 million hectares (375 million acres) fell under the control of the federal government. The Alaska Organic Act of 1884 recognized vaguely that Indians and Eskimos in the territory "shall not be disturbed in possession of lands actually in their use or occupation or now claimed by them." Otherwise, most of Alaska remained in the public domain.

Historically, all public land initially controlled by the federal government in a newly acquired area was opened to private ownership through one of a series of land transfer laws, such as the Homestead Act of 1863 or the Mining Law of 1872. In Alaska, however, the Department of the Interior imposed a ban on private preemption or settlement that remained in force for nearly a century. Thus began a spirited struggle for Alaska land that continues to the present, a struggle that has involved argument between the U.S. federal government and the Alaskan state government, between native groups and each government, between settler and native groups, and between developers and environmentalists. The opponents in these arguments have occasionally shifted in their alliances with others involved in the issue, but the scope of the struggle has grown dramatically and regularly.

The federal government chose to maintain its hold on Alaskan land until 1958, the year of statehood. On the eve of statehood, only 200,000 hectares (500,000 acres) was privately owned, with most private land located around Anchorage and Fairbanks—and most of it patented recently under the Homestead Act. The federal government had reserved a number of plots for military purposes and a large area north of the Yukon River for a Naval Petroleum Reserve.

The Alaska Statehood Act of 1958 ignored Indian and Eskimo claims and granted the state the right to select 42 million hectares (104 million acres) by 1983 for state use. Most of what has been selected is in the Alaska Range and the Yukon Valley, although the state also controls a large area on the North Slope.

In 1968, petroleum was discovered at Prudhoe Bay on the North Slope. At least partly to dispose of native claims to Prudhoe Bay and to accelerate petroleum development generally, Congress passed in 1971 the Alaskan Native Land Claims Settlement Act. This gave the state's Eskimos, Aleuts, and Indians nearly $1 billion and the rights to 18 million hectares (44 million acres) of land. Most of this land was given to profit-making regional development corporations, although about 40 percent went to serve 200 villages who maintain control of surface rights.

The transfer of land to the state and to native groups proceeded slowly amidst a tangle of conflicting claims. It was during this period that environmentalists became concerned at what they saw as the potential loss of a rich natural resource, the vast wilderness areas of Alaska. In 1978, the U.S. House of Representatives passed the Alaska National Interest Lands Conservation Act, which withdrew 58 million hectares (144 million acres)—over a third of the state—primarily for national parks and wildlife refuges. A threatened filibuster by Alaska Senator Gravel kept the Senate from acting on the bill, but President Carter used a little known 1906 Antiquities Act to create 22 million hectares (54 million acres) of Alaskan national monuments. Furthermore, he proposed another 22 million hectares for wildlife refuges and 4.5 million hectares (11 million acres) to be exempted from mining.

Outraged at these moves to set aside this land from possible development, many Alaskans argued that land withdrawal at the scale proposed would strangle the state's economy. The governor at that time, Jay Hammond, spoke for this view when he complained that "Alaska is being called upon at once to be the oil barrel for America and the national park for the world."

Congress passed a compromise Conservation Act in 1980. This designated 42 million hectares (104 million acres) for parks and refuges, including 23 million hectares (57 million acres) specified as wilderness (Figure 16-5). It also ex-

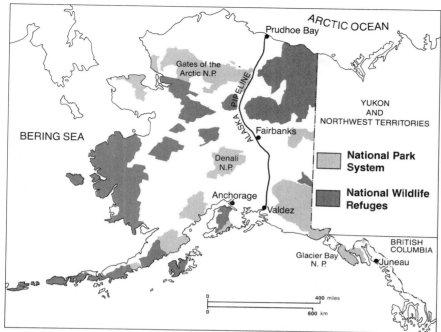

FIGURE 16-5

ALASKA'S NATIONAL PARKS AND WILDLIFE REFUGES. The state has
nearly a third of the country's entire national park and refuge acreage.

pedited the conveyance of land to native groups
and to the state, and it extended the transfer
deadline to 1994. Less than 20 percent of the
state's total to be transferred from the federal
government had actually been shifted when
transfers had been halted because of conflicting
claims.

The struggle for control of Alaskan lands is
not likely to be over because conflicting goals remain:

> Fundamentally, the issues have been ones of
> priority. The environmental movement argues
> that Alaska's lands belong primarily to all the peo-
> ple of the United States and require legal protec-
> tion to prevent undue development, particularly
> of mineral resources. The native corporations es-
> tablished by the Alaskan Native Land Claims Set-
> tlement Act of 1971 argue that their rights to land

> should take first priority because they represent
> aboriginal rights recognized in the Alaska Organic
> Act of 1884 and implicitly confirmed in the Alaska
> Statehood Act of 1958. The state, while recogniz-
> ing the paramount legal rights of the federal gov-
> ernment and the rights of the natives, argues that
> the recent massive federal withdrawals of land
> from individuals have seriously impinged upon
> the rights confirmed to Alaska by the Statehood
> Act.[4]

It seems reasonable to expect that even the extension of the transfer deadline to 1994 will not
be sufficient to permit resolution of the conflicting claims for use of Alaska's lands.

---

[4] Donald F. Lynch, David W. Lantis, and Roger W. Pearson, "Alaska: Land and Resource Issues," *Focus*, 31 (January–February, 1981), p. 2.

Additional Readings

Booth, C. W. *The Northwestern United States.* New York: Van Nostrand, 1971.

Dicken, Samuel N., and Emily F. Dicken. *Oregon Divided: A Regional Geography.* Portland: Oregon Historical Society, 1982.

Farley, A. L. *Atlas of British Columbia: People, Environment, and Resource Use.* Vancouver: University of British Columbia, 1979.

Gastil, Raymond. "The Pacific Northwest as a Cultural Region," *Pacific Northwest Quarterly,* 64 (1973), pp. 147–162.

Hull, Raymond, et al. *Vancouver's Past.* Seattle: University of Washington Press, 1974.

Meinig, Donald W. *The Great Columbia Plain: A Historical Georgraphy, 1805–1910.* Seattle: University of Washington Press, 1968.

Minghi, Julian V. (ed.). *People of the Living Land: Geography of Cultural Diversity in British Columbia.* Vancouver: Tantalus Research, 1972.

Robbins, William G., et al. (eds.). *Regionalism and the Pacific Northwest.* Corvallis: Oregon State University Press, 1983.

Robinson, J. Lewis, and Walter G. Hardwick. *British Columbia: One Hundred Years of Geographical Change.* Vancouver: Talon Books, 1973.

Shortridge, James R. "The Collapse of Frontier Farming in Alaska," *Annals of the Association of American Geographers,* 66 (1976), pp. 583–604.

Warren, James C. *Seattle: King County and Its Queen City.* Woodland Hills, Calif.: Windsor Publications, 1981.

CHAPTER SEVENTEEN

THE NORTHLANDS

The United States and Canada are in many ways creations of a frontier experience. The push westward remains part of recent history. Many still live who remember hearing from their parents of the often heroic struggle with the land that is such a large part of our settlement history.

The American historian Frederick Jackson Turner and many other students of American attitudes and customs who agreed with him have argued that because this frontier experience is still recent, it continues to have a noticeable impact on our society. The high value we attach to land ownership, by comparison with most Europeans, stems in part from the widespread availability of land and from the drive to possess land that helped push settlement westward. More land always seemed available over the next hill or beyond the next river. Those who lacked land believed that, through perseverance and diligence, land could be acquired; those who owned land could always reach for more.

The Canadian-American frontier is largely gone today. One can still find ads promising "government lands at $5 an acre," and some families do still move off from settled society and attempt to begin anew in areas that were heretofore unoccupied and unused. The ads are basically flimflam, and the infrequency of frontier settlement can be seen in the wide reporting it receives when it does occur. Although humans presumably have the technology to live anywhere on the earth's surface, those areas of Canada and the United States that can be occupied with moderate physical and economic effort are already staked out.

One vast region of North America remains sparsely settled; many tens of thousands of square miles are totally unoccupied. This is the North American Northlands (Figure 17-1). Extending as far south as the northern Great Lakes states and including the great majority of interior and northern Canada and Alaska, it is easily the largest of North America's regions. Generally difficult for human settlement, it is the inhospitable nature of the physical environment plus the

FIGURE 17-1
THE NORTHLANDS.

consequent thinness of settlement that gives the region its special character.

A HARSH ENVIRONMENT

A brief identification of the key features of the Northlands environment must include mention of its cold temperatures, long winters, thin soils, poor drainage, and low precipitation. It is the combination of these basic elements that limit broad human use of this region.

Climate

If Americans were asked to give a one-word definition of the nature of the Northlands, "cold" would probably be the most commonly used ad-

jective. This would certainly be a valid judgment. Average January temperatures in this region range from a high of about −7°C (20°F) along its southern Great Lakes margin to a full −40°C (−39°F) in parts of arctic Canada. Temperatures can reach −60°C (−78°F).

Not only are winter temperatures quite low across most of the region, but winters are also long. The frost-free period (the average time between the last frost in the spring and the first in the fall) is roughly 135 days at the southern margins of the area but little more than 14 days in length along parts of the Arctic Ocean borders of the two countries. Across most of the region, the frost-free period is less than 90 days. Because virtually all major food crops need a growing season of longer than 90 days, they can be grown in only a few small areas along the southern margins of the Northlands.

Summers, generally short and cool, can have surprisingly warm days. The region's climate is largely continental. Maritime moderation is significant only along the peripheries, mainly in the east and west. Thus, along with the low winter temperatures come brief but warm summers. It is an old joke among residents of the Yukon or the Northwest Territories that, when asked by visitors what they do in summer, the reply is "Well, if it falls on a weekend, we go on a picnic."

This dramatic seasonal variation in temperature results from the great shifts in length of day and the angle of incidence of the sun's rays. The earth's axis, that line connecting the two poles, tilts at an angle of 23½° from the plane of revolution around the sun. As the earth follows its annual path around the sun, the north pole is tilted toward the sun in our summer and away from it in our winter. Thus, it is dark everywhere north of 66½°N latitude, the Arctic Circle, for at least one day at midwinter, and there is at least a 24-hour period at midsummer without the sun setting. Moreover, during the winter, the sun, when it rises, remains low on the horizon. The sun's rays, striking the earth at a low angle, must cover a relatively large area. Much less heating results from this in comparison with the summer period, when the sun's rays are far more direct.

In midwinter, Fairbanks must make do with the light from a sun hanging just above the horizon for perhaps two hours. In midsummer, the sun sets briefly around midnight. Even southern sections of the region receive only six or seven daylight hours during the winter. For people accustomed to sleeping in darkness and being active during daylight, this seasonal fluctuation demands substantial psychological adjustments.

Precipitation amounts vary widely across the Northlands. Highest levels are found in the far southeast where both winter and summer storm systems move along the coast and combine to dump more than 100 centimeters (40 inches) of annual precipitation along the southern shore of Labrador. Precipitation levels drop rapidly toward the interior and north. Most of the Northwest Territories in Canada receive less than 25 centimeters (10 inches) annually; parts of the Arctic islands average less than 15 centimeters (6 inches). The cold air masses usually found over this area hold little moisture and, therefore, produce little precipitation. Also, the Arctic Ocean, covered by ice for all or much of the year, supplies little or no evaporation from the sea. The northern Arctic islands are the most arid parts of Canada and the United States. Snowfall amounts in the Northlands are consequently much lower than might be expected. Whereas parts of Labrador and interior Quebec may average more than 150 centimeters (60 inches) of snow a year, most of the Northwest Territories receive less than 60 centimeters (24 inches). The Arctic islands get only about 25 centimeters (10 inches) of snow annually, roughly the same amount as Washington, D.C.

Terrain and Soils

Despite the paucity of precipitation, little of the Northlands provides the appearance of a dry environment. In the summer, in fact, much of the region is covered with standing water, and the mosquito stories of the Northlands match those generated anywhere on the continent. This is due, in part, to the low levels of evaporation and evapotranspiration found in this cold climate. In the Arctic portions of the region, standing water is also supported by the widespread existence of *permafrost* (Figure 17-2). This is a subsurface layer of permanently frozen ground, which may be only a few meters thick, but which sometimes extends downward for more than 300 meters (1000 feet). As the surface layer thaws during the short summer to a depth of perhaps 1 meter (3 feet), water is held on the surface by the frozen layer underneath instead of penetrating the soil. This creates an extremely boggy, shifting surface. Construction in permafrost is difficult. Buildings must be placed on piles sunk deeply into the permafrost for sta-

An upland tundra in central Alaska. The vegetation of the tundra is often a rich mixture, with small flowers blooming in profusion in early summer.

bility, and roads must be repaired extensively each year to maintain any semblance of an even roadbed. Heat given off by buildings or absorbed from the sun's rays by road surfaces causes additional melt of the permafrost under the construction. Unless care is taken, a building constructed on permafrost can sink gradually into the ground, plunging deeper and deeper as it continues to melt the permafrost directly under it. Roughly half of Canada and most of Alaska are underlain by continuous or discontinuous permafrost.

MAJOR METROPOLITAN AREA POPULATIONS IN THE NORTHLANDS, 1981	
Sudbury, Ont.	149,923
Thunder Bay, Ont.	121,379
Duluth-Superior, Minn.-Wis.	93,745[a]
Fairbanks, Alaska	22,645[a]

[a] Indicates 1980 population.

Although considerable local terrain variation is found, much of the Northlands topography is either flat or gently rolling. The North Slope of Alaska and much of the central Canadian Arctic is a broad flat plain, as are the valleys of the Mackenzie River and the Yukon River in central Alaska.

Almost all of the area to the south and east of the coastal plains is a part of the Canadian Shield, North America's largest physiographic province. The Shield is underlain by very old, heavily metamorphosed rocks. Substantial local relief is found mainly along the southeastern edges of the Shield, especially just north of the St. Lawrence Valley. Nearly all of this 3 million square kilometer (1.1 million-square-mile) area was scoured and contoured repeatedly during the past million or so years of the Pleistocene (Ice Age), removing much of the soil from the Shield and depositing it in valleys or to the south. Drainage patterns were greatly disrupted; the landscape is dotted with thousands of lakes and

FIGURE 17-2

DISTRIBUTION OF TUNDRA, TAIGA, AND PERMAFROST. A harsh winter
and a short growing season dominate much of interior and northern Canada.
Most Canadians live south of the taiga.

bogs that mark former lakes and is drained by rivers following tortured, twisted pathways to their destinations. Fertile soil is confined to some of the old river valleys and to those lakes that have been filled by sedimentation and decayed vegetation. But even these areas are often acidic, poorly drained, and of low agricultural quality.

Vegetation

Most of the Northlands can be placed in two large, distinct vegetation areas. Stretching across the entire southern arc of the region is a coniferous forest called the boreal forest (or sometimes the taiga). Covering hundreds of thou-

sands of square miles, these closely ranked spruces, firs, pines, and tamaracks appear to blanket the landscape in a dark, almost black mass when seen from the air. Slow growing and never really tall, these trees decrease in number and height from south to north across the taiga. Around the Great Lakes, a mixture of pines and hardwoods predominate.

Passing just south of Hudson Bay, then angling northwest to the mouth of the Mackenzie River and across the northern edge of Alaska, is the tree line. It should be viewed as a transition zone rather than an actual line. North of it, climatic conditions are too harsh for treelike veg-

389

A view of the boreal forest, or taiga, along the Yukon River in the Yukon Territory. (Judith Currely/National Film Board—Photothèque.)

etation. Beyond lies the tundra, a region of lichens, grasses, mosses, and shrubs. At first glance, tundra vegetation appears homogeneous because it is all of the same low height. In fact, considerable variation exists based on temperature, precipitation, and soil differences. Still, to those not sensitive to these differences, the overwhelming visual impact of the tundra is of a broad, usually flat or rolling terrain devoid of any tall vegetation.

HUMAN OCCUPATION

Nearly all parts of the Northlands are sparsely populated with highest densities found along the southern margins. American Indians and Inuit (Eskimos) are important over much of the region. Inuit are the predominant population in most of the Arctic. American Indians are found mainly in the boreal forest area. In pre-European days, the Indians were hunters and fishers; their characteristically widespread population was generally associated with hunting societies. Although their economies were similar, there were substantial cultural differences between eastern and western groups.

By comparison, the Inuit culture was, and is, remarkably uniform across its broad distribution, which extends around the Arctic margins from Siberia across North America to Greenland. The most recent arrivals of America's pre-European population, the Inuit settled largely along the water margins of the Arctic. They depended on fishing and hunting for their livelihood, and developed an effective, self-sufficient life-style in that harsh environment.

The arrival of Europeans in the Northlands brought an end to much of this traditional native culture. Fur traders acquired many of their pelts from the Indians, and European goods entered the Indian economy as a result. Along with these trade goods came European diseases, liquor, and Christianity, all of which worked against the structure of the Indian culture. Although these changes did not reach many Inuit until comparatively recently, the results have been much the same. Where hunting and fishing continue, the motorboat, rifle, and snowmobile have usually replaced the kayak, bow, and dogsled.

Most Northlands Indians and Inuit no longer exist by hunting and fishing. They have moved in substantial numbers into towns or the economy of their villages has changed. Many urban places of the Northlands today have large native populations. They generally occupy the bottom rungs of the social and economic ladder, with far more than their share of unemployment, disease, and poverty. Those who are poorly educated and trained are less able to compete for the better paying urban jobs.

Both governments have had difficulty defining their posture in their dealings with the Northlands' natives. The Canadian government never adopted the "remove-them" approach that long characterized American attitudes toward Indians. Even in the United States, contact with Alaskan natives came sufficiently late to allow some softening in these harsh attitudes. The na-

A modern Eskimo village at Payne Bay, on the Ungava peninsula in Quebec. Electricity, gas heat, public schools, and prefabricated dwellings are all part of the rapid decline of the pre-European culture of the Eskimo population within the space of one or two generations. (Dunkin Bancroft/National Film Board—Phototèque.)

tive population remains scattered across the state, and the American government has recently approved the return of large blocks of Alaskan land to Indian and Inuit control (see Chapter 16). Canadian Inuit and Indian groups in the Northwest Territories and parts of Quebec claim aboriginal rights to the land, rights not surrendered by treaty as elsewhere in Canada. The federal government has agreed to negotiate settlements with native groups whose rights have not been extinguished by treaty, and Quebec paid Inuit groups around James Bay a substantial amount for part of the land in the development of the James Bay power complex.

Still, even the most benign government cannot easily solve the problem of how to deal with a population subgroup that is vastly different culturally and technologically inferior as compared to the dominant group. Even if it were possible to create the kind of isolation that would allow the continuance of traditional lifestyles, such a policy would refuse to Indians and Inuit the opportunity to integrate into the predominant economy of the country. To encourage wholesale change, on the other hand, would result in the destruction of much that is valuable in the traditional culture. Not surprisingly, both countries follow something of a middle course, providing services, such as basic education and health care, while also encouraging the maintenance of traditional culture. Although certainly not an unqualified success, this is probably a reasonable approach and has the support of most native leaders.

Early European Settlement and the Northern Economy

The Northlands offered little of interest to most Europeans who came to North America. Early settlers generally selected the far greater agricultural promise of the warmer climate and better soils to the south.

Where Northlands' settlement did occur, its focus was usually either extractive or military. French voyageurs, fur trappers and traders in search of animal pelts, pushed their canoes far beyond the agricultural settlements along the lower St. Lawrence as early as the middle of the seventeenth century. They extended French political control across the Great Lakes and northward into the headwaters of streams draining into Hudson Bay. The Hudson's Bay Company, an early British fur-trading company, established itself on the margins of Hudson Bay and then pushed south and west, thus blocking further French expansion westward. By the mid-eighteenth century, the Hudson's Bay Company, which had been granted a trade monopoly to the area by the British government, was in control of the entire boreal forest reaching from Hudson Bay westward to the Rocky Mountains, with further extension of influence into the Arctic as well. This vast extractive empire brought with it only a minimal number of small and widely scattered settlements.

The voyageurs and Hudson's Bay Company relied on the numerous lakes and streams of their area for transportation, and they located small forts at control points along the water routeways. At places where important streams met lakes, where streams ended and an overland portage began, or where rapids or falls were encountered, vessels had to be unloaded and their goods moved and reloaded onto other boats; these break-in-bulk points provided effective control of the entire water system. The sites of many early French forts are occupied today by important urban centers; the continuing break-in-bulk function remains one reason for their existence. Cities, such as Ottawa and Sault Ste. Marie in Canada and Chicago, Detroit, and Pittsburgh in the United States, have grown around the sites of former forts (Figure 17-3).

Logging

The boreal forest of the southern half of the Northlands contains the largest area of uncut forest remaining in North America. Until recently, the lumbering and the pulp and paper industries only nibbled at the edges of this vast forest. The area of the upper Great Lakes was logged on a massive, devastating scale during the

FIGURE 17-3

CANADA'S POPULATION DISTRIBUTION. In the cartogram the size of an
area is proportional to its population, not its geographic extent. Most Canadians
are urbanites, and about 30 percent live in one of the country's three largest
metropolitan areas—Toronto, Montreal, and Vancouver—while so few live in the
Northlands that it is hardly visible here.

late 1800s and the early twentieth century. Tree
removal was so complete that the area became
known as the Cutover Region. Because little reforestation was practiced at that time and because the cold climate of the boreal forest slows
regrowth, much of this area is only now recover-
ing its previous appearance. Repeated fires, which
were fueled by wood debris left from wasteful
logging, and a hardpan soil that developed after
all vegetation cover had been removed, further
slowed the reforestation process. Even today, the
Cutover Region is still not an important logging
area.

Lumber and, particularly, pulp and paper operations now dot the southern margin of the boreal forest from Quebec to Manitoba, and cutting is gradually moving northward. The rich
spruce forests south and southeast of Hudson
Bay are the prime source of raw material for paper mills, which are themselves located on some
of the plentiful water power sites along the
southern margins of the Canadian Shield, many
in the St. Lawrence Lowland. Quebec is today
the leader in Canadian pulp and paper production, and Canada is the world's number two exporter of pulp and paper products, most of which

393

goes to the United States. After motor vehicles, pulp and paper products are Canada's most important export in value, with forest products other than pulp and paper also ranking high.

Much of the capital for the Canadian lumbering and the pulp and paper industries as well as most other Canadian extractive industries has come from outside Canada. Until the end of World War I, the United Kingdom was the principal supplier of investment capital, but since then, an increasing majority has come from the United States (see Chapter 6). Because adequate development capital for large-scale growth was unavailable locally, foreign capital has aided Canada in many positive ways. However, outside ownership frequently has also meant outside control. This rankles many Canadians, who, with justification, have often felt unable to govern their own extraction policies. For example, after 1911, Canadian newsprint was admitted duty free into the United States, and American interests subsequently expanded greatly their Canadian operations. Many Canadians also believed that too much of Canada's raw materials were exported for manufacture elsewhere. They feared this was hurting internal manufacturing growth and limiting the number of high-skill, high-wage jobs available to Canadian workers. Finally, they felt that too much of the profits gained from Canadian operations were leaving the country.

In recent years, the Canadian government as well as some provincial governments have responded to these complaints. To be accepted, new foreign investment in Canada may require "partnership" with substantial Canadian money as well and with majority control placed in Canadian hands. The Canadian government also encourages the development of a capacity to manufacture internally more of Canada's industrial raw material product. In spite of this, Canada will undoubtedly remain the major supplier of raw materials to the United States—and its chief trade partner. Former President Nixon in one speech gave that distinction to Japan, much to the consternation of Canadians who have long

argued, again justifiably, that few Americans recognize the important role that Canada plays in the American economy.

The total value of Canadian exports in 1983 was $88 billion (Canadian currency); about 70 percent of that export was sent to the United States. Of these exports, approximately one-third were crude materials, another third were fabricated materials, and the final third, manufactured products. Imports totaled just over $75 billion, with the United States again claiming a 70 percent share. Manufactured goods accounted for nearly two-thirds of all imports. The statistics illustrate Canada's role as a raw material supplier and manufactured product buyer—a classic colonial relationship. The Canadian government intends to modify this relationship by ensuring that a greater share of Canada's raw materials become manufactured goods before leaving the country.

Mining and Petroleum

In 1983, the minerals industry contributed about 7 percent of the Canadian gross national product. The great majority of all production comes from the Northlands (see Figures 2-6 and 2-8). Also, the region's U.S. portion in the upper Great Lakes area is that country's leading source of iron ore plus a substantial contributor of copper. Alaskan North Slope petroleum has recently provided a large addition to the American energy supply. In short, mining and petroleum extraction together with lumbering comprise the bulk of economic activity in the Northlands.

As with logging, the accessible periphery of the Northlands became the region's first important mining districts. The Mesabi Range in northern Minnesota, an area of gently rolling elongated hills, along with neighboring areas in Minnesota, Wisconsin, and Michigan developed into America's chief source of iron ore late in the nineteenth century. The Great Lakes waterway provided an easy route for moving those ores to the industrial centers of the manufacturing core (see Chapter 5). Billions of tons of high-grade ore were moved by rail to Lake Superior ports and were loaded there onto large specialized lake ships that carried the ore to ports in northeastern Ohio, where it was then transferred to railroads for the trip to iron- and steel-processing plants in the Pittsburgh-Youngstown area. Today, most ore goes to the newer integrated iron and steel facilities at the southern end of Lake Michigan. Locks at Sault Ste. Marie that connect Lake Superior with the rest of the Great Lakes are the busiest in the world, largely as a result of this ore traffic.

Most of the high-quality iron ores are now gone from the Lake Superior mining district. Attention has turned to a lower-grade ore called taconite, also found in huge quantities in the district. The iron content of taconite, roughly 30 percent compared to perhaps double that for the richer ores, is so low that to ship the ore to the lower Great Lakes for processing is considered far too expensive. Thus, the bulk of the ore is reduced (and its iron content increased) through a process called beneficiation. The ore is ground into a fine powder, much of the rock removed, and the resulting material pressed into small pellets with a much higher iron content. This greatly lowers the cost of shipping taconite, thereby increasing its transferability. That is, its movement costs are sufficiently low relative to the value of the improved ore to make it profitable to ship. Without this reduction in bulk, the taconite might be ignored in favor of other sources of supply that, although more distant, offer an ore that is actually cheaper to move to market because of its higher ore content.

The great cost of shipping low-grade ores is the major factor in choosing to locate many smelting operations near the source of supply instead of near the market. For example, copper, which seldom represents as much as 5 percent of its ore and often less than 1 percent, is nearly always refined near the mine. The smelting and refining of ores is the major form of manufacturing employment in the Northlands, and the large smokestacks of the refineries are the central element of the skyline of some of the largest cities in the region.

These smokestacks, incidentally, also reflect one of the key problems with this and almost any mining company—environmental disruption. Lake Superior iron ore is mined by the open pit method. Until recently, the land was simply left in its disturbed condition to recover as it might. With the topsoil gone, sterile materials exposed on the surface, and normal drainage patterns destroyed, natural recovery is extremely slow, and abandoned mines remain as ugly scars on the landscape. Today, in some places, an attempt is being made to aid recovery by recontouring land, planting, and fertilizing. There have been some successes, but such efforts tend to be difficult and expensive.

Other environmental problems developed at the processing plants. A vast quantity of waste material, or slag, is created during ore processing. For example, one beneficiation plant at Silver Bay, Minnesota, on Lake Superior was dumping some 250,000 metric tons (270,000 tons) of waste material into the lake each day. The topography around many refining centers gradually changes as the waste is piled into great hills. Also, many kinds of refineries produce gaseous and liquid wastes that can be harmful. If one were to visit Sudbury, Ontario, during the mid-1960s, the first indication that this important mining and smelting center was near were the trees covered with brown leaves in midsummer. Substantial improvement is apparent now, but treatment of waste is usually expensive and often only partially successful.

The mining frontier pushed northward from the Great Lakes early in the 1900s. At first gold and silver were the principal minerals removed; then attention turned to such metals as copper, nickel, lead, and zinc. Sudbury, at the southern end of this great mining concentration along the Ontario-Quebec border, started first. With an urban area population of 150,000 people, it is the largest urban center in the Northlands. To the northeast of Sudbury, the Clay Belt district crosses the provincial border from Timmins to Val d'Or, with copper the leading mineral. The Chibougamau area, 350 kilometers (220 miles) farther northeast, is a newer mining region that produces a variety of metals, including lead, copper, and zinc. Together, these several mining districts comprise the major mining region in Canada.

Of the other metallic mining areas scattered across the Shield, the most notable is the Labrador Trough along the border between Labrador and Quebec. The existence of large bodies of high-grade iron ore in the region was first confirmed early in the twentieth century. Its inaccessibility coupled with the adequate supplies of the Lake Superior district slowed development of the area. The decline in the quality of the lat-

ter district's ores plus growing demand in Canada and the United States encouraged the exploitation of the Labrador Trough. A railroad, opened in 1954 at a cost of nearly $400 million, connected Schefferville, the early center of production, with Sept Iles, some 550 kilometers (340 miles) south on the St. Lawrence River. Today, Schefferville's mines have been closed, and the community is in decline. Other mines in the area, such as those at Wabush and Carol Lake, Newfoundland, and Gagnon-Fremont, Quebec, maintain the two provinces as the leaders in Canadian iron ore mining.

Smaller concentrations of mining activity are

Thunder Bay, Ontario, at the western end of Lake Superior, Canada's second largest port, is a major transshipment center for Canadian grain, ore, and coal. A similar facility for American iron ore is found a short distance southwest at Duluth-Superior. (Lakehead Harbour Commission.)

found in a number of other locations in Canada's Northlands, including mines in western Ontario (iron ore), Flin Flon on the Manitoba-Saskatchewan border (copper and zinc), Thompson (a leader in nickel production), and Yellowknife on Great Slave Lake (a center of gold mining). Substantial recent interest in gold mining and the opening of new mines in the Northwest Territories and northern Ontario propelled gold to a number 3 ranking in Canadian minerals production in 1983, behind copper and coal and just ahead of iron ore.

Canada is the world's third leading metal producer, behind the Soviet Union and the United States. On a per capita basis, Canada is surely in first place. As noted, much of that product is exported. One result is that the Canadian mining industry is particularly susceptible to international economic shifts, especially those felt by the American metals industry. For example, the recession in the U.S. metals industry in the early 1980s depressed the sale of Canadian copper by one-third, iron ore by over one-third, and nickel by nearly one-half between 1980 and 1983. The economic results were devastating to many of the single-industry mining communities of the Northlands.

Accessibility remains a major problem for Northlands mineral production; construction of railroads was necessary before large-scale mining could begin in many of these areas. Many other sites, particularly in the far north, still have not been fully explored geologically.

Until recently, little petroleum was produced in the Northlands. Canada's major proven resources of liquid petroleum are on the prairies of Alberta; American demands were met by domestic production elsewhere or by imported supplies. Recent declines in developed reserves, price increases, and increases in petroleum use have encouraged exploitation and development in both countries. In Alberta, a major initiative financed by both public and private funds is underway to develop the Athabasca tar sand petroleum reserves (see Chapter 12). Elsewhere, geologic conditions suggest that the best prospects for petroleum and natural gas discoveries are found along the mountain margins in the west to the Mackenzie River delta and in a broad band extending from Alaska north of the Brooks Range (the North Slope) across Canada's northern Arctic islands. Oil deposits have definitely been indicated in the Mackenzie Delta, and gas has been found on Arctic islands. Exploitation of these northern fields will be gradual, however, slowed by the high cost of developing a resource that is cheaper at other locales and by Canada's growing desire to husband its natural resources.

The United States, in contrast, has moved rapidly to develop its North Slope petroleum fields. Some oil producers paid well over $1 billion just for the right to search for oil in the region. The Alaska Pipeline, the most expensive privately financed construction project in world history, opened in the summer of 1977. America's fear of an inadequate energy supply resulted in the country's most dramatic technological adventure since the boom days of extraterrestrial exploration in the 1960s.

Transportation of crude petroleum was the principal problem involved in opening the North Slope fields. At first, many assumed that large tankers could carry the petroleum through waterways between Canada's Arctic islands to the Atlantic Ocean, and one specially reinforced ship did make the voyage. Canadian concern for the long-term ecological damage that an oil spill could cause in cold Arctic waters, coupled with the near impossibility of making the trip during any period other than the few warm months, forced abandonment of this plan. The only real alternative was to construct a pipeline. For a time a joint American-Canadian project was considered in which a pipeline extending from the North Slope perhaps as far as the lower Great Lakes was to be built. Differences over management and construction timetables plus the huge cost of construction led to its abandonment as well. A pipeline costing $8 billion and crossing central Alaska to the port of Valdez on the Pacific was finally built.

This section of the Alaskan petroleum pipeline on the North Slope about 200 kilometers (120 miles) south of Prudhoe Bay has been elevated above ground to help ensure that the heated petroleum does not melt the permafrost. (Alyeska Pipeline Service Company.)

Hydroelectricity

About 70 percent of all electrical power in Canada comes from hydroelectric facilities, mostly located along the southern margins of the Shield where streams crossing the Shield's hard rock drop onto the lowlands of southern Ontario and Quebec. Until recently, the Churchill Falls project in Labrador was the largest in the country. Quebec has finished an even larger project on the margin of James Bay. These projects, like so many other developments, have been extended northward onto the Shield. The electricity generated by them is both cheap and abundant. As a result, Quebec is a major center for aluminum smelting, a process that requires large amounts of electricity. The province has also contracted to sell surplus electricity to New York, Ontario, and several power companies in northern New

England. Hydro Quebec, the provincially owned electrical company, hopes to export $1.5 billion in hydroelectricity by the late 1980s, with 60 percent of that total going to the United States.

Transportation

In a region where isolation is so pervasive, transportation is quite important. The region's problems in developing transportation result from more than the difficult environment, however important that factor might be. The Northlands' sparse population and especially its lack of cities means that even if transportation routes could be constructed cheaply, they would be used relatively little. The large nodal growth necessary to support a well-integrated transport network is unlikely to develop in the region. Hence, it is difficult to justify high transportation expenditures unless a large deposit of a particularly needed mineral is available.

Today only the Mackenzie River is used to any extent for river transport to the north. Diesel fuel is the most carried freight. Few railroads penetrate far into the Northlands; they include: one to the Labrador Trough from Sept Îles; one which carries mostly export wheat from the Prairie Provinces to Churchill on Hudson Bay; one to the mining developments on Great Slave Lake; and one to James Bay in Quebec. Other rail lines serve the southern margins of the region, such as the mining districts on the Quebec-Ontario border. The Mackenzie Highway also reaches Great Slave Lake, and the Alaska Highway and some of its branches pass through the western margins of the region.

Taken together, these conventional routeways provide access to only a small part of the Northlands. Small ships serve some coastal towns on an occasional basis. For much of the region, however, the light airplane and its bush pilot is the only transportation link available. Close to a score of carriers operate a relatively dense pattern of scheduled routes in the north. Although expensive, some studies suggest that air transport may be only twice as expensive as truck transport on the winter highways, and it cuts development costs tremendously. No region of the world, with the possible exception of the Soviet Union's own northlands and the Australian outback, places greater reliance on the airplane for basic transportation.

Tourism

As North Americans have gained higher and higher incomes with an accompanying growth in leisure time, they have extended their need for recreation space. As nearby areas have become overused, their search has reached farther and farther away from the settled portions of the continent. No major national, provincial, or state park is today so isolated that it does not face the problem of overuse during peak tourist periods. The Northlands, the continent's largest empty space, is felt by many to have almost unlimited tourist potential. This belief is only partly justified.

For many in the large urban centers just south of the region's margins, a summer vacation in the North Woods has long been a standard part of life. Large sections of the southern boreal forest are heavily used by tourists. Northern Minnesota, Wisconsin, and Michigan are an easy day's drive from some of the largest U.S. cities. In Canada, Montrealers flock to the Laurentians a short way north. The jam of cars carrying weekend vacationers north of Toronto can extend for 300 kilometers (185 miles) on a fine Friday afternoon in summer.

To the north, in the central and northern boreal forest and in the tundra, tourist volumes are much smaller. The big game and sport fish found there are justifiably famous, attracting many hunters and fishermen who are willing to fly to isolated lodges in the area. The total number of visitors involved is small, however, and it is likely to remain so. For the average vacationer, these areas offer nothing that cannot be found closer to home with less effort and expense. Also, the fragile environment cannot support anything approaching heavy use. For example, this is an area where vehicle tire marks made on the tundra more than 40 years ago during World War

II are still visible, and many people fear that the traffic traveling the road constructed along the Alaska pipeline will have a detrimental impact on the environment.

A Nodal Settlement Pattern

Although the total regional population is not large, the great majority of people in the Northlands live in villages, towns, and cities (Figure 17-4). Agriculture, a major support of dispersed settlement elsewhere, is only locally important. The principal agricultural settlement areas are the Lake St. John-Saguenay Lowland of Quebec, the Clay Belt along the Quebec-Ontario border, and the Peace River District on the British Columbia-Alberta border. (Both of the latter two areas were settled in this century and represent the last significant examples of something that was once a key part of North American life— frontier agricultural settlement.) Nearly all of the larger cities are dominated by a single major economic activity and are located in the south. Many, such as Sudbury or Flin Flon, are mining and smelting towns. Others, such as Duluth or Thunder Bay, are primarily transportation centers. Chicoutimi's economy focuses on aluminum smelting. Most smaller towns in the boreal forest are similarly unifunctional.

In the far north, European development has resulted in few permanent settlements. Most whites in the area work, in some capacity or another, for the Canadian or U.S. governments or in resource exploitation. Far northern communities are extremely isolated, often with predominantly male populations, and the labor force frequently spends periods of weeks away from the communities for family visits and recreation. As with communities everywhere that are totally focused on minerals extraction, they are often short lived owing to resource depletion.

The Inuit constitute the far north's majority population. Most live in small villages along the coastline, although more and more of them are drifting into the towns of the area. Again, it is a nuclear settlement pattern; few people live away from these settlement nodes.

FIGURE 17-4

SETTLEMENT ZONES OF NORTHERN CANADA. Most of northern Canada
is a zone of recent European settlement, usually the result of resource extraction
activities. The Far North remains largely beyond the settlement frontier.

GOVERNMENT ATTITUDES AND NORTHLANDS DEVELOPMENT

The Northlands have presented both governments with a series of important and often per-
plexing problems. This is a vast area, largely unoccupied, and only now gradually being

exploited. At an earlier time, perhaps neither
country would have concerned itself with the integrity of these lands or their people. However,
times have changed. Today, the region, still
largely frontier, is controlled by industrial, urban economies that share an increasing concern

Many of the towns of the Northlands are mining centers, and their skylines are often dominated by the chimneys of the refineries that reduce the bulk of the product to lessen the cost of shipment. This is a view of Flin Flon, Manitoba. (Manitoba Government Photograph.)

for what their growth has meant to the natural environment.

Canadians, particularly, with the Northlands so much a part of their country, are concerned about the problems of development there. The government and many universities in Canada actively sponsor a wide range of research efforts to help learn more about the Northlands environment. As mentioned earlier, Canadians are tempering their desire for the wealth almost surely available in the region's resources with a concern for what extraction will do to its obviously fragile environment. The United States shares these concerns, although on a much smaller scale.

The native population in the Northlands compounds these problems. Both Indian and Inuit fertility rates are high. Even with a fairly high mortality rate and with some migration toward the margins of the region, rapid growth in numbers still is occurring. Unemployment rates among these groups are already high—often above 50 percent—and good jobs cannot be provided without substantial economic growth. Indians and Inuit are caught in a nearly impossible situation; they have learned to desire many of the comforts and products of a developed economy, but they seldom have the means of attaining those goals.

The necessity of focusing development on mineral resources coupled with the high cost of transportation to widely scattered locations, suggests that even a decision to encourage extractive industries would have only a small and short-lived impact on the economy of most far northern communities. Economic opportunities are restricted and will probably remain so.

Several writers interested in the political geography of places have suggested that it is useful to look at what can be called a country's "effective" national territory. They contend that whatever a country's boundaries might seem to be, it is really in control only of that land actually occupied and actively administered. Large areas of empty land are, by this measure, not firmly a part of a country's territory.

No one is threatening to seize the empty lands of either Canada or the United States, but at least some aspects of this idea can be seen in their approaches to the Northlands. The construction of improved transportation facilities, the development of military bases, and the study of development schemes for the region are in many ways all indicative of this concern. This region will become more and more integrated into the economy and society of the rest of the two countries. It will not be set aside as a massive preserve for things as they once were. In the final analysis, neither country views that as an acceptable alternative.

Additional Readings

Bradbury, John, and Isabelle St. Martin. "Winding Down in a Quebec Mining Town: A Case Study of Schefferville," *Canadian Geographer*, 27 (1983), pp. 191–206.

Bruemmer, Fred. *The Arctic*. Englewood Cliffs, N.J.: Prentice-Hall, 1974.

Canada, Department of Energy, Mines, and Resources. *Annual Report*. Ottawa: Department of Energy, Mines, and Resources, 1966/67– .

Dacks, Gunston. *A Choice of Futures: Politics in the Canadian North*. New York: Methuen, 1981.

Hamelin, Louis-Edmond. *Canadian Nordicity: It's Your North Too*. Montreal: Harvest House, 1979.

McGinnis, Joe. *Going to Extremes*. New York: Alfred A. Knopf, 1980.

McPhee, John. *Coming into the Country*. New York: Farrar, Straus & Giroux, 1977.

Mowat, Farley. *The Great Betrayal: Arctic Canada Now*. Boston: Little, Brown, 1976.

Vanderhill, Burke G. "The Passing of the Frontier Fringe in Western Canada," *The Geographical Review*, 72 (1982), pp. 200–217.

Wonders, William (ed.). *The North*. Toronto: University of Toronto Press, 1972.

CHAPTER EIGHTEEN

HAWAII

When we initially designed this text, we gave serious consideration to excluding any major discussion of Hawaii. After all, we reasoned, this is a book that attempts an overview of the regional geography of North America, and a group of islands in the middle of the Pacific Ocean are certainly not a part of North America. Other perhaps calmer and more reasonable heads prevailed however. They contended that because Hawaii was a part of the United States and because our North America really meant Canada and the United States, the book would be incomplete without a section on Hawaii.

Those who encouraged us to include Hawaii were, of course, correct, and we were likely to have reached the same decision after further reflection. What is most interesting is that we even considered ignoring it. That we, as geographers, considered not discussing Hawaii is a good indication of both the physical separation of Hawaii from the continent and the distinctiveness of its physical and cultural landscapes.

Technically, the Hawaiian archipelago is a string of islands and reefs 3300 kilometers (2050 miles) long that forms a broad arc in the mid-Pacific. The archipelago begins in the east with the island of Hawaii and ends almost at the International Date Line with a small speck in the ocean called Kure Atoll (Figure 18-1). Only the easternmost 650 kilometers (400 miles) of the state contains islands of any size as well as almost all the people. It is to this portion, then, which is usually considered as the actual "Hawaii," that we will limit our discussion. Among the hundred or so bits of land to the west, Midway Island, occupied by a military base 2500 kilometers (1500 miles) west of Honolulu, is easily the largest.

The eight main islands of Hawaii contain more than 99 percent of the state's land area and all but a handful of its people. The island of Hawaii, at 8150 square kilometers (4021 square miles), itself comprises nearly two-thirds of the state's total area. The smallest of the eight is Kahoolawe, at 125 square kilometers (45 square miles). It is uninhabited and used only as a mil-

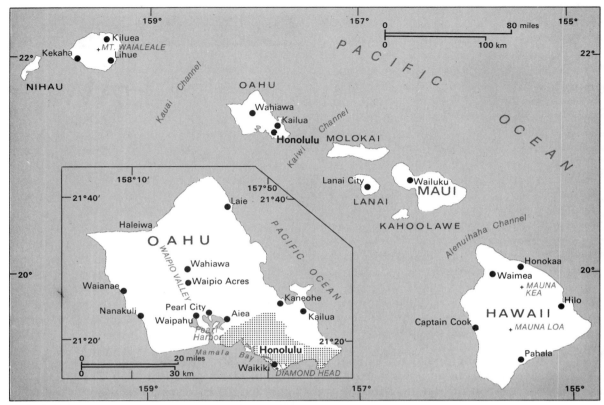

FIGURE 18-1
HAWAII.

itary bombing and firing range, to the annoyance of many residents in the state.

LOCATION AND PHYSICAL SETTING

Mid-Pacific

Hawaii is near the middle of the Pacific Ocean. Honolulu, the state capital, is 3850 kilometers (2400 miles) west of San Francisco, 6500 kilometers (4000 miles) east of Tokyo, and roughly 7300 kilometers (4500 miles) northeast of the

MAJOR METROPOLITAN AREA POPULATIONS IN HAWAII, 1980	
Honolulu	762,565
Hilo	37,017

Australian coast. The great swath of islands stretching across so much of the south and west Pacific are, with a few small and largely inconsequential exceptions, themselves well over 3500 kilometers (2175 miles) away. This might be viewed as a case of extreme isolation. Until the last couple of centuries, this was probably true; but as countries around the Pacific Basin began to communicate more with one another and to use the ocean's resources, these islands in the middle of the ocean became an important center of interaction. The strategic value of Hawaii's location is great in the affairs of the central and north Pacific.

Geology

The Hawaiian chain is merely the visible portion of a series of massive volcanoes. The ocean floor in this area is 4000 to 5000 meters (13,000 to 15,000 feet) below sea level. Hence, for a volcano to break the water's surface requires a mountain already approaching five kilometers (three miles) in height. If one wishes to be a little picky and ignore the oceans in determining the height of mountains, Mauna Kea, on the "Big Island" of Hawaii, may be the world's tallest. It stands 4528 meters (13,784 feet) above sea level. In addition, its base is some 5400 meters (18,000 feet) below sea level, for a total height of nearly 10,000 meters (32,000 feet), almost 1000 meters (3000 feet) taller than Mount Everest.

The kind of volcanic activity that created the islands and that continues there today has, for the most part, not been of the explosive type where large pieces of material are thrown great distances by violent explosions. Instead, the islands have been built up by repeated flows of lava from fissures. Volcanic cones resulting from explosive eruptions do exist on the islands. Diamond Head, the famous Honolulu landmark, is the largest at about 240 meters (810 feet). More common, however, are features formed from a gradual buildup of material as a sequence of lava flows piled one layer on top of another. The usual shape of volcanic mountains formed in this way is domelike, with the main feature being undulating slopes instead of steep cliffs. Such rolling highlands are found in many areas on the islands.

Several of the volcanos on the "Big Island" remain active. Mauna Loa pours out lava on the average of once every four years. A 1950 eruption covered some 100 square kilometers (36 square miles). Volcanic activity poses a constant threat to Hilo, the island's largest town. The most recent occurrence was in 1984. Another volcano, Kilauea, is usually active, but lava actually flows from it about once in every seven years. A 1960 flow from Kilauea covered 10 square kilometers (4 square miles), adding nearly a square mile to the island's size.

Hawaii is characteristically a state of rugged slopes and abrupt changes in elevation. This is the result of the erosion of the volcanic surfaces by moving water. Sea cliffs cut by waves form a spectacular edge to parts of the islands. Such cliffs on the northeast side of Molokai stand as much as 1150 meters (3600 feet) above the water and are among the world's highest; others on Kauai exceed 600 meters (2000 feet). Some small streams on the northeast side of the "Big Island" drop over such cliffs directly into the sea.

Stream erosion has heavily dissected many of the lava surfaces. Canyons lace many of the domes. The floor of Waimea Canyon on Kauai, is more than 800 meters (2500 feet) below the surface of the surrounding land. Waterfalls several hundred feet high are common on the islands. Hiilawe Falls, in Waipio Valley on Hawaii, has a drop of about 300 meters (950 feet). The Pali, on Oahu, is a line of cliffs where the headwaters of streams eroding from opposite sides of the island meet. Those flowing east have eroded the ridges separating them to cut a broad lowland; the westward-facing valleys are higher and remain separated by ridges. The result, viewed from the coast, is a jagged cliff that looks like a series of connected Vs.

One important result of this intense erosive action is a limited amount of level land on the islands. Kauai is particularly rugged, with the only lowlands formed as a thin coastal fringe, and even that does not exist in the northwest. Maui has a flat, narrow central portion separating mountainous extremities. Molokai is reasonably flat on its western end. Oahu has a broad central valley plus some sizable coastal lowlands. The island of Hawaii has only some limited coastal lava plains. The total area of these level sections is not great; flat land on the islands is limited and often quite valuable.

Climate

Hawaii's oceanic location has, obviously, a substantial impact on its climate. It is the ocean that fills the winds with the water that brush the islands' mountains. The ocean also moderates the islands' temperature extremes—Honolulu's record high of 31° C (88° F) is matched by a record low of only 13° C (57° F). The maritime in-

The natural landscape of Hawaii is dramatic and exotic when compared to the
rest of the United States. The erosion of igneous rocks in a warm environment
has created a topography of rugged cliffs covered with a varied vegetation. This
is on the "Garden Island" of Kauai. (Hawaii Visitors Bureau.)

fluence tends to delay seasons on the islands a
bit as well. Whereas most places in the United
States have their highest temperatures in July
and August and their lowest in January and
February, highs in Hawaii occur in September
and October and the yearly low may be delayed
until early March. This lag reflects slower changes
in the ocean's water temperatures.

The state has a definite tropical location. The

latitude of Honolulu, about 20° N, is the same as
Calcutta and Mexico City. As a result, there is
little change in the length of daylight or the angle of incidence of the sun's rays from one season to another. This factor plus the state's maritime position means that there is little seasonal
variation in temperature. Honolulu in January
averages 22° C (72° F); the average July temperature is 26° C (78° F). It is variations in precipi-

tation, not in temperature, that mark the major changes in season on the islands.

It has been suggested that Hawaii has only two seasons, a drier summer between May and October, and a moister winter from October to April. During the summer the islands are under the persistent influence of northeast trade winds. They approach Hawaii over cool waters located to the northeast and create characteristic Hawaiian weather—breezy, sunny with some clouds, warm but not hot. In winter, these trade winds disappear, sometimes for weeks, allowing "invasions" of storms from the north and northwest that bring greater amounts of moisture as well as winds from the south and west. These storms may bring the very heavy rainfall expected in tropical locations. Honolulu has received as much as 43 centimeters (17 inches) of rain in a single 24-hour period. Other Hawaiian weather stations have recorded 28 centimeters (11 inches) in an hour and 100 centimeters (40 inches) in a day, both of which rank near world records.

The topography of the islands creates extreme variations in precipitation from one location to another. As air masses are forced to rise up and over the islands' mountains, they cool, their moisture-carrying capacity is reduced and orographic precipitation results in a process similar to that found along the shores of the North Pacific Coast (see Chapter 16). Mount Waialeale, on Kauai, receives 1234 centimeters (486 inches) annually, making it one of the world's wettest spots, and Waimea, also on Kauai, receives about 50 centimeters (20 inches) annually—yet these two sites are only 25 kilometers (15 miles) apart. For another example, within the metropolitan area of Honolulu it is possible to live in a semiarid climate with less than 50 centimeters (20 inches) of rainfall annually or

A substantial ranching economy is found on many of the drier portions of the islands, especially on the Big Island of Hawaii. Despite the small size of the state, the pattern of concentrated private landholding results in the existence of a number of ranches each covering many thousands of acres. (Hawaii Visitors Bureau.)

on the margins of a rainforest drenched by 300 centimeters (120 inches) of precipitation a year. Unlike the Pacific Northwest, the greatest precipitation on the higher mountains in Hawaii occurs at fairly low elevations, usually between 600 and 1200 meters (2000 to 4000 feet). Air masses, tending to take the easiest path, push around these isolated mountains instead of up and over them. Hence, the lower northeastern flanks of Mauna Kea average well over 250 centimeters (100 inches) of precipitation yearly, with as much as 750 centimeters (295 inches) received in some areas; the opposite southwestern slopes average less than a tenth that amount. Leeward mountain slopes and leeward locations, in general, are much drier than windward slopes because the descending air warms and increases its moisture-carrying capacity.

Much of the volcanic soil is permeable. This allows water to percolate rapidly, draining beyond the reach of many plants. Thus, many areas of moderate to low precipitation are arid in appearance. Visitors to Hawaii are often surprised at the expanses of desert and semidesert that are present on most of the islands.

Biota

The isolation of the Hawaiian islands coupled with their generally temperate climate and great environmental variation has created a plant and bird community of vast diversity, the great majority of which are unique to the islands. There are several thousand plants native there and found naturally nowhere else, and 66 uniquely Hawaiian land birds have also been identified. Interestingly, there were no land mammals on the islands until humans arrived.

Unfortunately, island ecosystems are especially fragile, as numerous studies have shown. Many island species have developed in small, isolated locations in response to a peculiar ecological framework and to the absence of competition. These small niches can disappear easily under human influence; when they do, plant or animal species disappear with them. Of Hawaii's 66 unique birds, more than one-third are now extinct, and most of the rest are on the list of rare and endangered species. It is harder to determine the impact on the islands' plants. Still, by 1900, most native plant cover below 450 meters (1500 feet) had been destroyed. Much of that at higher elevations is also succumbing to human encroachment.

Simple human destruction may not be the primary way in which these native species are eliminated. Alien plants and animals introduced to the islands often do a far better job. For example, people brought with them rats, dogs, pigs, and goats. Rats took an especially heavy toll of native bird life by eating their eggs. Because few Hawaiian plants had thorns or any other protection from grazing animals, feral animals often did great damage to the vegetation. Also, plants by the hundreds were imported either intentionally or accidentally, and many native plants could not compete with them. Government attempts to protect endangered species have resulted in the creation of a number of sanctuaries on the islands, but the success of these efforts has not been great.

POPULATING THE ISLANDS

The Polynesians

The Polynesian settlement of Hawaii was a segment in one of humankind's most audacious periods of ocean voyaging. These people set out on repeated voyages in open canoes across broad oceanic expanses separating small island clusters. Settlers who came to Hawaii 1000 years ago, for example, are presumed to have come from the Marquesas, 4000 kilometers (2500 miles) to the southwest. There was some kind of pre-Polynesian population on the island, but it was probably absorbed by the newcomers. A second substantial wave of Polynesian migrants arrived 500 or 600 years ago.

The massive effort of these voyages, which matched anything accomplished by the Vikings, apparently became too great. As a result, Hawaii spent several hundred years in isolation after the second migration period. During the isolation, the Hawaiians solidified a complicated social organization in their insular paradise. Hereditary rulers held absolute sway over their populations and owned all of the land. By the late eighteenth century, when Europeans found the islands, the benign environment supported a large population that numbered about 300,000 natives.

Early European Impact

The first European to visit Hawaii, which he dubbed the Sandwich Islands, was Captain James Cook in 1778. Cook himself was killed on the shore of the Big Island, but news of his discovery spread rapidly after reaching Europe and North America; it was quickly recognized that the islands were the best location for a way station to exploit the trade developing between North America and Asia. In the 1820s, the whaling industry moved into the North Pacific and, for the next half-century, the islands became the principal rest and resupply center for whalers. About the same time, conservative Protestant missionaries came to the islands. Like most of the whalers, they were from New England, and, very successful in their missionary work, they were for decades a major influence on the islanders.

The effect of all of this on the Polynesians and their culture was catastrophic. It was an almost classic example of the negative consequences that are possible when a long-isolated, technologically inferior culture comes into sudden contact with more advanced and broadly traveled outsiders. As Hawaiians accepted aspects of European culture, their ability to live within their own environment declined. Famine, heretofore unknown, became a part of life for the native islanders. A whole series of infectious diseases, endemic among the visitors but previously unknown on the islands, came to Hawaii. Measles, leprosy, smallpox, syphilis, tuberculosis, and the like swept the islands. One outbreak of either cholera or bubonic plague reportedly halved the population in 1804.

Within three generations, the economic and social order of the Polynesian Hawaiians was destroyed. Their numbers, falling from 150,000 in 1804 to perhaps 75,000 in 1850, are fewer than 10,000 today (although State of Hawaii statistics indicate that an additional 17 percent of the state's population is part Polynesian).

The Asian Arrival

The first Hawaiian sugar plantation was established in 1837, although the islands did not become a substantial producer until after the middle of the century. Between then and the end of the century, Hawaii grew to the rank of a major world sugar exporter.

This development led to a great need for agricultural laborers. Native Hawaiians were used for a time, but they proved to be unwilling workers. Besides, their declining numbers provided nothing like the labor force needed. The plantation owners, who were Americans and Europeans, turned to Asia for the required work force. The first few hundred contract workers from China arrived in 1852. Japanese contract laborers began arriving in 1868; Filipinos in 1906. Between 1852 and 1930, 400,000 agricultural laborers, mostly Asian, were brought to Hawaii. Most were men, and many chose to leave the plantations at the end of their contract period; hence, it was necessary constantly to replenish the labor force.

In 1852, ethnic Hawaiians represented over 95 percent of the population of the islands. By 1900, they were less than 15 percent of the total, whereas nearly 75 percent were Oriental. Japanese were the largest single ethnic group, comprising more than 50 percent of the total population of just over 150,000.

Current Ethnic Patterns and the Second American Influence

After 1930, the mainland United States became the main source of new residents in Hawaii (Figure 18-2). During the 1970s, about 70 percent of the 234,000 people who migrated to Hawaii were from the continental United States. In 1910, only about 1 resident of Hawaii in 5 was of European ancestry (referred to in Hawaii as Cau-

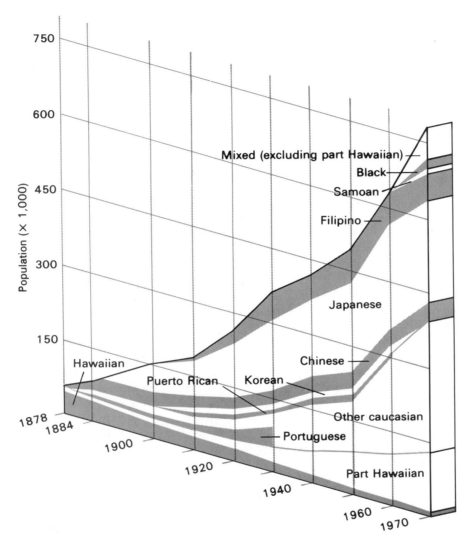

FIGURE 18-2
HAWAIIAN POPULATION BY ETHNIC GROUPS (from R. Warwick Armstrong, ed., *Atlas of Hawaii*. Honolulu: University of Hawaii Press, 1973). Diversity is the dominant theme in Hawaiian ethnicity.

casian). By 1980, Caucasians outnumbered Japanese 222,000 to 198,000 in the state's nonmilitary population. Nearly 40 percent of the state's total current population is Caucasian or part-Caucasian.

Modifications in U.S. immigration laws during the 1960s resulted in a much larger Asian share of total immigration into the country. Hawaii, close to Asia, has been an important destination for these migrants. About half of the new increase in the Hawaiian population resulting from immigration is currently due to Asian migrants, mostly from the Philippines. Net migration accounts for about 40 percent of the state's total annual population increase.

Implications of Ethnicity

Marriage between people of different ethnic groups is increasingly common in Hawaii. About 75 percent of the Caucasians still marry other Caucasians, and 60 percent of the Japanese still marry Japanese. A decade earlier, 81 percent of Caucasians and 75 percent of Japanese married within their ethnic group. For most of the other ethnic groups, about half or more of all marriages are outside the group. In mainland United States, social acceptance of such marriages has remained less common.

The Japanese and particularly the Chinese have done well in the Hawaiian economy. Educational standards among both groups have been high, and average Chinese income levels have surpassed Caucasian averages in recent decades. Newer immigrant groups, notably Filipinos, and native Hawaiians have comparatively low incomes and low-status occupations.

Although by no means a completely integrated society, Hawaiian residential integration easily surpasses that in other parts of the United States. No census tract anywhere in Hawaii is as much as 40 percent Chinese in population, and none is more than 70 percent Japanese. Within Honolulu, where most of the state's residents are found, a clear majority of tracts have a population mixture that includes at least 10 percent of each of the city's three major ethnic groups—

The ethnic diversity of the Hawaiian population is captured in this crowd scene in Honolulu. (Werner Stoy/Camera Hawaii.)

TABLE 18-1
Percentage Distribution of the Hawaiian Population

Location	Date					
	1831	*1878*	*1910*	*1940*	*1970*	*1980*
Oahu	23	35	43	61	82	79
Honolulu	10	24	27	42	42	38
Rest of Oahu	13	11	16	19	40	41
Other Islands	77	65	57	39	18	21

SOURCE: R. Warwick Armstrong (ed.), *Atlas of Hawaii*. Honolulu: University Press of Hawaii, 1973, p. 100; *1980 U.S. Census of Population*, Vol. 1, *Hawaii*. Washington, D.C.: U.S. Government Printing Office, 1983.

Caucasians, Japanese, and Chinese. The poorer ethnic groups are less well integrated.

Population Distribution

The total population of Hawaii fell from its pre-European peak to a low of 54,000 in 1876 before beginning to grow again. By the early 1920s, the state's population had reached pre-European levels. Over 400,000 service personnel, stationed on the islands during World War II, temporarily pushed the total to about 850,000. In 1980, the state had 965,000 residents. Because of immigration, Hawaii's annual rate of population growth is about 2.5 times the national average.

The pre-European population was spread across the islands, with the "Big Island" occupied by the largest number of people. Since European discovery, the islands' population has been concentrated on Oahu (Table 18-1). Honolulu, with its fine harbor, became the principal port city. The other islands also lost in absolute terms. Their postcontact maximum to 1970 of 165,000 was well below precontact estimates. That number had declined to 132,000 in 1960, although, by 1980, there had been a substantial increase to 193,000. The trend in outer islands population growth will probably continue as new tourist developments search out less crowded areas of the state.

POLITICAL INCORPORATION INTO THE UNITED STATES

The political history of Hawaii was turbulent during the 120 years after Cook's discovery. The various kingdoms of the islands were eliminated by a strong chief, Kamehameha, between 1785 and 1795. The missionaries' growing influence gradually made a sham of the authority of the Hawaiian rulers, and, during the nineteenth century, competing European political interests moved in to fill the resulting vacuum. From the 1820s to 1850s the French pushed their influence in the region. Much of the Hawaiian monarchy seemed to favor a relationship with Britain, however, and for a brief period in 1843 Britain annexed Hawaii. It was only the balance between the different powers operating in Hawaii that kept the islands independent during the nineteenth century.

The increasing role of Americans made it inevitable that, if it was to lose its political independence, Hawaii would be annexed by the United States. As American plantation owners increased in number and influence, their dissatisfaction with the Hawaiian government grew. In 1887, they forced the monarchy to accept an elected government controlled by the planters. The monarchy was overthrown completely in 1893, and the new revolutionary government immediately requested annexation by the United

States. Initially refused, they were finally accepted as a territory in 1898.

Unlike earlier acquisitions, no provision was made at the time of annexation for the eventual admission of Hawaii to statehood. Despite repeated requests, Congress long refused to grant the islands the status of a state. The reasons for this steadfast refusal were varied and never really clear, but they likely centered around concerns over the allegiance of the Asian population of Hawaii and the long distance separating the islands from the mainland. Not until 1959, after Alaska was admitted as a state and all possible economic arguments against Hawaiian statehood were gone, did it become the fiftieth state.

THE HAWAIIAN ECONOMY

Landownership

Roughly half of all land in Hawaii is government owned. This in itself is not usual because large proportions of all of the Western states and provinces are governmentally owned. What is unusual is that it is the state, not the federal government, that controls 80 percent of that land. Most of it is in the agriculturally less desirable portions of the islands, and the bulk is in forest reserves or conservation districts. Most federal lands are primarily in national parks on the "Big Island" and Maui or in military holdings on Oahu and Kahoolawe.

It is the unusual concentration of private lands in the hands of a relatively few major landowners that sets the state off from the rest of the country (Figure 18-3). Seven-eighths of all privately owned land in Hawaii is in the hands of only 39 owners; each owns 200 hectares (5000 acres) or more. Six different landowners each control more than 40,000 hectares (100,000 acres) out of a state total of about 1,040,000 hectares (2,570,000 acres) and the largest, the Bernice P. Bishop estate, holds more than one-sixth of all the privately owned land in the state. Smaller unit ownership of private land is most extensive on Oahu, but even there the larger owners con-

trol more than two-thirds of all privately owned land. Two of the islands, Lanai and Niihau, are each nearly entirely controlled by a single owner, and on all of the other islands (except Oahu) major landowners control about 90 percent of all privately held property.

Most of these large landholdings were created during the nineteenth century period of freewheeling exploitation on the islands. Land had previously been held entirely by the monarchies, not by commoners. This land passed into the hands of non-Hawaiian private owners during the political decline of the monarchy. With the deaths of the early owners, most estates have been given over to trusts to administer rather than passing directly to heirs. This has made it difficult to break up the ownership pattern.

This concentration of landownership repre-

FIGURE 18-3

PATTERNS OF LAND OWNERSHIP ON OAHU (from R. Warwick Armstrong, *Atlas of Hawaii.* Honolulu: University of Hawaii Press, 1973). Most of the land area is held either by the state or by a very small number of large land owners. One result is that land prices in Hawaii are extremely high.

sents a major problem for the state, especially in urban areas. The landowners have leased land for urban use. As the leases expire, they can be renewed at a much higher rate because the urban development provided by the lessors has greatly increased the value of the land. The security of landownership, so long a part of the American psyche, is, thus, effectively denied to many in Hawaii. Where smaller holdings are available and where population growth is rapid, most notably on Oahu, this has led to a tremendous increase in land prices.

In addition to high land values, other sometimes less direct problems result from the landownership pattern. High land costs have helped create pockets of high population density, again particularly on Oahu. Many major investors are unwilling to build on land they do not own.

Hawaii attempted to moderate the impact of large landholdings through the Land Reform Act of 1967. This act extended the state's power of eminent domain by allowing Hawaii's housing authority to condemn individual parcels of land, usually at the time of lease renewal. Title would be transferred to the tenants at a fair market price determined by arbitration or negotiation. The law was declared unconstitutional by lower courts, but in 1984 the U.S. Supreme Court overturned that ruling, thus paving the way for change.

Agriculture

Sugar, and later pineapples, fueled the Hawaiian economy for many decades after the 1860s. The economy remained primarily agricultural until World War II. In recent decades agriculture has continued to show modest gains in income, but its relative importance has declined greatly. Sugar's share of all income generated in the state fell from 20 percent in 1950 to 5 percent in 1980. The income from pineapples dropped from 16.5 to 2 percent of the total. Even diversified agriculture, such as livestock raising on the island of Hawaii, which has been the prime growth component of the agricultural industry, had its income share decline from 4.6 to about 2 percent. Total employment grew by

nearly 20 percent between 1970 and 1980, but employment in sugar and pineapple production declined by more than 17 and 25 percent, respectively. Only 1 Hawaiian worker in 30 is currently employed in agriculture. Once the kingpin of the Hawaiian economy, agriculture today ranks no better than fourth or fifth.

In some ways this emphasis on declining agricultural importance is misleading. First, even with the move of some plantations out of production in the face of declining profits, Hawaii continues to provide a substantial share of the world's sugar harvest. Hawaii once controlled most of the world's pineapple production but lost part of its market after World War II to newer producing areas; however, production remains at about 650,000 tons annually. Hawaii remains the world's largest supplier of pineapples.

Gross economic statistics for the share overwhelmingly emphasize the position of Oahu, where today more than 80 percent of the state's entire economy is concentrated. The role of agriculture remains great on the other islands. Both Lanai and Molokai depend on pineapples for much of their employment and income. Livestock and sugar form the backbone of the economy on the "Big Island," as does sugar and pineapples on Maui and Kauai. In short, only on Oahu has agriculture been replaced as the most important sector of the economy.

The Federal Government

As agriculture declined and lost its dominance over the Hawaiian economy, its place was first taken by the federal government. Over the past several decades, governmental expenditures have increased at a rate roughly comparable to the growth of the total economy, maintaining about a one-third share of all expenditures. Most of this has come from the military. The strategic value of Hawaii's location in the mid-Pacific is obvious. It is the headquarters for the U.S. Pacific Command, center of Pacific operations for all branches of the armed forces, and home of a number of major military bases. The military controls almost 25 percent of Oahu, including

Hawaii's apparently urban economy is largely a function of the domination by Honolulu of the state's economic structure. Urban functions, other than tourist services, remain relatively unimportant on all of the other islands. These pineapple fields are on the island of Maui. (Arnold Nowotny/USDA Soil Conservation Service.)

the land around Pearl Harbor, one of the finest natural harbors in the Pacific. Nearly 1 Hawaiian worker in 4 is an employee of the military. The 57,000 military personnel and their 64,000 dependents together represented 13 percent of Hawaii's population in 1980. The armed forces are also the largest civilian employer in the state.

The armed forces have long been a substantial part of the Hawaiian community, and they are generally well received. Still, the heavy de-

pendence on the military is often viewed as unfortunate, especially when periods of military cutbacks affect the local economy adversely. The relative importance of the military is likely to decline somewhat in the future, although it will almost surely remain a major presence.

Tourism

For decades, Hawaii represented an exotic, distant tropical paradise to most Americans. Many dreamed of a visit to its balmy shores. The airplane and cheap fares turned that dream into reality. The first regularly scheduled transpacific flights began in 1936. As late as 1951, however, nearly half of the 52,000 annual visitors to Hawaii reached there by ship. Then came a period of larger airplanes, lower fares, and a booming national economy. The number of visitors increased sixfold in every decade, to 320,000 in 1965 and 1,819,000 in 1971. By 1981, 3.9 million people (most of them tourists) visited the state. Hawaii's government and businesses encouraged these visits through the Hawaii Visitors Bureau, which widely advertised the tropical and exotic image of the islands. Tourism

These two photographs, both of Waikiki Beach and Diamond Head in Honolulu, were taken approximately 25 years apart, the first (below) in the early 1950s, the second (right) in the mid-1970s. Comparatively inexpensive air flights from the mainland opened Hawaii to the mass of American tourists. For most, Waikiki was the principal destination. (State of Hawaii Public Archives.)

(Werner Stoy/Camera Hawaii.)

became the principal growth sector of the economy, increasing its share of total island income from 4 percent in 1950 to 18 percent in 1970 and 28 percent in 1980. Tourism income now nearly matches federal government expenditures in Hawaii; if the pattern of growth continues, the tourist industry promises to become the kingpin of the state's economy.

To many Hawaiians, this influx of tourists is a mixed blessing. In 1952, only 2800 hotel units existed on the islands. By 1982, there were 58,000 units. Nearly 34,000 of the existing units were on Oahu, with over 29,000 crowded along the famous beaches of the Waikiki-Kahala coast in Honolulu. On a typical day 60,000 tourists flock to the high-rise hotels along the beach. Diamond Head, long the city's most famous landmark, suddenly found hotels shrouding its lower elevations. The congestion, decline in scenic quality, and possible pollution problems resulting from this crowded development are of great concern to many. Recent expansion has focused on the outer islands (the number of tourist rooms on Maui grew from 4000 to 11,600 between 1970 and 1980), and there is fear that without careful control these problems will spread throughout the state.

A second problem with the tourist economy is its unstable nature. The downturn in the national economy in the early and mid-1970s had a heavy impact on the Hawaiian tourist industry because many Americans chose to economize by cutting travel costs with a vacation closer to home. Also, although Hawaii has striven to develop a year-round tourist flow, the summer months remain the peak period; this results in higher unemployment during the winter.

Most visitors to Hawaii come from the mainland of the United States, especially the West Coast. A growing number of Asians, mostly Japanese, also visit the state. These latter visitors are causing some local concern. In an attempt to control the total tour package and, thus, increase profits, Japanese tour operations have purchased a number of hotels and other tourist facilities in Hawaii. Other Japanese firms, simply looking for a good investment, have bought a variety of Hawaiian businesses. All of this has caused some Hawaiians to urge control of such sales to foreigners, arguing that the state's economy could become Japanese controlled. Perhaps they can appreciate better the native Hawaiian reaction to the American influx of a century ago. Japanese tourists typically spend nearly twice as much daily as their American counterparts, and spend far more for gift and souvenir items. Thus, the growth in Asian tourism is especially important to Hawaii.

Transportation

The insular nature of Hawaii creates a transport situation far different from that of the mainland United States. First, it is entirely dependent on air and sea movement; overland transport is of only local importance. Nearly all visitors arrive by air. Hawaii imports much of its food, nearly all of its energy supplies and vehicles, and a host of other commodities. Its agricultural exports amount to about 11 million tons annually.

The Hawaiian economy is, thus, quite vulnerable to any breakdowns in transportation. The International Longshoreman's and Worker's Union, the islands' strongest union, has demonstrated that dependence on several occasions through strikes or threats of strike. It is Hawaii's dependence on shipping that has given the union its power.

HONOLULU

Honolulu dominates Hawaii. Nearly 80 percent of the state's population and a matching percentage of its economy is concentrated in the city and its suburbs. The second largest city outside Oahu, Hilo on Hawaii, although increasing its population at a more rapid rate than Honolulu, has only 37,000 residents. No other state, no other region discussed in this textbook, is so totally dominated by a single urban center.

Mountains, the ocean, and federally held land hug the city, crowding it into a series of narrow valleys and a fringe along the coast. Recent construction of two highways and tunnels have increased the city's connections with the northeast coast of the island (the Kilua-Kaneohe urbanized area on that coast has grown to over 100,000), and construction is also increasing along Oahu's central valley.

The problems of Honolulu are similar to those of other American cities, although somewhat magnified. Living costs are high, perhaps 20 percent above the national average. Among U.S. cities, only Anchorage, Alaska, matches the expense of living in Honolulu. Housing costs are also high, averaging over $120,000 for a single family home in 1980. With so much land held by estates, residential land for sale is scarce. One result is that only 41 percent of the population lives in owner-occupied housing. Housing is noticeably inadequate for low-income groups. The arrangement of the city, the many barriers provided by physical features and federal land, and the general high density of occupation all inhibit traffic flow and increase congestion. The city's leeward location leads to occasional periods of little wind, and a genuine atmospheric pollution problem is evident.

All of this does not particularly damn Honolulu as a city. Instead it emphasizes that it is a city with a set of problems and attractions much like that of any other American city. It may be in paradise, but it is a large city in paradise.

INTERISLAND DIVERSITY

The islands are part of the same state, they have similar geologic histories, and they are closely spaced in a vast ocean, yet each tends to have its own character. Oahu is densely populated and intensely used, and it offers a view of bustle and confusion common to urban America. The island of Hawaii, by comparison, offers an air of relative space and distance, with large ranches, high, barren volcanos, and large stretches of almost treeless land. Kauai is heavily eroded into a spectacular scenery of mountains, canyons, cliffs, and waterfalls. It is sometimes called the garden isle because of its lush tropical vegetation. Maui offers a contrast between the plantations of its central lowlands and the rugged mountains to either side. Niihau is a dry land of low hills that is populated by native Hawaiians somewhat isolated from the outside. Molokai is half pineapple plantations and half rugged mountains. Kahoolawe is barren, waterless, and bombed. Lanai is a blend of the others.

An Island Paradise?

It has become fashionable to talk of trouble in paradise when discussing Hawaii. Certainly the state has its share of problems. High living costs and high unemployment rates will continue to be difficult. The combined impact of the flood of tourists, of economic development, of change of many kinds in its physical environment has Hawaii groping for a solution. The economy is too closely tied to tourism and the military, as both are given to rapid fluctuation resulting from recession or changing military spending programs.

Still, these should not overshadow Hawaii's positive attributes. No part of America, and few parts of the world, can match its level of racial and ethnic assimilation. All people are not treated equally in Hawaii, but they are far closer to that goal than most of the rest of the world. Much of Hawaii remains a place of great beauty. Its climate is often superlative. Few other states in the country can match Hawaii's attempts at controlling growth to maintain the quality of their environment. Hawaii passed America's first statewide land-use program, and it has a statewide sign ordinance. Concern for tourist-oriented development on the outer islands has exerted a strong counterbalance to rapid growth. For many, Hawaii will remain an island paradise.

Additional Readings

Armstrong, R. Warick (ed.). *Atlas of Hawaii,* 2nd ed. Honolulu: University Press of Hawaii, 1983.

Carlquist, Sherwin John. *Hawaii, A Natural History,* 2nd ed. Honolulu: Pacific Tropical Botanical Garden, 1980.

Farrell, Bryan. *Hawaii: The Legend That Sells.* Honolulu: University Press of Hawaii, 1982.

Glick, Clarence E. *Sojourners and Settlers: Chinese Migrants in Hawaii.* Honolulu: University Press of Hawaii, 1980.

Kent, Noel J. *Hawaii: Islands Under the Influence.* New York: Monthly Review Press, 1983.

Lind, Andrew W. *Hawaii's People,* 4th ed. Honolulu: University Press of Hawaii, 1980.

Michener, James A. *Hawaii.* New York: Random House, 1959.

Morgan, Joseph. *Hawaii: A Geography.* Boulder, Colo.: Westview Press, 1983.

Ogawa, Dennis M. (ed.). *"Kodomo no tame ni"—For the Sake of the Children: The Japanese American Experience in Hawaii.* Honolulu: University Press of Hawaii, 1978.

GLOSSARY

Absolute Humidity: the mass of water vapor in the atmosphere per unit volume of space.

Accessibility: a locational characteristic that permits a place to be reached by the efforts of those at other places.

Accessibility Resource: a naturally occurring landscape feature that facilitates interaction between places.

Air Mass: a very large body of atmosphere defined by essentially similar horizontal air temperatures. Moisture conditions are also usually similar throughout the mass.

Alluvial Soils: soils deposited through the action of moving water. These soils lack horizons and are usually highly fertile.

Antebellum: before the war; in the United States, belonging to the period immediately prior to the Civil War.

Arete: a sharp, narrow mountain ridge. It often results from the erosive activity of alpine glaciers flowing in adjacent valleys.

Arroyo: a deep gully cut by a stream that flows only part of the year; a dry gulch. A term normally used only in desert areas.

Badlands: very irregular topography resulting from wind and water erosion of sedimentary rock.

Base Level: the lowest level to which a stream can erode its bed. The ultimate base level of all streams is, of course, the sea.

Batholith: a very large body of igneous rock, usually granite, that has been exposed by erosion of the overlying rock.

Bedrock: the solid rock that underlies all soil or other loose material; the rock material that breaks down eventually to form soil.

Bilingual: the ability to use either one of two languages, especially when speaking.

Biota: the animal and plant life of a region considered as a total ecological entity.

Boll Weevil: a small, greyish beetle of Mexico and the southeastern United States with destructive larvae that hatch in and damage cotton bolls.

Break-in-Bulk Point: commonly, a transfer point on a transport route where the mode of transport (or type of carrier) changes and where large-volume shipments are reduced in size. For example, goods may be unloaded from a ship and transferred to trucks at an ocean port.

Caprock: a strata of erosion-resistant sedimentary rock (usually limestone) found in arid areas. Caprock forms the top layer of most mesas and buttes.

Carrying Capacity: the number of people that an area can support given the quality of the natural environment and the level of technology of the population.

CBD: the central business district of an urban area, typically containing an intense concentration of office and retail activities.

Chinook: a warm, dry wind experienced along the eastern side of the Rocky Mountains in Canada and the United States. Most common in winter and spring, it can result in a rise in temperature of 20° C (35° F to 40° F) in a quarter of an hour.

Climax Vegetation: the vegetation that would exist in an area if growth had proceeded undisturbed for an extended period. This would be the "final" collection of plant types that presumably would remain forever or until the stable conditions were somehow disturbed.

Confluence: the place at which two streams flow together to form one larger stream

Coniferous: bearing cones; from the conifer family.

Continental Climate: the type of climate found in the interior of the major continents in the middle, or temperate, latitudes. It is characterized by a great seasonal variation in temperatures, four distinct seasons, and a relatively small annual precipitation.

Continental Divide: the line of high ground that separates the oceanic drainage basins of a continent. The river systems of a continent on opposite sides of a continental divide flow toward different oceans.

Coulee: a dry canyon eroded by Pleistocene floods that cut into the lava beds of the Columbia Plateau.

Conurbation: an extensive urban area formed when two or more cities, originally separate, coalesce to form a continuous metropolitan region.

Core Area: the portion of a country that contains its economic, political, intellectual, and cultural focus. It is often the center of creativity and change (see **Hearth**).

Crop-Lien System: a farm-financing scheme whereby money is loaned at the beginning of a growing season to pay for farming operations with the subsequent harvest used as collateral for the loan.

Culture: the accumulated habits, attitudes, and beliefs of a group of people that define their general behavior and way of life; the total set of learned activities of a people.

Culture Hearth: the area from which the culture of a group diffused (see **Hearth**).

Cut-and-Sew Industry: the manufacture of basic ready-to-wear clothing. Such facilities usually have a small fixed investment in the manufacturing facility.

Deciduous Forest: forests where the trees lose their leaves each year.

De Facto Segregation: the spatial and social separation of populations that occurs without legal sanction.

Degree Day: deviation of one degree temperature for one day from an arbitrary standard, usually the long-term average temperature for a location.

De Jure Segregation: the spatial and social separation of populations that occurs as a consequence of legal measures.

Demography: the systematic analysis of population.

Discriminatory Shipping Rates: a transportation charge levied in a manner that is inequitable to some shippers, primarily because of those shippers' location.

Dome: an uplifted area of sedimentary rocks with a downward dip in all directions. This is often caused by molten rock material pushing upward from below. The sediments have often eroded away, exposing the rocks that resulted when the molten material cooled.

Dry Farming: a type of farming practiced in semiarid or dry grassland areas without irrigation using such approaches as fallowing, maintaining a finely broken surface, and growing drought-tolerant crops.

Economies of Agglomeration: the economic advantages that accrue to an activity by locating close to other activities; benefits that follow from complementarity or shared public services.

Economies of Scale: savings achieved in production costs by larger enterprises because the cost of initial investment can be defrayed across a greater number of units produced.

Emergent Coastline: a shoreline resulting from a rise in land surface elevation relative to sea level.

Enclave: a tract or territory enclosed within another state or country.

Erratic: a boulder that has been carried from its source by a glacier and deposited as the glacier melted. Thus, the boulder is often of a different rock type from surrounding types.

Estuary: the broad lower course of a river that is encroached on by the sea and affected by the tides.

Evapotranspiration: the water lost from an area through the combined effects of evaporation from the ground surface and transpiration from the vegetation.

Exotic Stream: a stream found in an area that is too dry to have spawned such a flow. The flow originates in a moister section.

Extended Family: a family that includes three or more generations. Normally, that would include grandparents, their sons or daughters, and their children, as opposed to a **nuclear family,** which is only the married couple and their offspring.

Fall Line: the physiographic border between the piedmont and coastal plain regions. The name derives from the river rapids and falls that occur as the water flows from hard rocks of the higher piedmont onto the softer rocks of the coastal plain.

Fallow: agricultural land that is plowed or tilled but left unseeded during a growing season. Fallowing is usually done to conserve moisture.

Fault Block Mountain: a mountain mass created either by the uplift of land between faults or the subsidence of land outside the faults.

Fault Zone: a fault is a fracture in the earth's crust along which movement has occurred. The movement may be in any direction and involve material on either or both sides of the fracture. A "fault zone" is an area of numerous fractures.

Federation: a form of government in which powers and functions are divided between a central government and a number of political subdivisions that have a significant degree of political autonomy.

Feral Animal: a wild or untamed animal, especially one having reverted to such a state from domestication.

Fish Ladder: a series of shallow steps down which water is allowed to flow; designed to permit salmon to circumvent artificial barriers such as power dams as the salmon swim upstream to spawn.

Focality: the characteristic of a place that follows from its interconnections with more than one other place. When interaction within a region comes together at a place (i.e., when the movement focuses on that location), the place is said to possess "focality."

Functional Diversity: the characteristic of a place where a variety of different activities (economic, political, social) occur; most often associated with urban places.

Geomorphology: the study of the arrangement and form of the earth's crust and of the relationship between these physical features and the geologic structures beneath.

Ghetto: originally, the section of a European city to which Jews were restricted. Today, this is commonly defined as a section of a city occupied by members of a minority group who live there because of social restrictions on their residential choice.

Glacial Till: the mass of rocks and finely ground material carried by a glacier, then deposited when the ice melted. This creates an unstratified material of varying composition.

Great Circle Route: the shortest distance between two places on the earth's surface. The route follows a line described by the intersection of the surface with an imaginary plane passing through the earth's center.

Hearth: the source area of any innovation; the source area from which an idea, crop, artifact, or good is diffused to other areas.

Heavy Industry: manufacturing activities engaged in the conversion of large volumes of raw materials and partially processed materials into products of higher value. Hallmarks of this form of industry are considerable capital investment in large machinery, heavy energy consumption, and final products of relatively low value per unit weight (see **Light Industry**).

High-action Beach: a coastal area subject to greater

erosion (from ocean currents and wave movement) than deposition. The beach, therefore, is narrow.

Hinterland: the area tributary to a place and linked to that place through lines of exchange, or interaction.

Horizon: a distinct layer of soil encountered in vertical section.

Humus: partially decomposed organic soil material.

Hydrography: the study of the surface waters of the earth.

Hydroponics: the growing of plants, especially vegetables, in water containing essential mineral nutrients rather than in soil.

Igneous Rock: rock formed when molten (melted) materials harden.

Indentured Labor: work performed according to a binding contract between two parties. During the early colonial period in America, this often involved long periods of time and a total work commitment.

Inertia Costs of Location: costs born by an activity because it remains located at its original site, even though the distributions of supply and demand have changed.

Insular: either of an island or suggestive of the isolated condition of an island.

Intervening Opportunity: the existence of a closer, less expensive opportunity for obtaining a good or service, or for a migration destination. Such opportunities lessen the attractiveness of more distant places.

Intracoastal Waterway System: a waterway channel, maintained through dredging and sheltered for the most part by a series of linear offshore islands that extends from New York City to Florida's southern tip and from Brownsville, Texas, to the eastern end of Florida's panhandle.

Isohyet: a line on a map connecting points that receive equal annual precipitation.

Jurisdiction: the right and power to apply the law; the territorial range of legal authority or control.

Karst: an area possessing surface topography resulting from the underground solution of subsurface limestone or dolomite.

Kudzu: a vine native to China and Japan but imported into the United States; originally planted for decoration, for forage, or as a ground cover to control erosion. It now grows wild in many parts of the southeastern United States.

La Belle Province: the former nickname of the province of Quebec, still commonly used.

Lacustrine Plain: a nearly level land area that was formed as a lake bed.

Latitude: a measure of distance north or south of the equator. One degree of latitude equals approximately 110 kilometers (69 miles).

Leaching: a process of soil nutrient removal through the erosive movement and chemical action of water.

Legume: a plant, such as the soybean, that bears nitrogen-fixing bacteria on its roots and thereby increases soil nitrogen content.

Light Industry: manufacturing activities that use moderate amounts of partially processed materials to produce items of relatively high value per unit weight (see **Heavy Industry**).

Lignite: a low-grade, brownish coal of relatively poor heat-generating capacity.

Loess: a soil made up of small particles that were transported by the wind to their present location.

Longitude: a measure of distance east and west of a line drawn between the North and South Poles and passing through the Royal Observatory at Greenwich, England.

Maritime Climate: a climate strongly influenced by an oceanic environment, found on islands and the windward shores of continents. It is characterized by small daily and yearly temperature ranges and high relative humidity.

Mediterranean Climate: a climate characterized by moist, mild winters and hot, dry summers.

Metamorphic Rock: rock that has been physically altered by heat and/or pressure.

Metes and Bounds: a system of land survey that defines land parcels according to visible natural landscape features and distance. The resultant field pattern is usually very irregular in shape.

Metropolitan Coalescence: the merging of the urbanized areas of separate metropolitan regions; Megalopolis is an example of this process.

Monadnock: an isolated hill or mountain of resistant rock rising above an eroded lowland.

Moraine: the rocks and soil carried and deposited by a glacier. An "end moraine," either a ridge or low hill running perpendicular to the direction of ice movement, forms at the end of a glacier when the ice is melting.

Nuclear Family: see **Extended Family.**

Open Range: a cattle- or sheep-ranching area characterized by a general absence of fences.

Orographic Rainfall: precipitation that results when moist air is lifted over a topographic barrier such as a mountain range.

Outwash: rocky and sandy surface material deposited by meltwater that flowed from a glacier.

Overburden: material covering a mineral seam or bed that must be removed before the mineral can be removed in strip mining.

Permafrost: a permanently frozen layer of soil.

Physiographic Region: a portion of the earth's surface with a basically common topography and common morphology.

Plate Tectonics: geologic theory that the bending (folding) and breaking (faulting) of the solid surface of the earth results from the slow movement of large sections (plates) of that surface.

Platted Land: land that has been divided into surveyed lots.

Pleistocene: period in geologic history (basically the last 1 million years) when ice sheets covered large sections of the earth's land surface not now covered by glaciers.

Plural Society: a situation in which two or more culture groups occupy the same territory but maintain their separate cultural identities.

Polynodal: many centered.

Postindustrial: an economy that gains its basic character from economic activities developed primarily after manufacturing grew to predominance. Most notable would be quaternary economic patterns.

Precambrian Rock: the oldest rocks, generally more than 600 million years old.

Presidio: a military post (Spanish).

Primary Product: a product that is important as a raw material in developed economies; a product consumed in its primary (i.e., unprocessed) state (see **Staple Product**).

Primary Sector: that portion of a region's economy devoted to the extraction of basic materials (e.g., mining, lumbering, agriculture).

Pueblo: a type of Indian village constructed by some tribes in the southwestern United States; a large community dwelling, divided into many rooms, up to five stories high, and usually made of adobe; also, a Spanish word for town or village.

Quaternary Sector: that portion of a region's economy devoted to informational and idea-generating activities (e.g., basic research, universities and colleges, and news media).

Rail Gauge: the distance between the two rails of a railroad.

Rainshadow: an area of diminished precipitation on the lee (downwind) side of a mountain or mountain range.

Rangs: lines of long, narrow fields in French Canada, created to maximize access to rivers or roads for transportation.

Region: an area having some characteristic or characteristics that distinguish it from other areas; a territory of interest to people and for which one or more distinctive traits are used as the basis for its identity.

Resource: anything that is both naturally occurring and of use to humans.

Riparian Rights: the rights of water use possessed by a person owning land containing or bordering a water course or lake.

Riverine: located on or inhabiting the banks of or the area near a river or lake.

Scarp: also "escarpment"; a steep cliff or steep slope, formed either as a result of faulting or by the erosion of inclined rock strata.

Scots-Irish: the North American descendents of Protestants from Scotland who migrated to northern Ireland in the 1600s.

Secondary Sector: that portion of a region's economy devoted to the processing of basic materials extracted by the primary sector.

Second Home: a seasonally occupied dwelling that is not the primary resident of the owner. Such residences are usually found in areas with substantial opportunities for recreation or tourist activity.

Sedimentary Rock: rock formed by the hardening of material deposited in some process; most commonly sandstone, shale, and limestone.

Seigneuries: large land grants in French Canada possessing feudal privileges that were awarded by the kings of France to noblemen and the Church. These lands were then parceled out to individual settlers.

Sharecropping: a form of agricultural tenancy where the tenant pays for use of the land with a predetermined share of his crop rather than with a cash rent.

Shield: a broad area of very old rocks above sea level, usually characterized by thin, poor soils and low population densities.

Silage: fodder (livestock feed) prepared by storing and fermenting green forage plants in a silo.

Silo: usually a tall, cylindrical structure in which fodder (animal feed) is stored; may be a pit dug for the same purpose.

Sinkhole: crater formed when the roof of a cavern collapses; usually found in areas of limestone rock.

Site: features of a place related to the immediate environment on which the place is located (e.g., terrain, soil, subsurface geology, groundwater).

Situation: features of a place related to its location relative to other places (e.g., accessibility, hinterland quality).

Smog: mixture of particulate matter and chemical pollutants in the lower atmosphere, usually over urban areas.

SMSA: Standard Metropolitan Statistical Area. A statistical unit of one or more counties that focuses on one or more central cities larger than a specified size, or with a total population larger than a specified size; a reflection of urbanization.

Soluble: capable of being dissolved; in this case, the characteristic of soil minerals that leads them to be carried away in solution by water (see **Leaching**).

Space Economy: the locational pattern of economic activities and their interconnecting linkages.

Spatial Complementarity: the occurrence of location pairing such that items demanded by one place can be supplied by another.

Spatial Interaction: movement between locationally separate places.

Staple Product: a product that becomes a major component in trade because it is in steady demand; thus, a product that is basic to the economies of one or more major consuming populations (see **Primary Product**).

Temperature Inversion: an increase in temperature with height above the earth's surface, a reversal of the normal pattern.

Territory: a specific area or portion of the earth's surface, not to be confused with **Region.**

Tertiary Sector: that portion of a region's economy devoted to service activities (e.g., transportation, retail and wholesale operations, insurance).

Threshold: the minimum-sized market for an economic activity. The activity will not be successful until it can reach a population larger than this threshold size.

Time-Distance: a time measure of how far apart places are (how long does it take to travel from place A to place B?); this may be contrasted with other distance metrics such as geographic distance (how far is it?) and cost-distance (how much will it cost to get there?).

Township and Range: the rectangular system of land subdivision of much of the agriculturally settled United States west of the Appalachians; established by the Land Ordinance of 1785.

Transferability: the extent to which a good or service can be moved from one location to another; the relative capacity for spatial interaction.

Transhumance: the seasonal movement of people and animals in search of pasture. Commonly, winters are spent in snow-free lowlands and summers in the cooler uplands.

Tree Line: either the latitudinal or elevational limit of normal tree growth. Beyond this limit, closer to the poles or at higher or lower elevations, climatic conditions are too severe for such growth.

Tropics: technically, the area between the Tropic of Cancer (23½° N latitude) and the Tropic of Capricorn (23½° S latitude), characterized by the absence of a cold season. Often used to describe any area possessing what is considered to be a hot, humid climate.

Underemployment: a condition among a labor force such that a portion of the labor force could be eliminated without reducing the total output. Some individuals are working less than they are able or want to, or they are engaged in tasks that are not entirely productive.

Underpopulation: economically, a situation where an increase in the size of the labor force will result in an increase in per worker productivity.

Water Table: the level below the land surface at which the subsurface material is fully saturated with water. The depth of the water table reflects the minimum level to which wells must be drilled for water extraction.

Zoning: the public regulation of land and building use to control the character of a place.

MAP APPENDIX

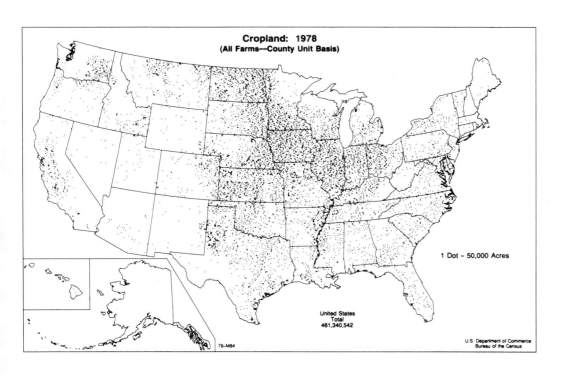

Cropland: 1978
(All Farms---County Unit Basis)

1 Dot – 50,000 Acres

United States
Total
461,340,542

78–M84

U.S. Department of Commerce
Bureau of the Census

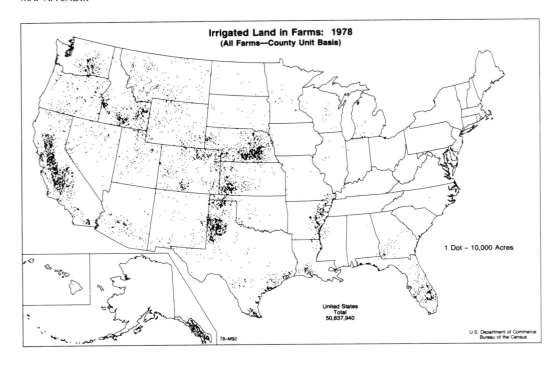

Irrigated Land in Farms: 1978
(All Farms—County Unit Basis)

1 Dot – 10,000 Acres

United States
Total
50,837,940

U.S. Department of Commerce
Bureau of the Census

78–M92

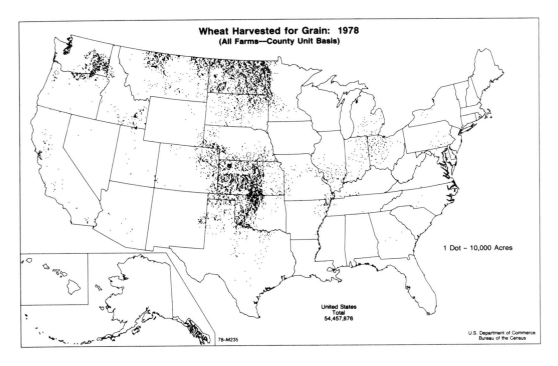

Wheat Harvested for Grain: 1978
(All Farms—County Unit Basis)

1 Dot – 10,000 Acres

United States
Total
54,457,876

U.S. Department of Commerce
Bureau of the Census

78–M235

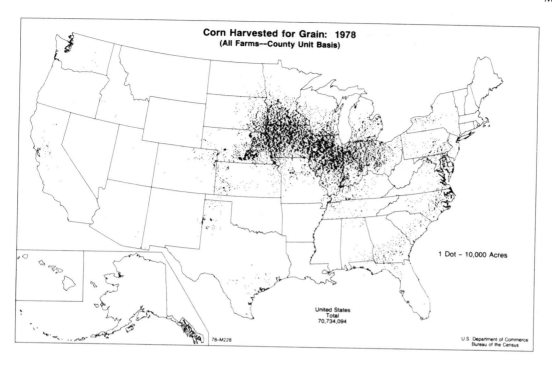

Corn Harvested for Grain: 1978
(All Farms--County Unit Basis)

1 Dot – 10,000 Acres

United States
Total
70,734,094

78-M228

U.S. Department of Commerce
Bureau of the Census

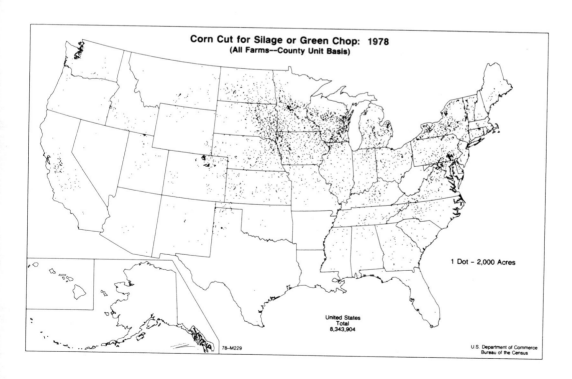

Corn Cut for Silage or Green Chop: 1978
(All Farms--County Unit Basis)

1 Dot – 2,000 Acres

United States
Total
8,343,904

78-M229

U.S. Department of Commerce
Bureau of the Census

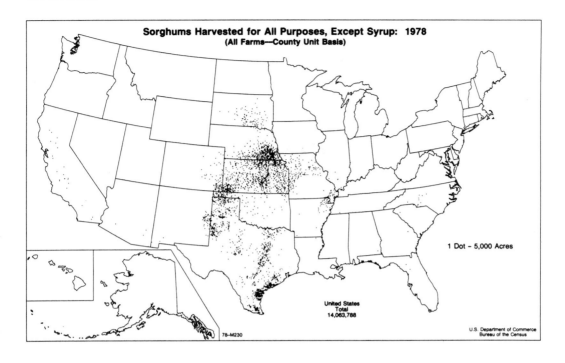

Sorghums Harvested for All Purposes, Except Syrup: 1978
(All Farms—County Unit Basis)

1 Dot – 5,000 Acres

United States
Total
14,063,788

78–M230

U.S. Department of Commerce
Bureau of the Census

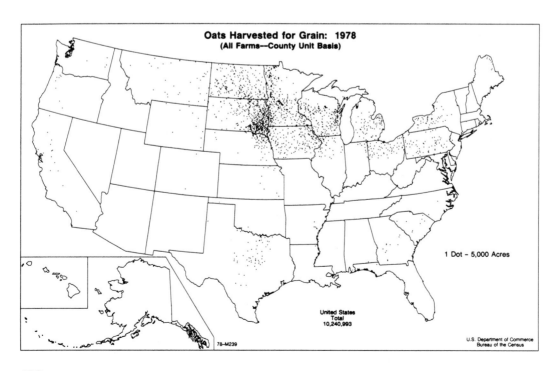

Oats Harvested for Grain: 1978
(All Farms—County Unit Basis)

1 Dot – 5,000 Acres

United States
Total
10,240,993

78–M239

U.S. Department of Commerce
Bureau of the Census

432

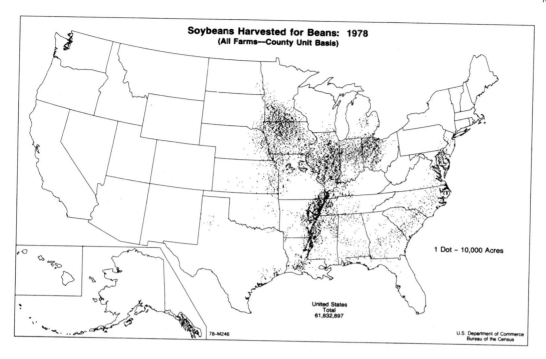

Soybeans Harvested for Beans: 1978
(All Farms—County Unit Basis)

1 Dot – 10,000 Acres

United States
Total
61,832,897

78–M246

U.S. Department of Commerce
Bureau of the Census

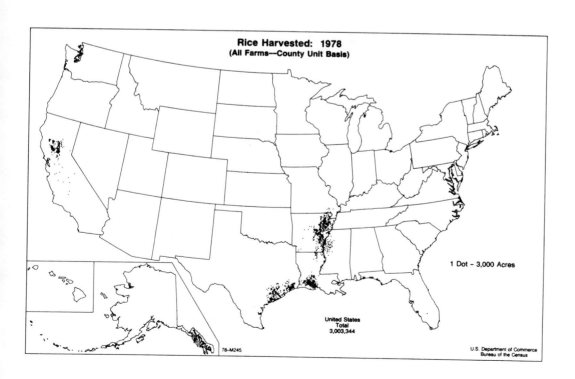

Rice Harvested: 1978
(All Farms—County Unit Basis)

1 Dot – 3,000 Acres

United States
Total
3,003,344

78–M245

U.S. Department of Commerce
Bureau of the Census

433

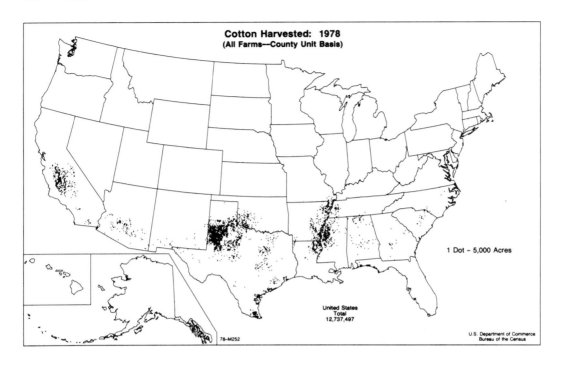

Cotton Harvested: 1978
(All Farms—County Unit Basis)

1 Dot – 5,000 Acres

United States
Total
12,737,497

78–M252

U.S. Department of Commerce
Bureau of the Census

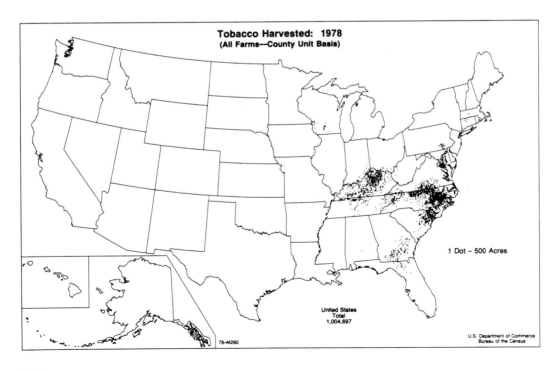

Tobacco Harvested: 1978
(All Farms—County Unit Basis)

1 Dot – 500 Acres

United States
Total
1,004,697

78–M260

U.S. Department of Commerce
Bureau of the Census

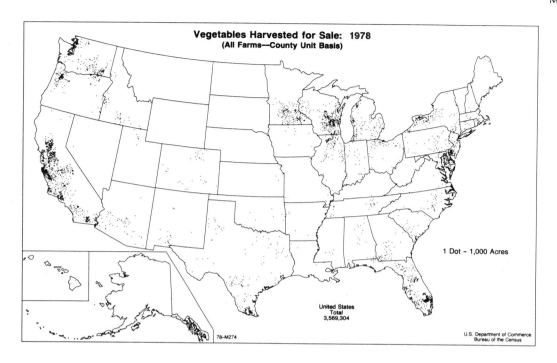

Vegetables Harvested for Sale: 1978
(All Farms---County Unit Basis)

1 Dot - 1,000 Acres

United States
Total
3,569,304

78–M274

U.S. Department of Commerce
Bureau of the Census

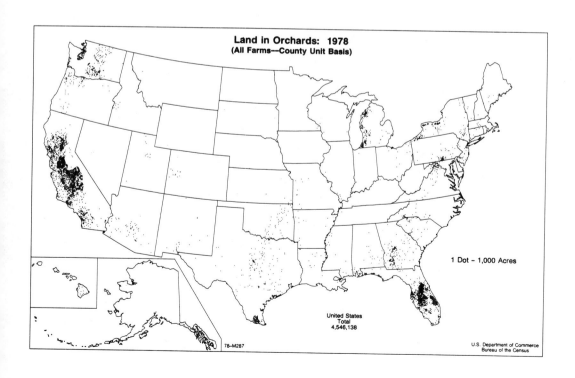

Land in Orchards: 1978
(All Farms---County Unit Basis)

1 Dot - 1,000 Acres

United States
Total
4,546,138

78–M287

U.S. Department of Commerce
Bureau of the Census

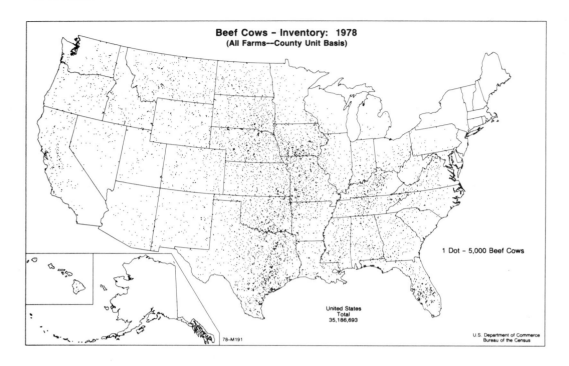

Beef Cows – Inventory: 1978
(All Farms—County Unit Basis)

1 Dot – 5,000 Beef Cows

United States
Total
35,186,693

U.S. Department of Commerce
Bureau of the Census

78–M191

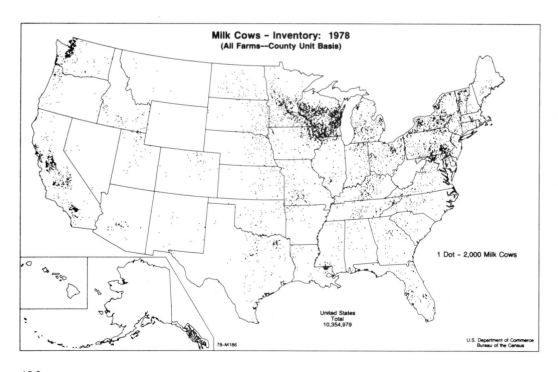

Milk Cows – Inventory: 1978
(All Farms—County Unit Basis)

1 Dot – 2,000 Milk Cows

United States
Total
10,354,979

U.S. Department of Commerce
Bureau of the Census

78–M186

436

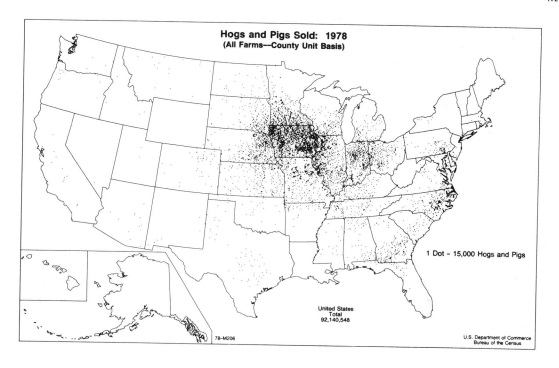

Hogs and Pigs Sold: 1978
(All Farms---County Unit Basis)

1 Dot – 15,000 Hogs and Pigs

United States
Total
92,140,548

78–M206

U.S. Department of Commerce
Bureau of the Census

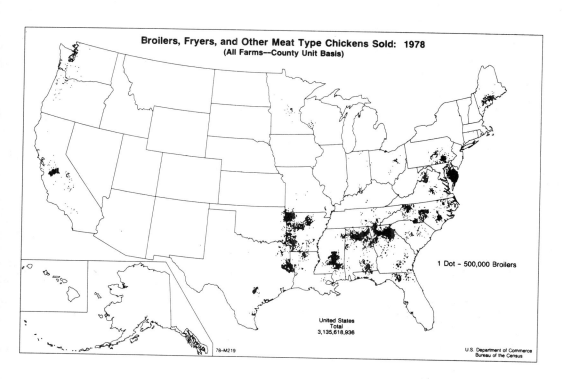

Broilers, Fryers, and Other Meat Type Chickens Sold: 1978
(All Farms---County Unit Basis)

1 Dot – 500,000 Broilers

United States
Total
3,135,618,936

78–M219

U.S. Department of Commerce
Bureau of the Census

INDEX